T0360725

Monopolizing Knowledge

In the nineteenth century, an ambitious new library and museum for Asian arts, sciences and natural history was established in the City of London, within the corporate headquarters of the East India Company. Funded with taxes from British India and run by the East India Company, this library-museum was located thousands of miles away from the taxpayers who supported it and the land from which it grew. Jessica Ratcliff documents how the growth of science at the Company depended upon its sweeping monopoly privileges and its ability to act as a sovereign state in British India. She explores how "Company science" became part of the cultural fabric of science in Britain and examines how it fed into Britain's dominance of science production within its empire, as well as Britain's rising preeminence on the scientific world stage. This title is part of the Flip it Open program and may also be available open access. Check our website Cambridge Core for details.

Jessica Ratcliff is a historian in the Department of Science and Technology Studies at Cornell University. She is the author of *The Transit of Venus Enterprise in Victorian Britain* (2008).

Science in History

Series Editors

Science in History is a major series of ambitious books on the history of the sciences from the mid eighteenth century through the mid twentieth century, highlighting work that interprets the sciences from perspectives drawn from across the discipline of history. The focus on the major epoch of global economic, industrial and social transformations is intended to encourage the use of sophisticated historical models to make sense of the ways in which the sciences have developed and changed. The series encourages the exploration of a wide range of scientific traditions and the interrelations between them. It particularly welcomes work that takes seriously the material practices of the sciences and is broad in geographical scope.

A full list of titles in the series can be found at: www.cambridge.org/sciencehistory

Monopolizing Knowledge

The East India Company and Britain's Second Scientific Revolution

Jessica Ratcliff

Cornell University

Shaftesbury Road, Cambridge CB2 8EA, United Kingdom

One Liberty Plaza, 20th Floor, New York, NY 10006, USA

477 Williamstown Road, Port Melbourne, VIC 3207, Australia

314–321, 3rd Floor, Plot 3, Splendor Forum, Jasola District Centre, New Delhi – 110025, India

103 Penang Road, #05–06/07, Visioncrest Commercial, Singapore 238467

Cambridge University Press is part of Cambridge University Press & Assessment, a department of the University of Cambridge.

We share the University's mission to contribute to society through the pursuit of education, learning and research at the highest international levels of excellence.

www.cambridge.org
Information on this title: www.cambridge.org/9781009379496

DOI: 10.1017/9781009379526

When citing this work, please include a reference to the DOI 10.1017/9781009379526

First published 2025

A catalogue record for this publication is available from the British Library.

A Cataloging-in-Publication data record for this book is available from the Library of Congress

ISBN 978-1-009-37949-6 Hardback

To the memory of Dr. Faidra Papanelopoulou (1978–2016)

Contents

Figures

Acknowledgments

Thanks, first of all, to Dan Schiller, and the Information in Society program at the Graduate School of Library and Information Science, University of Illinois at Urbana–Champaign, for giving me a postdoc that both saved my career and started me on a new research path. Dan's seminars on the political economy of information were the first steps that eventually led to this book. I also want to acknowledge the influence of Jim Bennett, whose brilliant work introduced me to the material culture of science, and whose lessons I will never stop learning from.

I want to sincerely thank the institutions that, through grants and fellowships, have financially supported this project: the National Maritime Museum London, the Max Planck Institute for the History of Science, the Singapore Ministry of Education, Cornell University and the Institute for Advanced Study, Princeton. I also want to give special thanks to Bronwen Bledsoe of the Cornell University Library, who managed to acquire funding to purchase access to an exorbitantly priced primary source database (Adam Matthew Digital's East India Company Online) after COVID-19 upended my research travel plans. Researching the East India Company generally requires expensive travel to London. Unfortunately, this new corporate-owned database of the British Library's East India Company records has not, as yet, made historical work of this kind any less expensive. This very situation – the political economy of access to these knowledge resources – is what this book is all about.

I am immensely grateful for the expertise, time and assistance of the archivists and librarians who made this work possible, including but not limited to: Margaret Makepeace and Antonia Moon at the British Library; Carole Atkinson at Cornell University Library; Sushma Jansari at the British Museum; Stephen Sinon at the New York Botanical Gardens Library; Divia Patel at the Victoria and Albert Museum; and Mark Glancey at the National Museum of Scotland. Thanks also to Lucy Rhymer and Rosa Martin at Cambridge University Press for all of the work they have done to help bring this book to print. Sally Evans-Darby was a fabulous copyeditor. Thanks also to Reshma Venkatachalapathy and her team at Integra Software Services in Pondicherry, who managed the book's production phase.

Several generations of graduate and undergraduate research assistants have contributed to this research by doing primary and secondary source surveys, organizing data and engaging with libraries and museums. My thanks and appreciation to: Christian Go (Yale–NUS College), Edwin Rose (Cambridge University), Samuel Schrivar, Minna Chow, Nnenna Ochuru, Milan Taylor and Skylar Xu (Cornell University).

I have also relied upon the support and feedback of many of my colleagues, who have been very generous with their time and ideas. Thanks to my colleagues at the Graduate School of Library and Information Science at the University of Illinois, Yale–NUS College (RIP!) and the Department of Science and Technology Studies at Cornell University. I would like to give special thanks to those who have taken the time to read and comment on draft versions of this book: Suman Seth, Nico Silins, Robert Travers, Anna Winterbottom, Malte Ziewitz and the anonymous reviewers of the book. In addition, I want to acknowledge those who have discussed and debated the project with me over the years and contributed to its final shape: Caroline Cornish, Felix Driver, Fredrik Albritton Jonsson, Taran Kang, Arun Kundnani, Odette Lineau, Owen Marshall, Simon Naylor, Aziz Rana, Simon Schaffer and Sujit Sivasundaram.

To Darcie ("Dar Dar") and Nicole ("Cole"), immense thanks for the all-important work that you do that has made this work possible. Nicole: without you, this book could not have been written.

Finally, to Nico and Ada (and Clover): thanks for the laughter, the hugs, the adventures and $\infty + 9$ other things that make it all worthwhile.

Introduction

The British East India Company is credited with great and terrible things. It is said to have had a direct hand in creating global capitalism, while at the same time contributing to modern forms of state.[1] "The corporation that changed the world" built an infrastructure of armies, ships, fortified port cities and a global financial network that moved vast resources between Britain and Asia.[2] The "original evil corporation" also forged a modern world economy in which imperialism and free markets went hand in hand.[3] The Company transformed the political and economic landscape of huge portions of South and Southeast Asia, brought the Chinese Empire into war and left some formerly affluent regions of the Indian subcontinent utterly impoverished. It gave shape to the modern sense of "Britishness" and was instrumental in the creation of the largest, most densely inhabited and possibly dirtiest city the world had yet seen: London c. 1830.

The Company also left its fingerprints all over the making of modern science. This book is about how, as an inextricable part of all the above, the Company shaped the global scientific order, and with lasting consequences. It focuses on the period between the Company's emergence as a territorial power in 1757 and the dissolution of the Company in 1858. This was a period in which the Company's empire expanded dramatically, and one in which its 150-year-old monopoly on trade with Asia would also slowly fall apart. This period is sometimes referred to as Britain's "second scientific revolution," in which scientific practices, professions, disciplines and theories also underwent radical change. The historical relationship between these two developments has been the subject of many studies. But the role of the East India Company in

[1] Stern, Philip J. *The Company-State: Corporate Sovereignty and the Early Modern Foundations of the British Empire in India*. Oxford University Press, 2011.

[2] Robins, Nick. *The Corporation That Changed the World: How the East India Company Shaped the Modern Multinational*. Pluto Press, 2012.

[3] Dalrymple, William. "Opinion: The Original Evil Corporation." *The New York Times*, September 4, 2019. See also Dalrymple, William. *The Anarchy: The East India Company, Corporate Violence, and the Pillage of an Empire*. Bloomsbury Publishing, 2019.

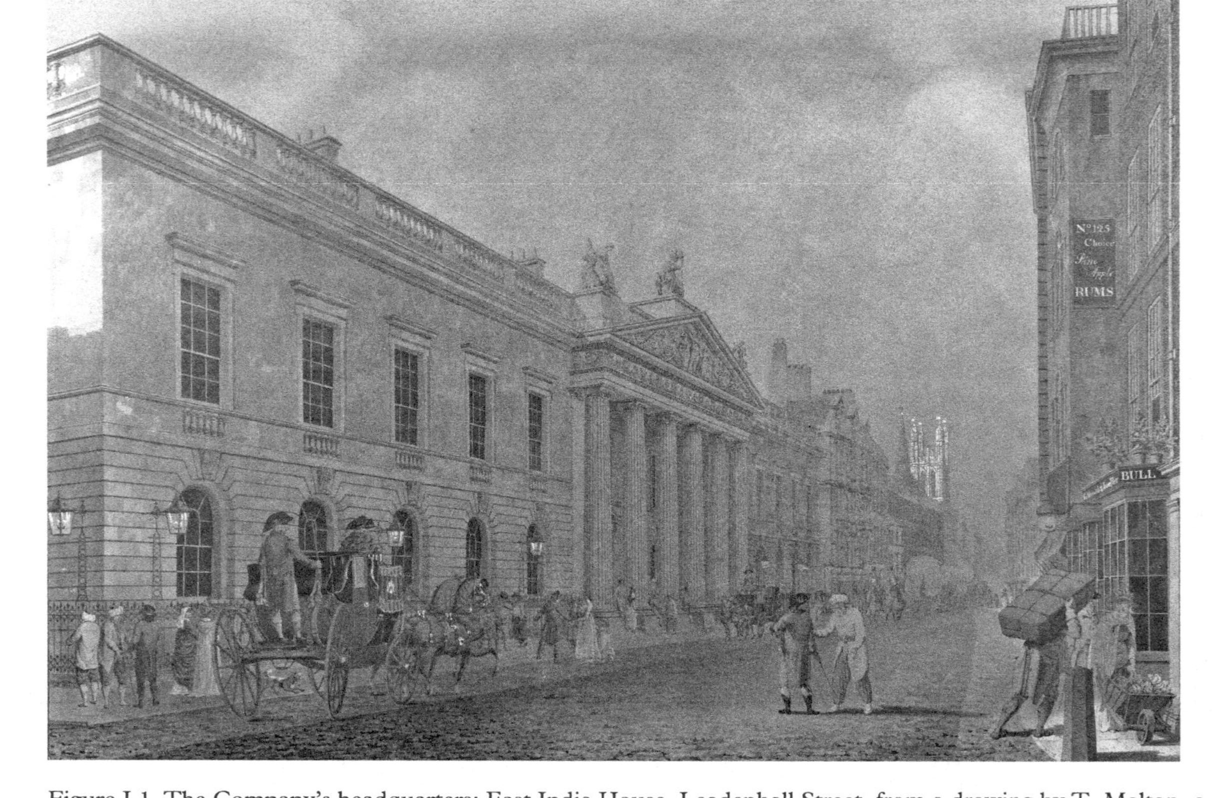

Figure I.1 The Company's headquarters: East India House, Leadenhall Street, from a drawing by T. Malton, c. 1800. Copyright British Library Board (asset WD 2460).

this history remains poorly understood. One aim of this book is to bring new clarity to that subject.

The Company's monopoly privileges – and its peculiar form of monopoly-based colonial capitalism – are key to understanding the particular dynamic connecting the growth of British science and the growth of the British Empire. From 1600 until 1813 (with some key gaps), the Company held a Crown-granted legal monopoly – one laid out in charter agreements between the Crown and the Company – on all British trade east of the Cape of Good Hope. The Company's monopoly deeply shaped British access to, and knowledge of, Asia's nature. In practice, the Company's control over access and movement and information was far from complete. But it was formidable and would, as we will see, play a significant role in shaping the cultures and practices of science in both Britain and the colonies. Especially in the period covered here, colonial and capitalist expansion coincided with an unprecedented information boom back in Europe.[4] Many new libraries, museums, botanical gardens and similar institutions for the management of knowledge resources were established in this period, and the expansion of data-intensive empirical practices is often taken as a hallmark of the so-called second scientific revolution.[5]

Another aim of this book is therefore to situate the history of Britain's second scientific revolution within a longer history of the global regulation of trade in knowledge resources. In order to do so, I have focused in particular on the history of the Company's practices of accumulation, management and production of what might today be called scientific information: manuscripts, books, specimens of natural history, technologies, antiquities, works of art and craft and so on.[6] As we will see, the Company's decision to establish a library and museum at its

[4] For example, in Peter Burke's *Social History of Knowledge*, he concludes that "The amount of new knowledge gathered or collected in ... 1750–1850, was staggering, especially the knowledge collected by Europeans about the fauna, flora, geography and history of other parts of the world." Burke, Peter. *A Social History of Knowledge II: From the Encyclopaedia to Wikipedia*. Polity Press, 2012, p. 12. The classic work on this is Cannon, Susan Faye. *Science in Culture: The Early Victorian Period*. Science History Publications, 1978. I address this historiography in Ratcliff, Jessica. "The Great (Data) Divergence: Global History of Science within Global Economic History." In *Global Scientific Practice in an Age of Revolutions, 1750–1850*, edited by Patrick Manning and Daniel Rood. University of Pittsburgh Press, 2016.

[5] See, for example, Sheets-Pyenson, Susan. *Cathedrals of Science: The Development of Colonial Natural History Museums during the Late Nineteenth Century*. McGill-Queen's University Press, 1988; Bennett, Tony. *Pasts beyond Memory: Evolution, Museums, Colonialism*. Taylor & Francis Group, 2004; MacKenzie, John M. *Museums and Empire: Natural History, Human Cultures and Colonial Identities*. Manchester University Press, 2009.

[6] Lissa Roberts has argued that global or transnational histories of science need to renew analytical focus on the "concept, processes and management of accumulation." The

headquarters in London, as well as the founding of two Company colleges in Britain, would mark a major change to the way the Company managed information, and it was both tied to wider changes in British science and a consequence of very particular political and economic changes affecting the accessibility and cost of accumulation. The Company's collections were accumulated with the monopoly advantage, and its library, museum and colleges were supported by the Company's tax revenues from the people of India. Those collections are now the property of the people of Britain, divided primarily among the British Museum, Kew Gardens, the British Library, the UK Natural History Museum and the Victoria and Albert Museum (see Figures I.2 and I.3).

A third overall aim of this project has been to provide a view of how scientific practices and cultures are defined and limited by political economic change. This book aims, in other words, to situate a key period in the history of science in Britain within economic history in such a way that we gain a better understanding of how the growth of science has occurred internal to, as part of, wider political economic change.[7] The essence of

Figure I.2 Syntypes of *Paludina lecythoides*, once in the East India Company's museum, now owned by the Natural History Museum, London. Copyright Trustees of the Natural History Museum, London (1842.9.30.47–48, SYNTYPES, Paludina lecythoides Benson, 1842).

current problem, according to Roberts, is that accumulation still tends to be "investigated separately within the fields of political economy and history of science" when in fact accumulation is a key point of coextension across information, capital and many other sources of value. Roberts, Lissa. "Accumulation and Management in Global Historical Perspective: An Introduction." *History of Science* 52, no. 3 (September 1, 2014): 227–246, p. 228.

[7] Elsewhere, I have argued that both the history of science and economic history would benefit from closer engagement: Ratcliff, Jessica. "The Great (Data) Divergence"; Ratcliff, Jessica. "Travancore's Magnetic Crusade: Geomagnetism and the Geography of Scientific Production in a Princely State." *British Journal for the History of Science; Norwich* 49, no. 3 (September 2016): 325–352.

Figure I.3 A circa first-century CE reliquary casket excavated from a Buddhist *stupa* in Bimaran, Afghanistan by the East India Company agent Charles Masson in c. 1833–1838. Now in the British Museum (1900,0209.1). Copyright Trustees of the British Museum.

that perspective was captured by contemporary observers Karl Marx and Friedrich Engels in an 1845 digression on the relationship between science and commerce: "But where would natural science be without industry and commerce? Even 'pure' natural science is provided with an aim, as with its material, only through trade and industry, through the sensuous [i.e. physical] activity of men."[8]

At just around the time that Marx and Engels wrote their now-famous early draft of their materialist philosophy of history, the Company was busy installing new museum galleries within its headquarters at India House (see Figures I.1 and I.4). Where a vast pay office had once been, curators were now busy at work on crateloads of fossils extracted from northern India, carefully cleaning and making plaster casts of the finds

[8] Marx, Karl and Friedrich Engels. *The German Ideology* (written 1845–1846, published 1935). Reprinted in Tucker, Robert C. *The Marx–Engels Reader*, W. W. Norton, 1978, p. 171.

Figure I.4 A view of a new gallery of the East India Company's museum at India House, 1858. From the *London Journal*. Courtesy of the New York Public Library Digital Collections (1858–03-06. https://digital collections.nypl.org/items/045309c0-041 f-0134-b03 f-00505686a51c).

before mounting them for public display. A few floors up, another set of offices had been emptied of desks and papers and replaced with glazed cases that would be filled with stuffed birds from Java. An adjacent room was where the Company's naturalist worked to mount insect collections, pinning beetles to felt boards.

But a corporate-imperial museum is not what Marx and Engels had in their sights. Their comment had a very specific target: a rival materialist talking about "secrets of nature" being revealed only to the eye of the chemist and the physicist, through their specialist instrumentation and experience. Marx and Engels found this type of materialism entirely misguided. They asserted that the real "material basis" of those chemists' knowledge did not begin and end with the ability to perceive, or even with the specially trained eye and a carefully constructed experiment. One must also, they claimed, take equal account of the laboratory, the building it is housed in, the source of oil and candles for lighting, the sewers, the streets, in fact the entire town in which the laboratory sat, not to mention the food that sustained the scientist, the clothing that kept them warm and so on. The material basis of science, Marx and Engels argued, must be recognized as the entire human/natural world within which those scientists were working, the whole mode of production. To bring the point home, their thought experiment continued, imagine if all the "unceasing . . . labor and creation . . . were interrupted only for a year." What, then, would the chemist be able to perceive? They would find "not only an enormous change in the natural world . . . the whole world of men and his own perceptive faculty, nay his own existence, [would be] missing."[9]

For the purposes of this Introduction, what is important about this extremely broad view of the material basis of science is how it describes science as coextensive with and fundamentally shaped by economic activity, but *not* in a way that suggests that *therefore* science is shaped or determined by the strategies of economically powerful actors or of economic self- (or corporate) interest. Instead, economic change and scientific change are coextensive because they are the material culture of science writ large, or the "matter at hand," as Gideon Freudenthal and Peter McLaughlin have described it (paraphrasing Boris Hessen and Henryk Grossman), that determines the conditions of possibility or the horizons of enquiry within which scientific change unfolds.[10] We thus may similarly ask, what would have happened to British science if its imperial connections had been cut off? Would London, Oxford and Cambridge still have emerged as world centers of scientific production by the start of the twentieth century?

Despite (or maybe because of) the fact that the Company was one of the largest and most important employers in nineteenth-century London, its social and cultural impact in Britain has, until recently, remained "strangely invisible," as John McAleer puts it.[11] This could be said for our understanding of the Company's place in the history of science in Britain as well. Several decades ago, David Arnold highlighted the fact that, because of the peculiarities of the Company-state model, "on the subcontinent the Company and its servants enjoyed a near monopoly

[9] Marx and Engels. *The German Ideology*, p. 17.

[10] Freudenthal, Gideon and Peter McLaughlin. "Classical Marxist Historiography of Science: The Hessen-Grossmann-Thesis." In *The Social and Economic Roots of the Scientific Revolution: Texts by Boris Hessen and Henryk Grossmann*, edited by Gideon Freudenthal and Peter McLaughlin. Boston Studies in the Philosophy of Science. Springer, 2009, pp. 1–40.

[11] McAleer, John. "Exhibiting 'The Strangest of All Empires': The East India Company, East India House, and Britain's Asian Empire." In *The MacKenzie Moment and Imperial History: Essays in Honour of John M. MacKenzie*, edited by Stephanie Barczewski and Martin Farr. Springer International Publishing, 2019, p. 26.

over Western scientific activity."[12] But this critical topic of the interaction between scientific practice and the Company's monopoly has not yet been examined in any detail.[13] McAleer suggests the invisibility of the Company as a cultural force is due to the oversized role it plays as a political economic force. Likewise, for the case of the absence of the Company within the history of science, there seems to have been an inclination among historians to assume that, at the institutional level within the Company, science could only have ever played a very minor role. For example, for Arnold, the Company's "scientific monopoly" translated into a "laissez-faire" approach that led to only very minor support for science, education or improvement schemes.[14] Naturalists and orientalists working under the Company in Asia are thus often depicted as having succeeded *despite* the lack of support or encouragement offered by the Court of Directors. Even histories dealing with the Company's library and museum in London have largely followed the same line of argument. For example, in his major new study of collecting and the Company, Arthur MacGregor downplays the significance of the political, economic and institutional context for science under the Company by focusing on the personal drive and passion of the individuals associated with the Company and those who built up the collections.[15] Theodore Binnema, too, takes a similar perspective in his history of science at the Hudson Bay Company, arguing that, while certain aspects of the Hudson Bay Company's institutional structure were undoubtedly important, the real driver of scientific culture at the Company was down to

[12] David Arnold also suggests that the Company may have denied requests by Charles Lyell and Charles Darwin, both of whom had expressed interest in traveling to India. Arnold, David. *New Cambridge History of India: Science, Technology and Medicine in Colonial India*. Cambridge University Press, 2000, p. 20.

[13] For example, in two major new works, one on "science and the state" and the other on "science and empire," the Company only appears in passing. Gascoigne, John. *Science and the State: From the Scientific Revolution to World War II*. Cambridge University Press, 2019. What is here called "Company science" also only appears in passing in Goss, Andrew. *The Routledge Handbook of Science and Empire*. Routledge, 2023.

[14] Similarly, Gascoigne concludes that the Company, as a whole, was largely "unresponsive" to the various proposals for scientific or technological projects put before the Court of Directors. Gascoigne, John. *Science in the Service of Empire: Joseph Banks, the British State and the Uses of Science in the Age of Revolution*. Cambridge University Press, 1998, p. 144.

[15] One of his central arguments is that the science and collecting done by employees under the Company were driven by a very different set of interests and values from that of the Company, and as a whole it functioned largely independent from (and despite) the vast political economic machinery within which it was embedded. MacGregor, Arthur. *Company Curiosities: Nature, Culture and the East India Company, 1600–1874*. Reaktion Books, 2018, pp. 8–9.

individual officers' motivations and interests.[16] Although Bernard Cohn famously argues that "the establishment of British hegemony in India was also a conquest of knowledge," he doesn't relate much of that process to the institutional or regulatory powers of the Company, whose patronage of science he describes as "haphazard," "filled with false starts" and generally "halted when the bookkeepers in Leadenhall Street became aware of the potential costs, and their ill effects on the balance sheet."[17] From this perspective, the Company's library and museum have generally been interpreted as, most fundamentally, an exercise in imperial image-making or, as Maya Jasanoff put it, "self-advertisement."[18]

To be sure, all of these perspectives capture elements of a complex picture of the Company's situation with respect to science in Britain and British India. The point is, however, that since the Court of Directors sat at the head of a company-state that monopolized all trade (including information) and travel (including scholars and scientists) between Britain and Asia, whatever the Court of Directors' forms of engagement with science (or not) will have had a significant impact on cultures of science in both Britain and British India. I hope to show that not only was there a much richer and more consequential engagement with science within the Company but also that its imprint upon the shape of scientific culture in Britain was much more significant than we have so far assumed.

Recent work has begun to uncover how, even beyond its impact on Britain's political economy, the Company was deeply intertwined in British culture and society.[19] Its employees included specialized

[16] Binnema, Theodore. *Enlightened Zeal: The Hudson's Bay Company and Scientific Networks, 1670–1870*. University of Toronto Press, 2014, p. 11.

[17] Cohn refers to the library and museum at India House as a collection of "what would become the popular relics of British conquest of India." Cohn, Bernard S. *Colonialism and Its Forms of Knowledge: The British in India*. Princeton University Press, 1996, pp. 87, 104.

[18] Jasanoff, Maya. *Edge of Empire: Lives, Culture, and Conquest in the East, 1750–1850*. Knopf Doubleday Publishing Group, 2007, p. 104. It's worth noting that similar perspectives have sometimes been taken on the Company's colleges and botanical gardens in British India. For example, Adrian Thomas argues that the scientific interests of the early directors of the Botanical Gardens in Calcutta had to disguise their research in language of "public utility" to appease the Directors. Thomas, Adrian P. "The Establishment of Calcutta Botanic Garden: Plant Transfer, Science and the East India Company, 1786–1806." *Journal of the Royal Asiatic Society* 16, no. 2 (2006): 165–177, p. 174.

[19] See especially Finn, Margot and Kate Smith. *The East India Company at Home, 1757–1857*. UCL Press, 2018; Makepeace, Margaret. *The East India Company's London Workers: Management of the Warehouse Labourers, 1800–1858*. Boydell Press, 2010; Saville-Smith, Kay J. *Provincial Society and Empire: The Cumbrian Counties and the East Indies, 1680–1829*. Boydell Press, 2018; Quilley, Geoffrey. *British Art and the East India Company*. Boydell Press, 2020. Also see Ehrlich, Joshua. *The East India Company and the Politics of Knowledge*. Cambridge University Press, 2023.

knowledge workers – librarians, curators, naturalists, surgeons, hydrographers, professors – who worked for the Company but whose professional worlds also intersected with other societies and institutions of science and education around Britain. The sheer size of the Company ensured that, even within the civic, amateur culture of science in nineteenth-century Britain, it was an important source not only of knowledge resources but also of labor and infrastructural support, and an uncountable number of other matters at hand.[20] And the Company's internal practices of information management and knowledge production have a complex, varied history, as Huw Bowen, Christopher Bayly and others have shown. This study is indebted to this scholarship and grapples with the British side, and the science dimension, of what Christopher Bayly called the colonial information order.[21]

In what follows, I use the term "knowledge resources" to describe the huge variety of materials being accumulated in the private, corporate and public collections that were growing at this time. The term reflects the fact that such collections were most often presented as aiming to produce useful knowledge, very broadly defined. In fact, the materials being accumulated were put into all kinds of different uses, such as being sold for cash or being given as gifts or bribes. And significant amounts were apparently never used for anything while in the Company's possession, remaining unpacked in warehouses or cellars for decades.

This book also uses the term "Company science" to capture several distinctions. First, I use it to indicate the narrower focus of this book on natural, philosophical and historical sciences at India House. This study has not attempted to encompass the accumulation and management of resources related to accounting, finances, governance and corporate management; that is, what T. R. Malthus, in his lectures at the Company's college, called "the branch of the science of a statesman or legislator."[22] Importantly, "Company science" at India House was connected to, and embedded within, the rest of the Company administration, and in this book I have tried to follow many of the connections between

[20] Equally important, as Geoffrey Quilley has argued for a different cultural setting, given the sheer size of the Company as a political and economic force in Britain and British India, even the ways in which the Company did *not* engage, or declined to support, or remained silent, or was inconsistent and impulsive would ultimately have an impact on the cultures and practices of science. See Quilley. *British Art and the East India Company*.

[21] It is not a global history of science, for all the reasons laid out in Sivasundaram, Sujit. "Sciences and the Global: On Methods, Questions, and Theory." *Isis* 101, no. 1 (March 2010): 146–158; Bayly, *Empire and Information*; Watt, James. *British Orientalisms, 1759–1835*. Cambridge University Press, 2019; Finn and Smith. *The East India Company at Home, 1757–1857*; Makepeace. *The East India Company's London Workers*; Quilley. *British Art and the East India Company*.

[22] Quoted in Gascoigne. *Science and the State*. p. 90.

the library-museum and the "science of a statesman" that occupied the Court of Directors and many other committees. But this work does not come close to dealing with the many places beyond the library-museum and colleges in which "science," understood in its broad premodern forms, can be found at work within India House.[23]

I also use "Company science" to refer to this study's focus on a subset of the much broader domain of colonial science. Many important recent studies have examined the rich, multicultural context within which colonial science was practiced in the Asian territories under Company influence or control.[24] The term "colonial science" has been defined in different ways: in terms of methods (i.e. "field science," "data collection"); in terms of use or aims (i.e. information produced and deployed as part of the ideological fabric of imperialism); or simply as shorthand for the complex world of scientific production and consumption within colonized territories.[25] Here, "Company science" refers to a particular colonial relationship of knowledge ownership and management: projects directly funded by the Company and material considered to be directly owned by the Company, and therefore

[23] The most in-depth study of the administrative practices at India House is Bowen. *The Business of Empire*. For the case of British India, see Bayly. *Empire and Information*. For two excellent examples of following natural knowledge through administrative practices, see Colpitts, George. "Knowing Nature in the Business Records of the Hudson's Bay Company, 1670–1840." *Business History* 59, no. 7 (October 3, 2017): 1054–1080; Bellenoit, Hayden J. *The Formation of the Colonial State in India: Scribes, Paper and Taxes, 1760–1860*. Routledge, 2017, chapter 4.

[24] For scholarship on science under the Company, I am especially indebted to Arnold, David. *Colonizing the Body: State Medicine and Epidemic Disease in Nineteenth-Century India*. University of California Press, 1993; Damodaran, Vinita, Anna Winterbottom and Alan Lester, eds. *The East India Company and the Natural World*. Palgrave Macmillan, 2015; Edney, Matthew H. *Mapping an Empire: The Geographical Construction of British India, 1765–1843*. University of Chicago Press, 1997; Grove, Richard H. *Green Imperialism: Colonial Expansion, Tropical Island Edens and the Origins of Environmentalism, 1600–1860*. Cambridge University Press, 1996; Ogborn, Miles. *Indian Ink Script and Print in the Making of the English East India Company*. University of Chicago Press, 2007; Winterbottom, Anna. *Hybrid Knowledge in the Early East India Company World*. Palgrave Macmillan, 2015. For the case of science and colonial corporations, an excellent summary is in Winterbottom, Anna. "Science." In *The Corporation as a Protagonist in Global History, c. 1550–1750*, edited by William A. Pettigrew and David Veever. Brill, 2019, pp. 232–254. Also see Harris, Steven J. "Long-Distance Corporations, Big Sciences, and the Geography of Knowledge." *Configurations* 6, no. 2 (1998): 269–304. For the Dutch case, see Cook, Harold J. *Matters of Exchange: Commerce, Medicine, and Science in the Dutch Golden Age*. Yale University Press, 2007. For science at the Hudson's Bay Company, see Binnema. *Enlightened Zeal*. A classic study of science as part of the political and economic apparatus of colonial pre-revolutionary France is McClellan (III), James Edward and François Regourd. *The Colonial Machine: French Science and Overseas Expansion in the Old Regime*. Brepols, 2011.

[25] A useful discussion on the definitions of colonial science is in Binnema. *Enlightened Zeal*, pp. 17–18. Also see Menon, Minakshi. "Indigenous Knowledges and Colonial Sciences in South Asia." *South Asian History and Culture* 13, no. 1 (January 2, 2022): 1–18.

institutionally part of the colonial state. Company science at India House is connected to, and dependent upon, Company science in the colonial governments in British India, the district offices and the moving military frontier at the edge of the Company's territory. It was constructed between Asia and the home country, but, as part of the colonial political economy, accumulation and management became concentrated back in Britain.

In Part I, I show that the Company's forms of engagement with scientific knowledge production changed significantly in response to wider political economic changes. The Company moved, over the course of 150 years, from largely outsourcing scientific and orientalist expertise needs to, by around 1800, attempting to centralize control over the accumulation and management, within its London headquarters, of a wide range of this kind of knowledge and expertise. Part II follows the expansion and impact of the Company's institutions of science in Britain up to the Rebellion of 1857 and the abrupt abolition of the Company. I argue that the Company's monopoly privileges, and their decay over this period, would leave a distinct impression on the shape of the Company's science. Company resources would feed directly into the growth of Britain's second scientific revolution in several key ways. Perhaps most importantly, as the Company itself became nationalized, and as its library and museum collections became Crown property, Company science fed directly and materially into Britain's growing public museum movement. As we will see, in this period in which the basic categories of "science" and "empire" began to take on distinctly modern terms, part of those developments involved the absorption of "Company science" into "public science."[26]

By clarifying the place of the East India Company in the history of science in Britain in this period, I hope to have opened up a new perspective on the long, varied and critically important historical connections between markets, states and modes of knowledge production. The Company played a specific and peculiar role in how and why, by the end of the nineteenth century, Britain and a few other regions of Europe had obtained a firm grip on much of the global business of knowledge production and management.

[26] Mark Harrison unpacks the "science/empire" issue in Harrison, Mark. "Science and the British Empire." *Isis* 96, no. 1 (March 2005): 56–63, p. 56. As Caroline Cornish and Felix Driver have recently argued with respect to the Royal Botanical Gardens at Kew, state science itself was changed as the Company was abolished and absorbed into the state. Cornish, Caroline and Felix Driver. "'Specimens Distributed': The Circulation of Objects from Kew's Museum of Economic Botany, 1847–1914." *Journal of the History of Collections* 32, no. 2 (August 8, 2020): 327–340. For two sustained arguments about the interdependence of states and corporations in the making of modern science, see Beckert, Sven. *Empire of Cotton: A Global History*. Vintage Books, 2015; Mazzucato, Mariana. *The Entrepreneurial State: Debunking Public vs. Private Sector Myths*. Public Affairs, 2015.

Part I

The Making of Company Science, 1600–1813

1 Science under the Company before Company Science

A Mathematical Lecturer for the City of London

One of the least-celebrated consequences of Spain's failed attempt to invade Britain in 1588 was the establishment of the first public lecture series on the mathematical sciences in England. The new "Mathematical Lecturer to the City of London" post was filled by Thomas Hood, an instrument maker and author. The son of a member of the Merchant Taylors' Guild, Hood stood at the center of the overlapping worlds of commerce and the sciences in early modern London. The guild, primarily involved in importing and exporting cloth, was then one of the wealthiest groups of incorporated tradesmen in the city. Hood likely was educated at the Merchant Taylors' School (*f.* 1561), which, in a period when numeracy was very rare, regularly taught arithmetic. He then went on to become the author of a series of textbooks on mathematics, astronomy and navigation. Tasked with helping to bring the riches of long-distance trade better within England's reach, his lectures and publications introduced and popularized the use of terrestrial and celestial globes. In 1598, he invented a calculating instrument, later known as the Hood sector, that reduced the laborious work of calculating logarithms (useful for both commerce and navigation) to simple addition and subtraction.[1]

Hood gave his first lecture on November 4, 1588 in the Gracechurch Street house of the wealthy guild member Thomas Smythe. Smythe was prominent among a consortium of City merchants who, with the formal support of the Privy Council, had joined in the funding of the new post. Smythe also had colonial ambitions: with his friend Sir Walter Raleigh he was involved in the Virginia Company, and was also a founder of the Muscovy Company and a shareholder in the Somers Isles (Bermuda) Company. And, in 1600, when Queen Elizabeth first granted an

[1] In that same year, Galileo invented a similar instrument. Highton, Hester. "Thomas Hood (bap. 1566, d. 1620)." In *Oxford Dictionary of National Biography*. Oxford University Press, 2004. See also Johnston, Stephen. "Mathematical Practitioners and Instruments in Elizabethan England." *Annals of Science* 48 (1991): 319–344.

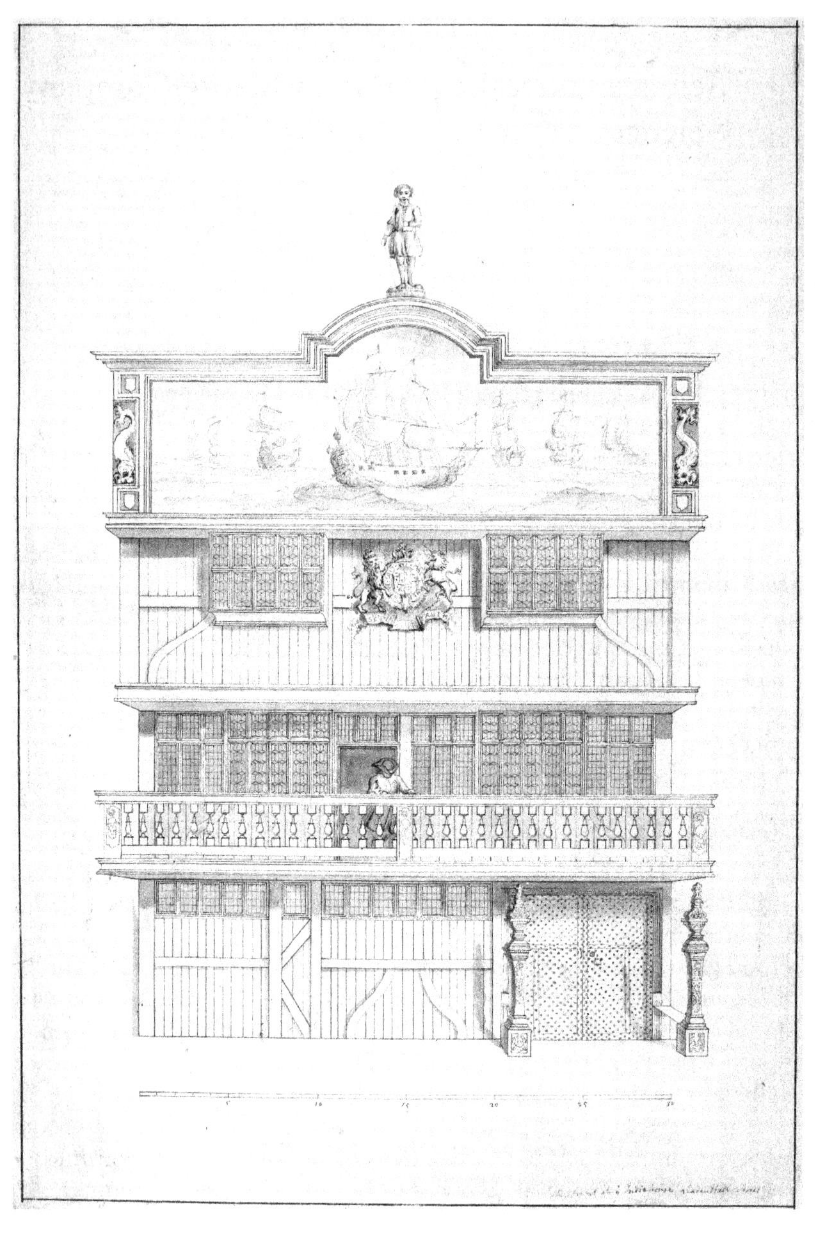

Figure 1.1 Drawing of the old India House on Leadenhall Street, 1628–1746, by George Vertue. Copyright British Library Board (asset WD 1341).

exclusive charter to the newly formed "Company of Merchants of London, Trading to the East," Smythe would be elected the first governor of the East India Company, a position he would keep (except for a few years) until 1623.

London-based merchants, mariners and traders, including those involved in the early East India Company, played a critical role in the development of the sciences in early modern Britain.[2] But while long-distance trade and attempts at colonization had always engaged with many branches of science, the structures that connected science to commerce would take various forms.

This chapter introduces the early East India Company and its modes of engaging with the sciences before the mid eighteenth century. Two aspects of science and the early modern Company are emphasized. First, before 1757, the Company generally contracted out many of the navigational, historical, medical, mathematical and other areas of technical expertise that supported and were supported by overseas trade. As an institution, the Company did directly own and manage a vast amount of information related to logistics, regulations and accounting. However, although the Company also depended upon technical and scientific expertise, it did not directly fund, manage or organize the other branches of science upon which its operations depended. There were exceptions: the Hood lectureship is one of several important instances in which the Company or its close associates would directly fund or otherwise support research or education in the sciences. Thus, in this period, and following a general pattern of early modern "contractor states," science generally grew and developed *under* the Company, if not *at* the Company. And, as the second part of the chapter explains, science *under* the Company found space to grow by way of the peculiar structure and organization of Company trading. The Company's allowance for malfeasance under the so-called private trade would be especially important to the growth of the curiosity and manuscript trade between Britain and Asia in this period.

[2] For example, the first English translation of Euclid's *Elements of Geometrie* (1570) was produced by haberdasher and alderman (and later mayor of London) Henry Billingsley. Six years later, the first English treatment of that keystone of the scientific revolution, Copernicus's *De Revolutionibus* (1543), was authored by Thomas Digges, a Member of Parliament and army officer. Digges's discussion of Copernicus's new heliocentric theory came not in a scholarly text but in an appendix to a widely circulated and popular mariner's almanac, *A Prognostication Everlasting* (1576). See Taylor, Eva Germaine Rimington. *The Mathematical Practitioners of Tudor and Stuart England.* Institute of Navigation at the Cambridge University Press, 1970; Hadden, Richard W. *On the Shoulders of Merchants: Exchange and the Mathematical Conception of Nature in Early Modern Europe.* SUNY Press, 1994.

Figure 1.2 East India House after a re-facing in 1726. *Illustrated London News*, August 30, 1890. Courtesy of Cornell University Library.

Science and the Company before 1757

The Company was organized with the aim of entering the lucrative maritime spice and drug trade, then centered in Sumatra, Java and neighboring islands. Much earlier in the 1500s, first the Portuguese and then the Dutch had established regular and profitable ocean commerce with several kingdoms and port towns along the Indonesian archipelago, and as this trade flourished it threatened the profits of the English Levant Company's overland Mediterranean trade in spice and other commodities from the East. The charter granted by Elizabeth in 1600 to the "Company of Merchants Trading to the East" gave this group a monopoly on English trade beyond the Cape of Good Hope (as it was then known) but only for a limited time; the monopoly was to be renegotiated every twenty years. On the Company's third voyage in 1607/8 a treaty was negotiated with the Mughal emperor, granting the Company trading rights at the port of Surat. This was not, however, an exclusive treaty and the Portuguese effectively dominated the Surat trade with Europe. This would change in 1619 when, after the Company's fleet defeated the Portuguese, the Mughal court granted the Company exclusive trading rights. By the 1620s, the East India Company had given up on the spice trade but had begun to discover that the European market for Indian textiles had unplumbed depths.

The Company was divided into the Court of Committees and the General Court, or Court of Proprietors. The General Court was made up of shareholders with holdings above a certain threshold. The Court of Committees (later Court of Directors) was made up of a governor (later chairman), a deputy-governor and twenty-four directors, each of whom was the head of an administrative committee. The governor and the directors were elected by ballot in the General Court.

Until well into the seventeenth century, Company investors were intensely aware of their relative insignificance and inexperience on the stage of inter-oceanic commerce, especially to any region beyond the transatlantic circuit. The informational foundations of England's early attempts at transoceanic trade thus began in large part as a process of capturing and translating sources from their Iberian and Dutch trading rivals (a practice that, as we will see, would continue over several centuries). When assessing the opportunities and dangers in investing in trade to the East, early investors turned to the encyclopedic and compiling "cosmographers" such as the chaplain and geographer Richard Hakluyt and the astrologer-mathematician and antiquary John Dee. Hakluyt had managed to acquire rare and valuable Portuguese, Spanish and Dutch travel accounts.[3] The Iberian powers closely guarded much of their information, but recent English naval successes against the Portuguese had provided information-rich plunder, and Hakluyt also gained access to Spanish and French sources while chaplain to an English diplomatic mission to Paris in 1585. In 1601, after being appointed as an advisor to the newly formed East India Company, Hakluyt produced an English translation of a Portuguese manuscript compendium of "the different and astounding routes by which in times gone pepper and spice came from India to our parts ... up to the year 1555."[4] The *Discoveries of the World*, as Hakluyt titled it, gave vital information on not only sea routes but also friendly and hostile ports, as well as legal-historical documentation relating to rights of conquest and discovery. "The work," announced Hakluyt, "though small in bulk

[3] On Hakluyt, see Carey, Daniel and Claire Jowitt, eds. *Richard Hakluyt and Travel Writing in Early Modern Europe*. Routledge, 2016.

[4] Hakluyt, Richard. *The Discoveries of the World [...]*, 1601. Recent reprint: Galvano, António. *Discoveries of the World: From Their First Original Unto the Year of Our Lord 1555*. Cambridge University Press, 2010.

containeth so much rare and profitable matter, as I know not where to seeke the like, within so narrow and street [straight] a compass."[5]

In seeking material to translate and compile, Hakluyt sometimes turned to the library of John Dee. Dee was a great "informer" of his day, an advisor to the Crown and colonial adventurers, and his library was rumored to be one of the largest in England, especially rich in accounts of travel and exploration.[6] From the 1570s onwards, Dee had been using his library to promote English economic and territorial expansion, drawing on his antiquarian collecting to make various arguments for British extraterritorial claims. His most ambitious work, *The Brytysh Monarchy*, was the first to use the term "British Impire," and it also made the argument (based on Welsh folklore about transatlantic voyages) for Elizabeth to be titled "Queen of the New World."[7]

The collecting, translating, copying and publishing activities of antiquarian geographers such as Dee and Hakluyt were the routes through which the would-be English colonists and adventurers gathered both critical intelligence and an ideological-legal justification for their projects in the earliest years of English colonialism. While profiting from granting the Crown and various colonial companies access to their private collections, both Dee and Hakluyt also argued that the English nation desperately needed its own repository. Hakluyt argued in 1587, for example, that the English Crown should also "collect in orderly fashion the maritime records of our own countrymen, now lying scattered and neglected, and ... bring them to the light of day in a worthy guise, to the end that posterity ... may at last be inspired to seize the opportunity offered to them of playing a worthy part."[8] Neither the English Crown nor the companies themselves attempted to form a centralized repository for information until nearly 200 years later.

Instead, as the colonial companies – the Virginia Company, the East India Company and the Hudson's Bay Company, to name a few – expanded their operations over the next century, this model of relying on the technical and informational resources of advisors would remain. From its earliest days, Company servants (i.e. employees) were engaged in a wide range of early modern sciences, but they tended to collect and (sometimes) publish as individuals, not for the Company. Wherever

[5] Hakluyt. *The Discoveries of the World [...]*, dedication.
[6] Hill, Christopher. *Intellectual Origins of the English Revolution – Revisited*. Clarendon Press, 1997, p. 18.
[7] Armitage, David. *The Ideological Origins of the British Empire*. Cambridge University Press, 2000, pp. 46–47.
[8] "Hakluyt, Richard (1552?–1616), Geographer." In *Oxford Dictionary of National Biography*. https://doi.org/10.1093/ref:odnb/11892.

Company-hired ships landed, captains and crew voraciously sought information that might help them get home with a profitable cargo or provide future advantage over their European trading rivals.

By the early eighteenth century, the Company managed to establish a more solid presence in Asia, still only in a string of fortified ports on the subcontinent and in the Malay archipelago. Backed by a growing military, and as the political and economic stakes of the Company's monopoly trade continued to grow, the Company became more deeply entangled in domains of knowledge upon which its operations always depended: knowledge of the science of navigation and of maritime defense, of fortification and surveying, and of the societies and natures within which its businesses were located. European foreigners now had to forge relationships with locals – both political or religious elites and workers of many kinds – through which it might be possible to gain insight into areas useful to the colonial project. In history, medicine, botany, agriculture, manufacturing, natural history, arts and crafts, linguistics, law, and many other fields, manuscripts and informants (such as local doctors, teachers and guides) were sought and their knowledge sometimes copied, appropriated, hybridized or even repressed by the foreigners. After a series of wars in 1757, the Company became a dominant territorial power, stepping in to take the reins of the Mughal Empire's centuries-old systems of governance. It was now a mammoth task of both appropriation and invention for Company servants to gain even a partial understanding of the land, languages, laws, and religious and civil structures in the societies that it was purportedly now governing. All the while, at the level of the Company's efforts to know itself and manage the many distant moving parts of which it was constituted, within the headquarters at India House a vast paper trail of colonial expansion was growing continuously.

But that bureaucracy – unlike, as we will see, in other European colonial administrations – did not yet extend to formally organizing, managing and producing what we would now call "scientific" or "historical" knowledge. Beyond its correspondence, accounting and finance records, and the (increasingly important) judicial and legislative records, the Company itself did not directly engage in collecting or natural knowledge resource storage and handling. Instead, until the late eighteenth century, the directors "outsourced," or contracted out, much of the knowledge required for long-distance trade. Beyond the account books and correspondence, the one policy the Company did have that related to archives and information (possibly on Hakluyt's advice) was that each ship captain had to deposit a ship's logbook – where daily recordings of distance logs, observations for latitude and longitude (and, after 1791, chronometer readings), and other measures and comments would be entered in a standard form – with the directors of the Company.[9]

Cartography and geography, for example, were initially managed not by any Company office but by the ship owners and, even more, by the captains hired separately for each voyage.[10] Generally, it was the ship captains who maintained their own chart collections as part of their set of navigational instruments. Captains, in turn, relied upon London's thriving commercial market in navigational knowledge throughout the seventeenth century and well into the 1750s. The Thameside Chartmakers, a branch of the Draper's Company, supplied both the captains hired by the Company and those of the Royal Navy with much of the charts, maps and plans used in commercial exploration and navigation.[11] Until well into the eighteenth century in Britain, hydrographical information, like charts and maps in general, were not produced, managed or controlled by the state. London was also a leading European center for the manufacture of the practical mathematical instruments depended upon by astronomers, navigators and surveyors hired by the Company.[12] The same outsourcing pattern held for the Company's surgeons and naturalists, who were required to purchase their own medicines, books and instruments. In England, Company surgeons decided what medicines to bring and purchased them with their own funds from any apothecary they chose. Similarly, Company writers (as the entry-level positions were

[9] Miller, David Philip. "Longitude Networks on Land and Sea: The East India Company and Longitude Measurement 'in the Wild', 1770–1840." In *Navigational Enterprises in Europe and Its Empires, 1730–1850,* edited by Rebekah Higgitt and Richard Dunn. Palgrave Macmillan, 2016, pp. 223–247; May, William Edward. "The Log-Books Used by Ships of the East India Company." *The Journal of Navigation* 27, no. 1 (January 1974): 116–118.

[10] The classic work on the shipping arm of the Company is Sutton, Jean. *Lords of the East: The East India Company and Its Ships*. Conway Maritime Press, 1981. Also see Chaudhuri, K. N. "The English East India Company's Shipping (c. 1670–1760)." In *Ships, Sailors and Spices: East India Companies and Their Shipping in the 16th, 17th and 18th Centuries,* edited by Jaap Bruijn and Femme Gaastra. NEHA, 1993, p. 64.

[11] Verner, Coolie. "John Seller and the Chart Trade in Seventeenth-Century England." In *The Compleat Plattmaker: Essays on Chart, Map, and Globe Making in England in the Seventeenth and Eighteenth Centuries*, edited by Norman Thrower and Joseph William. University of California Press, 1978, pp. 127–158. A detailed study of the business of mapmaking is in Pedley, Mary Sponberg. *The Commerce of Cartography: Making and Marketing Maps in Eighteenth-Century France and England*. University of Chicago Press, 2005. For the Dutch comparison, see Sutton, Elizabeth A. *Capitalism and Cartography in the Dutch Golden Age*. University of Chicago Press, 2015.

[12] See Bennett, Jim and Rebekah Higgitt. "London 1600–1800: Communities of Natural Knowledge and Artificial Practice." *The British Journal for the History of Science* 52, no. 2 (June 2019): 183–196; Baker, Alexi. "'Scientific' Instruments and Networks of Craft and Commerce in Early Modern London." In *Cities and Solidarities: Urban Communities in Pre-Modern Europe*, edited by Justin Colson and Arie van Steensel. Taylor & Francis, 2017, pp. 245–274.

called) or factors (merchants) wishing to learn the foreign languages or other skills useful for trade and diplomacy were on their own until the late eighteenth century, when outgoing writers were granted a "munshi's allowance" to hire tutors once they arrived in India.

At times the Company did directly fund or patronize experiments, publications or expeditions. For example, the directors gave free passage to the future Astronomer Royal Edmund Halley on his expedition to St. Helena to produce a chart of the southern stars in 1676.[13] Early in the seventeenth century, the Company, jointly with the Muscovy Company, funded expeditions in search of the Northwest Passage, and in the eighteenth century it supported the 1761 transit of Venus expeditions. The directors also gave periodic support to the Royal Society, for example by donating to its collections.[14] Generally, however, there was little in the way of Company-owned or produced science. Instead, the knowledge and technical skill that underpinned maritime commerce was something the Company rented or hired via the employment of individuals who themselves owned and possessed the relevant skills, knowledge and technology. To be clear: the Company was also itself the most important "contractee" of the Crown; in effect, the Company's monopoly was a way for the Crown to contract out the project of long-distance trade and colonization. The Company managed England's militarized trade with Asia and it, in turn, provided the Crown with revenue from import and export taxes, as well as lump-sum payments and loans from the Company.

None of the Company's imperial rivals contracted out so much of their information management. The medical, nautical, commercial, cartographic and archival information in the Iberian empires was highly centralized. Portugal dominated the Eastern trade for the entire sixteenth and much of the seventeenth centuries. From its first overseas expansion in North Africa in the early 1400s, Portugal had by 1550 established a disparate network of maritime trading ports that stretched from Japan through Southeast Asia to Goa, the Caribbean and South America. The *Casa da India* was its headquarters in Lisbon and the *Armazém da Guiné e Índias* was the shipyard where all training, shipbuilding and management of maritime supplies, including maps and instruments, was organized.

[13] See Grove, A. T. "St Helena as a Microcosm of the East India Company World." In *The East India Company and the Natural World*, edited by Vinita Damodaran, Anna Winterbottom and Alan Lester. Palgrave Studies in World Environmental History. Palgrave Macmillan, 2015, pp. 249–269.

[14] See, for example, Brown, Samuel and James Petiver. "An Account of Part of a Collection of Curious Plants and Drugs, Lately Given to the Royal Society by the East India Company." *Philosophical Transactions (1683–1775)* 22 (1700): 579–594.

Navigational knowledge was directed by the *cosmógrafo-mor* (chief cosmographer) and a group of pilots and scholars. The chief cosmographer's duties, according to a 1592 *Regiment*, included examining and rating makers of nautical instruments and charts and "authenticating" all charts, globes and maps. His office was also in charge of training future pilots in mathematics, astronomy and cosmography. All navigational information was kept in the strictest secrecy, including the officially sanctioned map for use by the pilots, the *Padrão Real*. The position was kept in this form until 1779, when it was completely reformed.[15]

In Spain, the Casa de Contratación (House of Trade) in Seville had since the early 1500s been the center of administration and information collection and production, including navigational, medical and natural philosophical works. The charts and surveys that went into the production of the *Padrón Real*, Spain's version of the *Padrão Real*, were kept under equally strict rules of secrecy. In the 1580s, there was a push to reform and extend the medical, navigational, geographical and cartographical information being collected at the Casa. In stark contrast to the case of Britain, in Spain all drugs used and dispensed in the colonies had to be tested and approved by a royal apothecary.[16] By the 1580s the Council of the Indies had developed a systematic and relatively homogeneous process of information gathering, based around what came to be known as the *relaciones* (geographical accounts). The *relaciones* were the returned answers to a standard-issue questionnaire produced by the Council of the Indies. By 1730 the questionnaire, sent to all parts of Spanish America, had grown to 435 questions. The *relaciones* were being collected with the intention of producing, at some point, edited authoritative editions. But, as Daniela Bleichmar argues, when the replies arrived from across the Atlantic, they were generally put directly into the state archives, where they remained, and "failed to become the basis of government action."[17]

France, whose empire was second in size only to Spain during the seventeenth century, had begun to organize many branches of science and medicine hierarchically under the state from the time of Louis XIV. France's "colonial machine," as James McClellan and François Regourd have called it, became, during the eighteenth century, inseparable from

[15] Bleichmar, Daniela, ed. *Science in the Spanish and Portuguese Empires, 1500–1800*. Stanford University Press, 2008, pp. 39–40. Also see Barrera-Osorio, Antonio. *Experiencing Nature: The Spanish American Empire and the Early Scientific Revolution*. University of Texas Press, 2006.

[16] Barrera, Antonio. "Local Herbs, Global Medicines: Commerce, Knowledge, and Commodities in Spanish America." In *Merchants and Marvels: Commerce, Science, and Art in Early Modern Europe*, edited by Pamela H. Smith and Paula Findlen. Routledge, 2002, pp. 163–182.

[17] Bleichmar. *Science in the Spanish and Portuguese Empires*, pp. 13–15.

the state's institutions of science.[18] Under the king and the Ministry of the Navy and the Colonies, the Depôt des Carts et Plans and the Observatoire Royale managed astronomy and cartography for the colonial fleets. Royal naval hospitals and medical schools developed techniques and trained surgeons that were sent to the colonies. The Jardin du Roi and other state-run gardens and agricultural societies were clearing houses for botany and natural history. Under the Académie des Sciences, the network of scientific correspondence reached the colonies and supported the publication and exchange of information. Science and empire were part of one vast state enterprise, in stark contrast to the case of Britain.

The Dutch Republic's Verenigde Oost-Indische Compagnie (VOC) was, in contrast, similar in many ways to Britain and the East India Company, although the VOC was more closely tied to the state. Another major difference was that the VOC built and owned all of its ships, whereas after the 1660s the Company generally hired out ships for individual voyages; only in India after 1800 did the Company start to build some of its own ships (and these mostly stayed in Asia). Also, while the Dutch Secretariat at The Hague was in many ways the center of navigational and trade information management, the Dutch also established a major hydrographic office and map seller in Batavia (Jakarta) in 1650.[19] Otherwise regarded as one of the least secretive or protective trading companies, the VOC did sometimes petition the state for patents of protection on maps, plans and other resources related to the Eastern trade. Generally, Dutch cartography and other colonial publications circulated widely, especially in England.[20]

The flipside of the East India Company's outsourcing of knowledge management was that those who the Company hired were largely free to profit from knowledge or information gained while under the employ of the Company. By the early decades of the eighteenth century, there was a robust market in travel accounts and histories of English (and increasingly Scottish) seamen and traders. Captains regularly published accounts of voyages, including diaries, routes and charts, hoping to defray

[18] McClellan and Regourd. *The Colonial Machine*. A useful summary of their argument for the "colonial machine" metaphor is in McClellan, James E. and François Regourd. "The Colonial Machine: French Science and Colonization in the *Ancien Regime*." *Osiris* 15 (2000): 31–50.

[19] Zuidervaart, Huib J. and Rob H. van Gent. "'A Bare Outpost of Learned European Culture on the Edge of the Jungles of Java': Johan Maurits Mohr (1716–1775) and the Emergence of Instrumental and Institutional Science in Dutch Colonial Indonesia." *Isis* 95, no. 1 (2004): 1–33.

[20] Davids, Karel. "Public Knowledge and Common Secrets. Secrecy and Its Limits in the Early-Modern Netherlands." *Early Science and Medicine* 10, no. 3 (2005): 411–427. Also see Sutton, Elizabeth A. *Capitalism and Cartography in the Dutch Golden Age*. University of Chicago Press, 2015.

some of the costs of their voyages in this way.[21] For the same reason, returned surgeons printed herbals, natural histories or their own travel accounts. Often the Company, or a group of directors, would contribute to the publication by agreeing in advance to purchase (i.e. subscribe to) a certain number of books.[22]

This model of decentralized resource management at the Company was not limited to the domain of knowledge; it appears to fit neatly within a well-established structural form that was especially prominent in the early modern English state. Some historians have suggested that this model was the precursor to the modern nation state. Arguing against earlier understanding of early modern states as unorganized and ineffective, and relatively unimportant compared to the pace of individual enterprise, John Brewer, for example, shows that states were in fact highly adept at the central function of raising armies. But this ability, in turn, depended upon being able to procure food, clothing, transport and weaponry, all of which, in *its* turn, required collaboration with private enterprise. The Royal Navy, for example, was in the period a major purchaser and consumer of goods but was not itself a producer. Instead, it contracted out to private firms everything from victualling to shipbuilding to gunsmithing.[23] The contractor mode of state enterprise may also be applied to the provisioning of information and knowledge at the early modern Company. It captures the knowledge management practices of the Company up to at least the Seven Years' War. As we will see in the next sections, that model began to change toward the end of the eighteenth century, as the Company began to accumulate its own stores of knowledge resources. The timing of this shift can also be aligned with a broader historical pattern away from the contractor model and toward the centralization of such functions under state offices.[24]

[21] On the cultural and scientific influence of travel literature in the eighteenth century, see Mary Louise Pratt's classic *Imperial Eyes: Travel Writing and Transculturation*. Routledge, 2007. Also see Driver, Felix and Luciana Martins. *Tropical Visions in an Age of Empire*. University of Chicago Press, 2005; Thell, Anne M. *Minds in Motion: Imagining Empiricism in Eighteenth-Century British Travel Literature*. Bucknell University Press, 2017.

[22] Winterbottom, Anna. "Producing and Using the Historical Relation of Ceylon: Robert Knox, the East India Company and the Royal Society." *The British Journal for the History of Science* 42, no. 4 (December 2009): 515–538.

[23] See, for example, Harding, Richard and Sergio Solbes Ferri. *The Contractor State and Its Implications, 1659–1815*. Universidad de Las Palmas de Gran Canaria, 2012; Sanchez, Rafael Torres. *Military Entrepreneurs and the Spanish Contractor State in the Eighteenth Century*. Oxford University Press, 2016. This more recent work builds on John Brewer's introduction of the idea of the fiscal-military state and the role of contractors in Brewer, John. *The Sinews of Power: War, Money, and the English State, 1688–1783*. Knopf, 1989.

[24] Harling, Philip and Peter Mandler. "From 'Fiscal-Military' State to Laissez-Faire State, 1760–1850." *Journal of British Studies* 32, no. 1 (1993): 44–70.

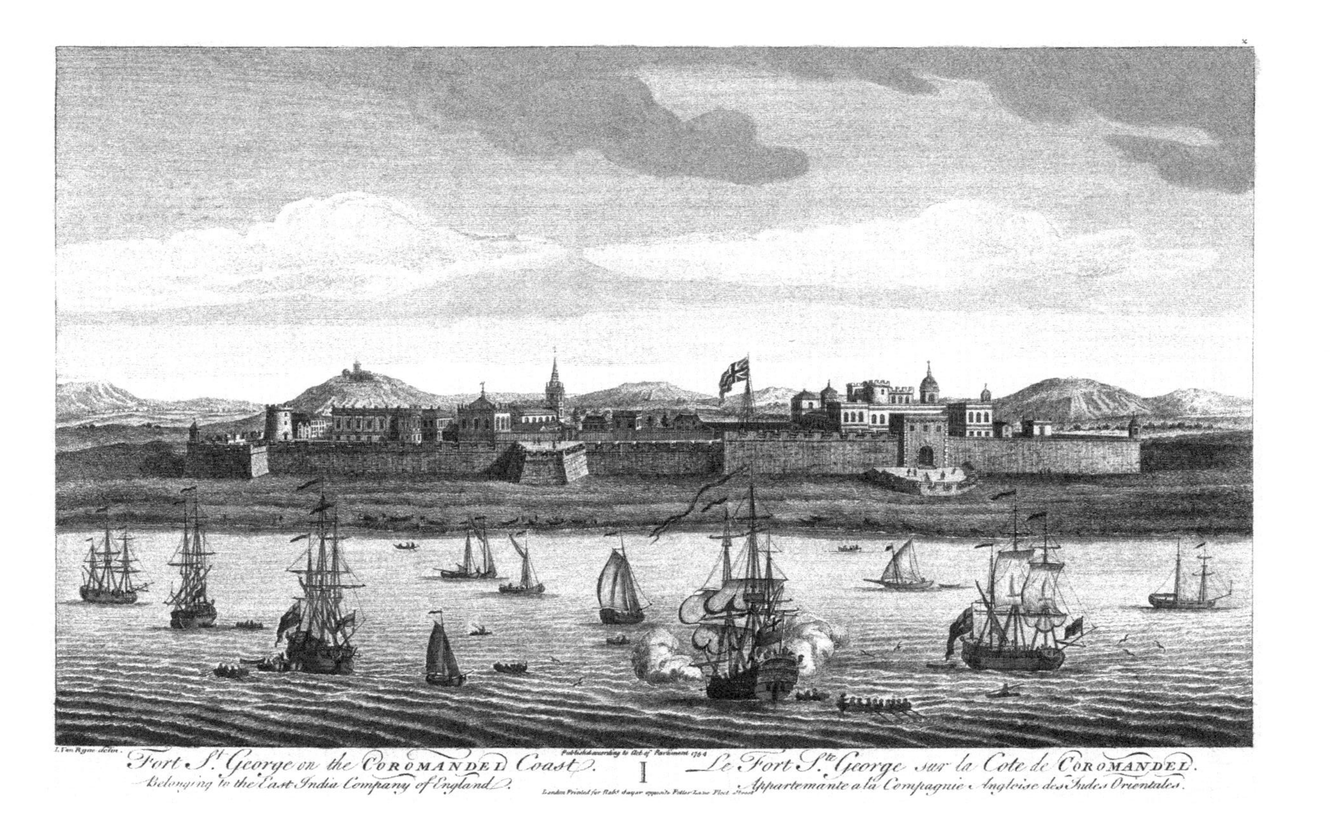

Figure 1.3 Fort St. George, Madras in the mid eighteenth century. From a print by Jan Van Ryne (1712–1760); photo by Ken Welsh/Universal History Archive/Universal Images Group via Getty Images.

Collecting and the Internal Free Trade

Long-distance trade and associated new commodity regimes were essential to a wide range of practitioners, practices, institutions and cultures that came to define the new sciences of early modern Europe. Histories of collecting and natural history under the Company in the early modern period have often focused on individual collectors and their networks of correspondence and exchange. For good reason, many historians have organized their examinations of seventeenth- and eighteenth-century knowledge and empire around a series of case studies of individual collectors or naturalists. In the early modern period, it was precisely at the level of individual agency that science under the Company was primarily organized (or unorganized) at this time. Emily Erikson has argued that it was this decentralized organization of the Company that fostered a robust information exchange in general: "When the English Company had a decentralized organizational structure, which is to say that significant autonomy lay in the hands of employees, social networks encouraged the transmission of local information and led to the incorporation of more ports and goods into the English trade network."[25] It was also within this decentralized mode of practice that a vibrant culture of "collecting" and natural history grew in Britain and its colonies in Asia.

Underneath the umbrella of the East India Company's monopoly on all trade east of the Cape of Good Hope, there was a vibrant ecosystem of private trade conducted by individuals for their own profit. Some were illegal "interlopers"; that is, British individuals conducting trade between Britain and Asia without permission of the Company. But the bulk of the private trade, which has been estimated to be anywhere from 10 percent to 50 percent of the total British Eastern trade, was legal and under license from the Company. Ship captains and other officers were a great beneficiary of the legal private trade, or, as Erikson calls it, the "internal free trade market" under the Company monopoly.[26] The vast majority of individual income from a voyage would come from this. In 1740, for example, according to K. N. Chaudhuri, a commander's salary might be £10, but he might be expected to make between £5,000 and £10,000

[25] Erikson, Emily. *Between Monopoly and Free Trade: The English East India Company, 1600–1757*. Princeton University Press, 2014, p. 107. Erikson and her collaborators further argue that the exchange and management of information at the level of individual traders in pursuit of private trading opportunities was critical to the Company's extraordinary longevity. See Erikson, Emily and Peter Bearman. "Malfeasance and the Foundations for Global Trade: The Structure of English Trade in the East Indies, 1601–1833." *American Journal of Sociology* 112, no. 1 (2006): 195–230; Erikson, Emily and Sampsa Samila. "Networks, Institutions, and Uncertainty: Information Exchange in Early-Modern Markets." *The Journal of Economic History*, no. 4 (2018): 1034.

[26] See Erikson. *Between Monopoly and Free Trade*.

with his allotted space for his own capital goods if in command of a large ship.[27] Some individuals were licensed to run private trading firms in the inter-Asian trade (this, for example, is how the majority of the opium trade with China worked). Critically, agents and servants of the Company were also, until the 1760s, allowed to accept personal gifts in the course of their duties. Within all ranks of the Company, from surgeons to governors, collecting in one way or another had long been common. In the period before territorial and wartime expansion under the Company, much of this collecting would be conducted on a small scale, for personal use, as piece-by-piece sale or barter, or – in the cities and towns in the East – by visiting the local markets or bazaars. Thus factors, governors and other servants based in the colonies were allowed to engage in a certain amount of direct trade between Asia and Britain; cargo holds, for example, contained space reserved for the "private trade," which also had its own warehouse back in East London.

Together with the practice of contracting out knowledge management (described in more detail in Chapter 3), malfeasance encouraged a robust growth of collecting *under* the Company, if not really *at* the Company. Importantly, this is not to say the Company did not attempt to monopolize aspects of the knowledge trade or control the direction of information flows or of technology. As a corporation of shareholders, vulnerable to stock-price dips related to rumors of poor crops or lost cargoes, those involved in core businesses of shipping and trade had always tried in various ways to control the means of private communication to and from Asia, as evident in the many rules and regulations regarding personal correspondence. Miles Ogborn, Huw Bowen and others have documented the persistent belief among many directors in London that a shadowy, subterranean network was at work in their territories, attempting to undermine the authority and interests of India House.[28] Company employees had a great deal of autonomy relative to other trading companies, and the threat of competition from other British groups was real. Bowen's study of the arms and instruments trade between Britain and the subcontinent shows, for example, that the Company couldn't even stop English companies from providing arms to its own enemies.[29] Thus, as we will see, the Company certainly severely restricted who could travel to and collect information within its territories. Some governors were also notorious for trying to censor the press in India, shutting down English-language presses in Calcutta and Bombay, and even attempting to keep

[27] Chaudhuri. "The East India Company's Shipping (c. 1670–1760)," p. 66.
[28] Ogborn. *Indian Ink*, p. 98; Bowen. *The Business of Empire*.
[29] Bowen, Huw V. "Trading with the Enemy: British Private Trade and Supply of Arms to India, 1750–1850." In Harding and Ferri. *The Contractor State*, pp. 32–53.

printing technology out of the hands of local rulers, most famously in explicitly forbidding its servants to provide printing presses.[30] Even if, relative to other European empires, there were less restrictions on information and communication, the Company was very far from overtly pursuing a "free market" in knowledge.

Unlike textiles, spices, raw materials and other commodities, manuscripts and curiosities were not sold directly by the Company at the auction rooms within India House. Instead, this material passed directly into private hands. From the Americas to China, European traders could easily supplement their primary trade with natural and artificial curiosities as well as highly valued drawings and paintings of local flora, fauna and scenes produced specifically for European collectors.[31] Apothecaries, alchemists, physicians and gardeners were key purchasers of exotica from abroad. Already by the mid seventeenth century, the curiosity market was so well established that a regular global trade could form around popular items for cabinets or *wunderkammer*, such as the "most rare and precious Commodity" of the "Teeth or Horns of the fishes called Sea-Unicorns" (narwhals), which according to one captain in 1656 were sought across Europe for the "Closets of the Curious."[32] By the mid seventeenth century, London was awash in natural and artificial curiosities.[33] Fascinating evidence of its scope is found in a remarkable archive of records of the purchases made by the merchant and Barbados investor William Courten (1642–1702). Courten bought items from more than eighty individuals, many of them trading within walking distance of his rooms in London's Middle Temple.

Out of this commerce would emerge the collection that would become the first public museum in England: the Tradescant's museum. It was put together by London gardeners and plant merchants. John Tradescant and his son formed their collection by way of their status as semiofficial buyers for the royal gardens. They formed the Musaeum Tradescantium, which from the 1630s was open for viewing at a building christened "The Ark" in Lambeth. The museum was just one of many sites where seventeenth-century Londoners might come into contact with displays of collections

[30] Nair, Savithri Preetha. "'... Of Real Use to the People': The Tanjore Printing Press and the Spread of Useful Knowledge." *The Indian Economic and Social History Review* 48, no. 4 (2011): 497–529.

[31] See, for example, Fan, Fa-ti. *British Naturalists in Qing China: Science, Empire, and Cultural Encounter*. Harvard University Press, 2004.

[32] Rochefort, Charles-César, comte de. *The History of the Caribby-Islands ... With a Caribbian Vocabulary / Rendered into English by John Davies ....* London, 1666, pp. 116–118.

[33] For an overview, see, for example, Arnold, Ken. *Cabinets for the Curious: Looking Back at Early English Museums*. Ashgate, 2006.

or exotica from abroad. Traveling shows, outdoor exhibitions and even some public houses, such as Don Saltero's coffee-house, all put on exhibitions of curiosities, usually for a small price.[34] According to its Royal Charter, when the Royal Society was established in 1662, a key reason for organizing such an institution was the pressing need for a central repository or "storehouse" of information resources.[35] And the scope and scale of these early modern museums, society collections and public exhibitions were matched and often exceeded by those of colonialists and plantation owners (such as Hans Sloane) or the Company servants (such as Elihu Yale and Josiah Child) who made huge personal fortunes in India.[36]

In Asia, collections were also expanding (and sometimes dissolving) in the context of increasing European presence. It was essential for would-be traders to bring out to the East items from Europe or elsewhere that could be offered as gifts in exchange for gaining commercial preferences and trading rights or bartered for other items. The social conventions of commercial diplomacy required the exchange of gifts and presents; the gift exchange was fundamental to the formation of British trading networks within Asia.[37] It was therefore essential for Company agents to cultivate an understanding of the collecting interests of the local elite, who were as avid collectors of curiosities and exotica as their European counterparts.[38] The factors at the Company's ports in early seventeenth-century India reported constantly on the kind and quantity of gifts required. "Something or other, though not worth two shillings, must be

[34] Altick, Richard Daniel. *The Shows of London*. Harvard University Press, 1978; Stewart, Larry. "Other Centres of Calculation, Or, Where the Royal Society Didn't Count: Commerce, Coffee-Houses and Natural Philosophy in Early Modern London." *British Journal for the History of Science* 32, no. 2 (1999): 133–153; Qureshi, Sadiah. *Peoples on Parade: Exhibitions, Empire, and Anthropology in Nineteenth Century Britain*. University of Chicago Press, 2011.

[35] Thomas, Jennifer. "Compiling 'God's Great Book [of] Universal Nature': The Royal Society's Collecting Strategies." *Journal of the History of Collections* 23, no. 1 (May 1, 2011): 1–13.

[36] Terrall, Mary and Adriana Craciun, eds. *Curious Encounters: Voyaging, Collecting, and Making Knowledge in the Long Eighteenth Century*. University of Toronto Press, 2019; Delbourgo, James. *Collecting the World: Hans Sloane and the Origins of the British Museum*. The Belknap Press of Harvard University Press, 2017.

[37] See, for example, Subrahmanyam, Sanjay. "Frank Submissions: The Company and the Mughals between Sir Thomas Roe and Sir William Norris." In *The Worlds of the East India Company*, edited by H. V. Bowen, Lincoln Margarette and Nigel Rigby. D. S. Brewer, 2002, pp. 69–96.

[38] See, for example, Winterbottom. *Hybrid Knowledge*, pp. 27–28 for examples from the subcontinent. Also see Subrahmanyam, Sanjay. *Europe's India: Words, People, Empires, 1500–1800*. Harvard University Press, 2017. For the court of Siam's collection of European artifacts, see Hodges, Ian. "Western Science in Siam." *Osiris* 13 (1998): 80–95.

presented every eight days," writes the chief factor at Ajmere:[39] "The Great Mogul was exceedingly delighted with anything strange .... Rich gloves, embroidered caps, purses, looking and drinking glasses, curious pictures, knives, striking clocks ... if [you have] a jack to roast meat on, I think he would like it, or any toy of new invention."

The governor of Surat requested a long list of items to be used as gifts, including "two suits of armour, swords" and live animals, preferably "mastiffs, greyhounds, spaniels, and little dogs."[40] Sir Thomas Roe asked for "pictures well-wrought, those of France, Germany, Flanders, &c. being fittest for that purpose." On another occasion, the Company sent as a gift for a Mughal ruler "a coach and horses, with a coachman who had been in the service of the Bishop of Lichfield, to drive the coach."[41] Likewise, Company servants also brought home a constant stream of gifts of rare or valuable curiosities for the royals of England. Jewels, artworks and live animals were especially popular among the royalty.[42] Such exchanges had long been a fundamental aspect of diplomacy and commerce.

*

Like all European trading empires in the early modern era, the successful expansion of Company influence in Asia depended upon a great deal of social, scientific and technical expertise. Unlike many of its rivals, however, much of the knowledge and expertise upon which the English East India Company depended was highly decentralized, with the production and management of natural and technical knowledge generally contracted out to the surgeons, ship captains and factors hired by the Company. The Company's unique formalization of malfeasance – the "internal free trade" – further supported a system whereby a great deal of scholarship and collecting was done *under* the Company, but not *by* the Company. To be sure, some of the manuscripts, curiosities and works of art acquired by Britons in Asia – whether gifts, purchases or plunder – had found their way to the Company's headquarters in India House since the seventeenth century. Although no evidence describing a cabinet or *wunderkammer* at India House has yet been found, a few references to curiosities and a more substantial collection do exist. For example, in the spring of 1667, during his monthlong tour of England, Prince Cosimo III

[39] This and the remaining examples are described in Sainsbury, W. Noel, ed. *Calendar of State Papers Colonial, East Indies, China and Japan, Volume 2, 1513–1616*. Her Majesty's Stationery Office, 1864, pp. vii–lxxvii.

[40] Sainsbury. *Calendar of State Papers Colonial*, pp. vii–lxxvii.

[41] Sainsbury. *Calendar of State Papers Colonial*, pp. vii–lxxvii.

[42] See Grigson, Caroline. *Menagerie: The History of Exotic Animals in England, 1100–1837*. Oxford University Press, 2016.

of Tuscany visited India House, which was "full of rare and curious things, both animal and vegetable ... which came from India, and are kept here to gratify the curiosity of the public."[43] There is also evidence, as Anna Winterbottom has shown, of interaction during this period between the Company's collections and the early Royal Society of London (est. 1663), which had its own cabinet, and which offered financial encouragement to Company servants who would collect for the Royal Society's repository.[44] In these and other ways, even while collecting under the Company was restricted to private trade, India House was one of the sites where privately collected materials, as well as gifts offered to the directors or the Company itself, were stored and displayed. In the later eighteenth century, however, as the Company gained in political and economic standing on the subcontinent, the nature of the India House collections would begin to change.

Clearly, the opening of the library and museum at India House in 1801 did not by any means mark the start of collecting by Company servants or the first accumulation of curiosities and other artifacts at India House. What it did mark, however, as we will see, is a key moment in a structural reorganization of the sciences within the Company; in other words, a new relationship between the Company as an institution and the knowledge and expertise upon which its operations depended. But the beginning of that reorganization, as we will see in the next chapter, would be rooted first and foremost in a changed political economy within North India, after the Company's first major defeat of the Mughals and the subsequent destabilization of the old networks of education, scholarship and expertise that had grown up around the Mughal courts.

[43] Magalotti, Lorenzo. *Travels of Cosmo the Third, Grand Duke of Tuscany, through the England during the Reign of King Charles the Second (1669)* .... J. Mawman, 1821.

[44] Winterbottom. *Hybrid Knowledge*, p. 28.

2 The Roots of Company Science in Asia

Tipu Sultan's Plunder

In an early phase of the third Anglo-Mysore wars in September 1790, the Mysorean army under the command of Tipu Sultan very nearly overwhelmed a contingent of the Company's forces. The Company's army was forced to quickly retreat to Coimbatore, leaving roughly 500 of their own dead and many of their supplies behind. Among these, Tipu's men found trunk-loads of books, instruments and manuscripts. Those that were deemed valuable or useful were moved to the palace library at Seringapatam, where they were carefully shelved. Some manuscript works of science were considered so valuable as to be rebound in red Moroccan leather decorated with gold tooling.

Thus it was that when, a few years later, the royal palace at Seringapatam was stormed by British troops and Tipu's famous library was seized by the prize agents, some Company officers found their own lost journals beautifully preserved in the palace collections. Francis Buchanan (later Buchanan Hamilton), a naturalist and surveyor who in the spring of 1800 was sent out on a mission to conduct a survey of the Company's latest territorial acquisition, was one such officer who discovered his work had been preserved at the Royal Library. When he arrived in Seringapatam, he was greeted with a notebook of his own that he had lost over a decade earlier. As Buchanan explains in a note he later wrote on the frontispiece.

> These notes were taken by me at the Botanical Garden Edinburgh in summer 1780 [actually 1781]. In a voyage to India in 1785 Mr. Boiswell, then my mate and who remained in the country, had by mistake put them up in his trunk and lost them at the affair near Satimangulum where they were taken by Tippoo and by him bound up in their present form. At the taking of Seringapatam they fell into the hands of Major Ogg who has restored them to me.[1]

[1] Reprinted in Watson, Mark F. and Henry J. Noltie. "Career, Collections, Reports and Publications of Dr Francis Buchanan (Later Hamilton), 1762–1829: Natural History Studies in Nepal, Burma (Myanmar), Bangladesh and India. Part 1." *Annals of Science* 73, no. 4 (October 1, 2016): 392–424, p. 408.

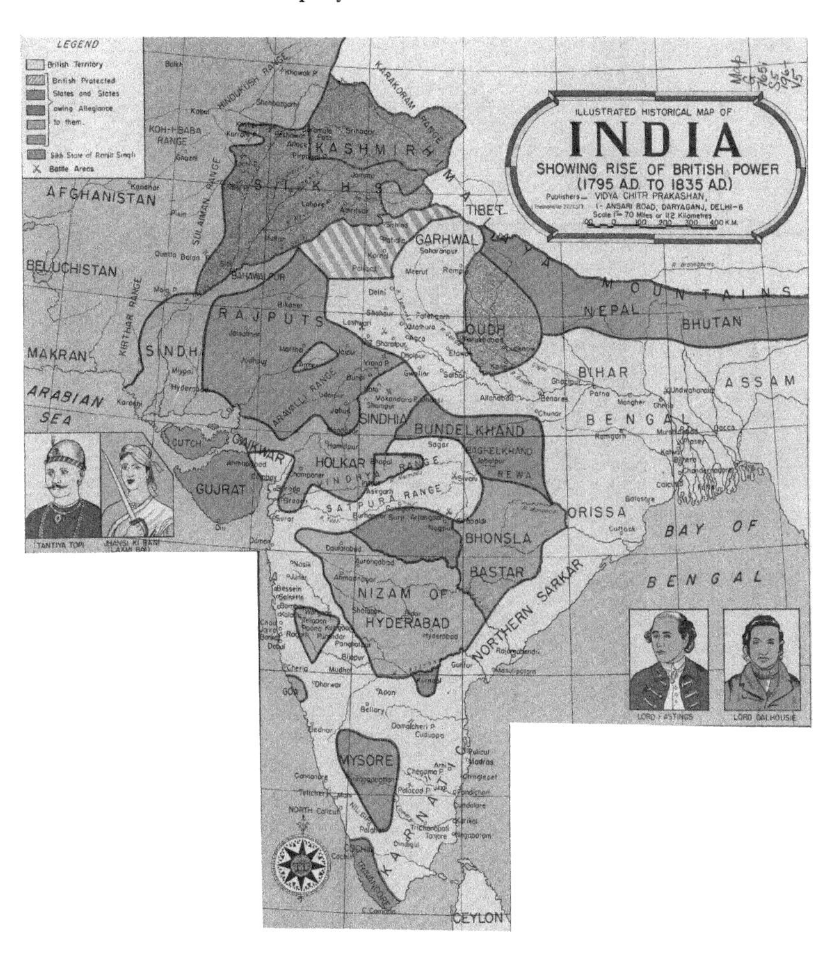

Figure 2.1 Map of East India Company territory in Asia, 1795–1835. Image courtesy of Vidya Chitr Prakashan, New Delhi.

Some years later, Buchanan would return to Britain with not only his journal (which eventually ended up back at the library of the Royal Botanic Garden Edinburgh) but also a major collection of specimens and records collected during his survey of Mysore. And now it was Tipu's personal writings and favorite books – including one of his copies of the Quran and a journal in which the sultan recorded his dreams – that would be rebound and carefully preserved in his enemy's library. Eventually, the majority of Tipu's library would also, after a bitter internal struggle between the directors and the governor-general, be sent back to India House (see Figure 2.2).

Figure 2.2 The Old Court House and Writer's Building, Calcutta. By William Daniell, 1787.

This mutual capturing and recapturing of book collections between Mysore and the British is just one small example of the ways in which the capture of cultural property, including knowledge resources, was always, on all sides, a part of wartime plundering and looting. For centuries, Mughal powers on the subcontinent had, through war and other means, accumulated great collections.[2] By the mid eighteenth century, however, patterns of accumulation would begin to shift toward European collections. This was a period in which, as recent studies have shown, in the wealthy, increasingly cosmopolitan towns on the subcontinent, foreigners were able to study and collect only through establishing relationships with locals willing to be intermediaries and knowledge brokers.[3]

In this chapter, I trace some of the wider political and economic changes that would allow foreigners, and particularly the British, to increasingly access and engage with the existing world of collecting, education and the sciences on the subcontinent. The result would be a slowing of the growth of resources in Indian centers such as Seringapatam and an acceleration of the growth of individual European-owned collections. The chapter begins by exploring changes in the patterns of accumulation that accompanied the conquest of Bengal. Here, I focus on the early careers of several Company servants who would eventually bring significant collections to Britain: Robert Orme, Alexander Dalrymple and Charles Wilkins. Each of these individuals would play an important role in the establishment of Company science back in Britain. And each, in their modes and methods of acquiring collections of knowledge resources from Asia, illustrate the debt that the growth of British resources would owe, in this period, to two major factors: wartime conventions of looting and plundering and (in consequence of the wartime upheaval) deepening social and political interaction between the foreign and local elite.[4] The final section of this chapter follows the Company's first steps toward moving from contracted-out to Company-owned science, with new spaces and

[2] Gahtan, Maia, Alessandro Nova and Eva-Maria Troelenberg, eds. *Collecting and Empires: The Impact of Empires on Collections and Museums from Antiquity to the Present*. Harvey Miller Publishers, 2019. For the Mughal and Indian context, see in this volume Koch, Ebba. "Jahangir's Hazelnut and Shah Jahan's Chini Khana: The Collections of the Mughal Emperors" and Guha-Thakurta, Tapati. "The Object Flows of Empire: Cross-Cultural Collecting in Early Colonial India," pp. 134–161; 218–237.

[3] Schaffer, Simon, Lissa L. Roberts, Kapil Raj and James Delbourgo. *The Brokered World: Go-betweens and Global Intelligence, 1770–1820*. Science History Publications, 2009; Raj, Kapil. *Relocating Modern Science: Circulation and the Construction of Knowledge in South Asia and Europe, 1650–1900*. Springer, 2007.

[4] Sandholtz, Wayne. *Prohibiting Plunder: How Norms Change*. Oxford University Press, 2007.

institutions of knowledge management being established in the wake of major land reforms in the 1790s.

Disaster Orientalism and Private Accumulation

In the India House library and museum's carefully kept day books, November 20, 1801 marks the first material to be deposited in the new collections: John Corse's presentation of "three Elephant heads with several detached parts intended to illustrate the natural history of those animals, so far as relates to their curious mode of Dentition."[5] Three days later Charles Wilkins, the librarian, deposited a copy of Corse's 1789 *Philosophical Transactions* essay on "Asiatic Elephants." On that day, Wilkins also deposited one of his own works, a catalog of "Sanskrita manuscripts presented to the Royal Society by Sir William and Lady Jones."[6] Into the stores on that day also came a "Persian manuscript traced on Oil Paper" with the English title "Mogul History" and six brass statues of Hindu deities.[7] These were presented to the library by John Roberts, one of the directors of the Company.

A week later, the first large deposit was recorded: fifty-seven volumes of printed material relating to the history of the Company; twenty-three volumes of manuscripts on the subject of India; thirty-seven rolls of maps and plans; thirty-five books of maps, plans and views; and four portfolios of maps.[8] This would turn out to be a large part of the collections gathered by Robert Orme who had been hired as the Company's historiographer in 1769. The Company's India House collections started, like most large collections, by absorbing and reordering other existing collections, particularly those privately held collections discussed in the previous chapter. And Orme's material was itself a collection of collections. Initially, his material was purchased at bazaars in India where local antiquities traders were often dealing in broken-up family collections. His collection also contained manuscript copies of Company records. And a great deal of his books, maps and manuscripts were acquired by "right of

[5] Sivasundaram, Sujit. "Trading Knowledge: The East India Company's Elephants in India and Britain." *The Historical Journal* 48, no. 1 (March 2005): 27–63.

[6] Jones's manuscripts would eventually be deposited at the India Office in 1872 and are now in the British Library. See Lawrence, Jonathan. "Building a Library: The Arabic and Persian Manuscript Collection of Sir William Jones." *Journal of the Royal Asiatic Society* 31, no. 1 (January 2021): 1–70.

[7] "*Siva* mounted on his bull; *Bhawam* standing upon a lion; *Ganesa* the God of Prudence and [Policy] and the offspring of Siva; *Hanuman* the monkey who attended *Rama* in his war against *Ravana* the Tyrant of Ceylon; a female figure supposed to be the Goddess *Saraswate*; Another figure supposed to be *Ganga*." BL MSS EUR F/303 1 (November 23, 1801).

[8] BL MSS EUR F303/1 (November 24 to December 2, 1801).

conquest"; that is, plundering, in wars with the Mughals, the French and the Spanish. Orme's collection, as we will see, was also an individual project (i.e. not a Company project) and was gathered (even when taken from various state archives) and used as his own private property. Eventually it would become an important founding collection for the India House library, and its origins and movements – like those of close contemporaries discussed later – illustrate some of the key steps in the Company's transition from a renter or contractor of knowledge resources to a manager and, eventually, producer of the same.

Robert Orme was born in the kingdom of Travancore (now in Kerala) in 1728, joining the Company at the age of fifteen when he first returned to Fort St. George from his English boarding school years at Harrow. This was the beginning of a period of near-constant territorial skirmishes between the French Company and the English Company, going back to the War of Austrian Succession (1740–1748), and engaging a range of Indian states that aligned themselves with either France or Britain. Orme rose steadily through the ranks of the Company from individual writer to factor to, by 1755, senior member of the Madras Council, the top political body in the presidency. Amidst the regular work of a Company writer and factor, such as managing correspondence and negotiating trade agreements, Orme had also been working his way through the largely unorganized records held at the Madras presidency. By the mid eighteenth century, the Company's operations at Fort William and Fort St. George had been regularized long enough that within the fort a large, largely unexamined archive of Company records had also accumulated. It was this collection that Orme had been probing and copying. Building up his own private version of the corporate archives, he employed local scribes to trace copies of manuscripts on oil paper.[9] Orme had also been collecting manuscripts from many other sources since joining the Company in 1742.[10]

During the early 1750s, Orme shared his collecting and archival passion with another young officer of the Madras administration, the Scotsman Alexander Dalrymple. Dalrymple had joined the Company at the age of fifteen, working first in the factory stores, then under the assay-master at the mint, and then on to various junior positions

[9] Delgoda, Sinharaja Tammita. "'Nabob, Historian and Orientalist.' Robert Orme: The Life and Career of an East India Company Servant (1728–1801)." *Journal of the Royal Asiatic Society* 2, no. 3 (1992): 363–376.

[10] Mantena, Rama Sundari. *The Origins of Modern Historiography in India: Antiquarianism and Philology, 1780–1880.* Palgrave Macmillan, 2012, pp. 39–41. Also see Srinivasachariar, C. S. "Robert Orme and Colin Mackenzie: Two Early Collectors of Manuscripts and Records." *Proceedings of the Indian Historical Records Commission*, Calcutta 6 (January 1924): 84, 89–91.

within governing committees. Meanwhile, as a clerk and secretary, he also had access to the Company records. As he tells it, he had been obsessed with the idea of an as yet undiscovered southern continent since his youth, and had set his sights on following "Magalhanes [Magellan] and Columbus"; thus, once in the Company's employment, "the desire of information" led him to seek out and copy Company records of eastward voyages.[11]

Orme, meanwhile, had risen to the rank of senior administrator in Madras when the Seven Years' War broke out. The decades between the Seven Years' War and the start of the American Revolution in 1772 were the years of the making of what would become the infamous nabob fortunes in India. In 1756, at the outbreak of the war in Europe, long-simmering hostilities between the Company outpost-town of Calcutta and the French-backed nawab of Bengal, Siraj-ud-Daulah, boiled over into armed conflict. The nawab's forces successfully captured Calcutta. Robert Clive was sent as commander of the Madras forces to recapture it. Clive's forces retook Calcutta and marched on to a surprisingly decisive victory over the nawab's forces, and, in the first of a series of direct political interventions, the Company installed one of Siraj-ud-Daulah's generals, Mir Jafar, as the new nawab. This was the beginning of a complex process that would result in the Company becoming politically and economically intertwined with the existing Mughal state. Perhaps most importantly, the Company had now secured significant revenue rights over the vast and wealthy region of present-day Bengal, Bihar and Orissa. After formally taking over the administration of Bengal, over the next several decades the Company increased land taxes, began promoting the cultivation of poppy (for the Eastern opium trade) instead of grains and effectively established control over the grain markets. The attempt to increase revenues from Bengal was in part a response to the Company's deteriorating finances, which had been hit hard by the cost of army expansion and the popularity of cheap smuggled Dutch tea in Britain and the Americas, undercutting their main line of trade revenue.[12] Meanwhile, in Bengal, food shortages beginning in 1768 were exacerbated by a poor annual monsoon season in 1769, and by 1770 the region

[11] Dalrymple, Alexander. *An Historical Collection of the Several Voyages and Discoveries in the South Pacific Ocean*. 1770, p. xxiii. On Dalrymple as a publisher and hydrographer, see the dissertation: Cook, Andrew. "Alexander Dalrymple (1737–1808), Hydrographer to the East India Company and the Admiralty, as Publisher: A Catalogue of Books and Charts." St. Andrews, 1993 (vol. 1) and 2012 (vol. 2).

[12] On the political and economic struggles of the Company in Bengal in this period, see Bowen, H. V. *Revenue and Reform: The Indian Problem in British Politics 1757–1773*. Cambridge University Press, 2002.

was suffering a severe famine, which was followed by an epidemic of smallpox. Up to 10 million, or 30 percent of the population, were killed. During the famine period, even while the Company's overall financial situation continued to deteriorate, the Company's revenues from the region steadily increased.[13] So too did the individual wealth of many Company servants.

Clive became a patron of Orme, and, when he returned to London in 1769, so did Orme, bringing with him a massive private library. Dalrymple, meanwhile, continued to collect hydrographical material from the Madras archives. In 1759, he had obtained permission and support to attempt to discover a new route to China through the Molucca Islands and New Guinea. He would conduct three voyages between 1759 and 1764. Meanwhile, the Asian fronts of the Seven Years' War had progressed. Not only had the Company established itself as the dominant European power in the northern Indian subcontinent and gained territorial control of a large portion of Bengal but it also now held Manila in the Spanish Philippines. The Admiralty and the Company had jointly invaded Manila, and the army plundered the city, invading private homes and burning, inadvertently, much of the state archives. During these voyages, and especially during his time in Manila in 1764, Dalrymple continued to amass his own personal collection.[14] As he later explained: "My peregrinations were of use even in this pursuit [i.e. building a collection]. I acquired amongst the Spaniards, some very valuable papers, and intimations from Spanish Writers, many of whose works [I] also procured."[15]

In this period, beyond the British-ruled territories, many native kingdoms extended their patronage to European naturalists, surgeons and engineers who had managed to gain a reputation on the subcontinent, and at this time Company servants were generally free to offer their employment to local British-aligned rulers. In the northeast, gardens and cabinets of curiosity were flourishing in Lucknow, capital of Awadh

[13] Damodaran, Vinita. "The East India Company, Famine and Ecological Conditions in Eighteenth-Century Bengal." In *The East India Company and the Natural World*, edited by Vinita Damodaran, Anna Winterbottom and Alan Lester. Palgrave Macmillan, 2015, pp. 80–101; Arnold, David. "Hunger in the Garden of Plenty: The Bengal Famine of 1770." In *Dreadful Visitations: Confronting Natural Catastrophe in the Age of Enlightenment*, edited by Alessa Jones. Routledge, 1999, pp. 81–111. Also see Damodaran, Vinita. "Famine in Bengal: A Comparison of the 1770 Famine in Bengal and the 1897 Famine in Chotanagpur." *The Medieval History Journal* 10, no. 1–2 (October 1, 2006): 143–181.

[14] David Routledge argues that in this period a great rupture in the historical record of the Philippines is produced, contributing to its ongoing lack of visibility today. Routledge, David. "The History of the Philippine Islands in the Late Eighteenth Century: Problems and Prospects." *Philippine Studies* 23, no. 1–2 (1975): 36–52.

[15] Dalrymple. *An Historical Collection*, p. xxiii.

(Oudh), under the rule of Nawab Asaf-ud-Daula (ruled 1775–1795).[16] The Oudh Royal Library was particularly famous.[17] The Serampore gardens started out in 1771 as a public garden supported by local rulers.[18] In the south, the nawab of Arcot in the Carnatic (in which Madras was situated) employed in the late 1760s the first of Linnaeus's "disciples" on the subcontinent. This Danish botanist, John Gerard Koenig, in turn became a prominent naturalist and collector and, via the nawab, supported a new generation of Company naturalist-collectors in southern India: James Anderson, physician general at Fort St. George in the 1770s; the physician and collector Patrick Russell; and botanist William Roxburgh (see Figure 2.3).[19] And, adjacent to the Carnatic, in the southern kingdom of Mysore, Sultan Hyder Ali was investing heavily in the growth of engineering, arts and sciences. Through his French allies, he imported European weaponry, instruments, works of art and literature and French military engineers.[20] Koenig eventually became salaried as a botanist by the Company as well after doing survey work in Siam and the Malay Peninsula. Russell would succeed Koenig as Company botanist in the Carnatic and made collections for the Madras government.[21] When Russell returned to Britain in 1790 (leaving his collections with the Madras government), Roxburgh, who had been experimenting with the cultivation of pepper, sugar-cane, coffee and other valuable commodities, was hired to replace him. Four years later, Roxburgh then moved up to Calcutta to direct the Company's new botanical gardens (more on that in the final section of this chapter).

In the same period, a small number of Company servants were becoming immensely wealthy from their personal dealings in India, and an

[16] Jasanoff, Maya. "Collectors of Empire: Objects, Conquests and Imperial Self-Fashioning." *Past & Present* 184 (2004): 109–135.

[17] Sprenger, Aloys. *A Catalogue of the Arabic, Persian and Hindu'sta'ny Manuscripts, of the Libraries of the King of Oudh*. Thomas, 1854.

[18] Sivasundaram, Sujit. "'A Christian Benares': Orientalism, Science and the Serampore Mission of Bengal." *The Indian Economic & Social History Review* 44, no. 2 (April 2007): 111–145.

[19] On Roxburgh, see Menon, Minakshi. "Medicine, Money, and the Making of the East India Company State: William Roxburgh in Madras, c. 1790." In *Histories of Medicine and Healing in the Indian Ocean World: The Medieval and Early Modern Period*, edited by Anna Winterbottom and Facil Tesfaye. Palgrave Macmillan, 2016. pp. 151–178; Desmond, Ray. *The European Discovery of the Indian Flora*. Royal Botanic Gardens, 1992.

[20] On the "proto-modernization" of Mysore under his rule, see Yazdani, Kaveh. "Haidar 'Ali and Tipu Sultan: Mysore's Eighteenth-Century Rulers in Transition." *Itinerario* 38, no. 2 (2014): 101–120.

[21] Desmond. *The European Discovery of the Indian Flora*; "Russell, Patrick (1727–1805), Physician and Naturalist." In *Oxford Dictionary of National Biography*. https://doi.org/10.1093/ref:odnb/24334.

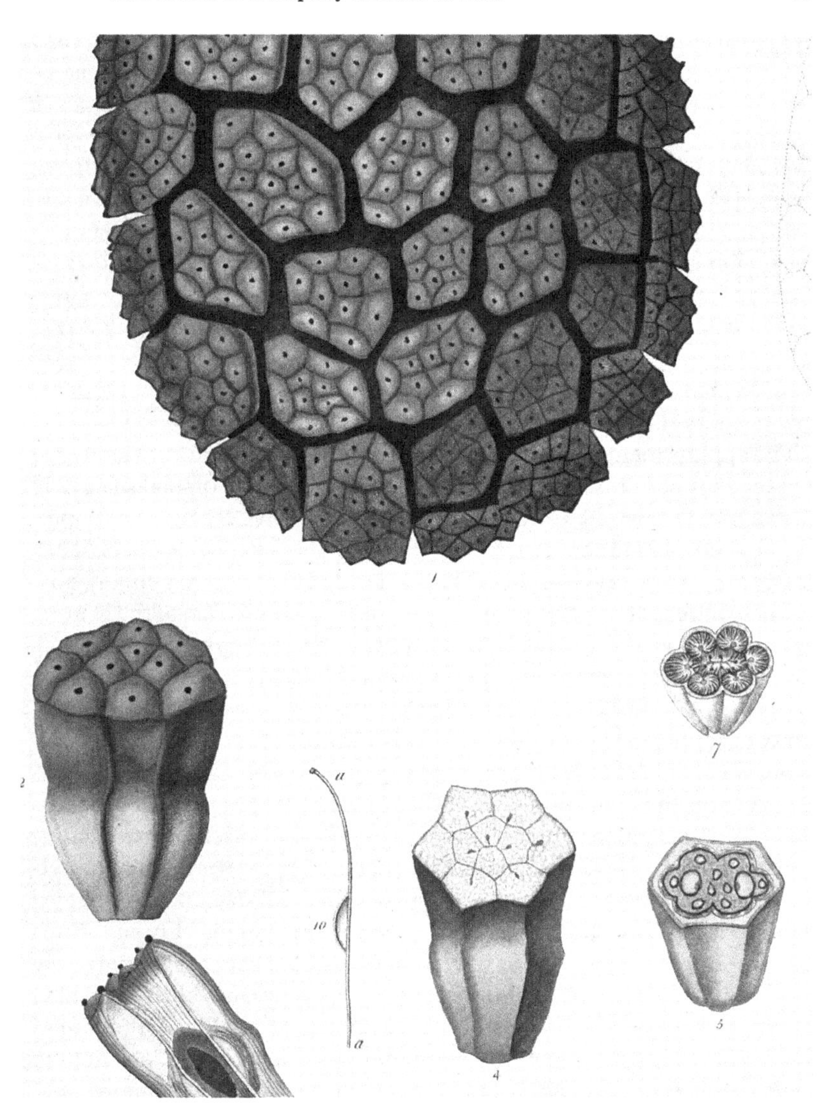

Figure 2.3 *Pandanus Odorifer* (*Pandanus Odorifissimus*), known for its aromatic oil, detail from Roxburgh, William. *Plants of the Coast of Coromandel: Selected from Drawings and Descriptions Presented to the Hon. Court of Directors of the East India Company*, vol. 1. London, 1795, plate 96. From the Biodiversity Heritage Library (www.biodiver sitylibrary.org/item/9711).

increasing number of these "nabobs" were now back in Britain with their riches on full display. At the same time, drought and famine were ravaging Bengal. The Company was in deep financial trouble and was at times in the early 1770s unable to pay the annual customs duties on its imports in London, which in turn was a significant blow to the income of the British government.[22] These and other factors combined to push the Crown to seek increased control over the Company. When the disastrous state of the Company's first decade of territorial rule became fully known in Britain, Parliament acted to impose for the first time a layer of direct Crown control over the Company's activities with the Regulating Act of 1773. The Act banned private trade and imposed a new organization on the Company's administration in India, with a single governor-general now heading the Government of India from Calcutta.

The appointment of Warren Hastings as the first governor-general after the Regulating Act is often taken as a turning point in colonial science in British India under the Company. Hastings promoted a version of imperial rule according to the "native" or "natural" laws already in place. Thus, as Hastings sometimes argued, the route to prosperous Company governance of Bengal must be guided by increasing knowledge of the history, laws and resources of the region, by way of the work of orientalists, surveyors and naturalists. In this mode, Hastings positioned himself as a translator of India for the British and as a "liberal" (i.e. generous) protector and promoter of the arts and culture.[23] With various forms of encouragement from Hastings, a new generation of Company servants also took on the collection and study of Indian language, history, geography and natural resources. These early British orientalists, such as Nathaniel Halhed, William Jones and Charles Wilkins, turned their interest to Sanskrit, Bengali and other languages that had so far been neglected relative to Persian, the language of state for the Mughal Empire. Such changes should not be read, as they sometimes have been, as primarily the result of Hastings's own qualities, or of "Enlightenment values" reaching British India by way of Hastings's patronage. Equally, if not more, important were the wider transformations in the British position in Bengal, which opened new opportunities for knowledge exchange, collecting and scholarship. For one thing, the new offices and administrative positions created in response to expanding British control over territorial revenue also created new sites and situations for British interaction with local scholars and administrators of the Mughal courts.[24] Furthermore,

[22] Bowen. *Revenue and Reform*, p. 39.
[23] Marshall, P. J. "The Making of an Imperial Icon: The Case of Warren Hastings." *The Journal of Imperial and Commonwealth History* 27, no. 3 (1999): 1–16.
[24] See, for example, Travers. *Empires of Complaints*.

Hastings's generation of scholars were perfectly poised to take advantage of the radical cultural and economic upheaval left by the famine.

It is that conjunction of genuine *amateur* orientalism with the brutality of the Company's expansion at the time that is critical to understanding the growth of British orientalism in the period. Before 1757, it had been common for Company servants to complain that the local administrative and learned elite were uninterested in sharing their knowledge. Hastings claimed there had long been a "jealous prejudice" against interlocution with the British, which led them to "guard" their knowledge from foreigners.[25] By the early 1770s, however, with their student numbers decimated and many formerly wealthy patrons now unable to support them, the local intellectual elite increasingly turned to the British for employment. *Pandits* (experts in Persian) and *munshis* (experts in Sanskrit and Hindu jurisprudence) in Bengal at this time were often from long lines of families that had served as scribes, accountants and translators for the Mughal elites.[26] In the kingdoms of southern India, scribal elites played a similar role, and in the late eighteenth and early nineteenth centuries more and more had, out of necessity, shifted to working for the British. There was plenty of work of this kind to be had: in a partial step toward directly supporting officer education, all young men entering the India service in Calcutta or Madras were now granted a "*munshi*'s allowance" with which to hire a teacher for language instruction in their first few years of employment.[27]

Native experts on local laws and religions were also increasingly hired directly by the presidency governments to compile and translate the existing laws and statutes in practice in different regions. Jones, for example, was given approval in 1788 to hire ten native scholars and writers, on competitive salaries, to produce a digest of Hindu laws of inheritance and contracts. And, as Jones explained to the Supreme Council, salaries for these positions would have to be high in order to attract qualified applicants. After the project was underway, one Company administrator noted that it was remarkable Jones had managed to hire enough qualified and willing people, interpreting the change as

[25] Wilkins, Charles and Charles Anthon, eds. *The Bhăgavăt-Gēētā, or, Dialogues of Krēēshnă and Ărjŏŏn: In Eighteen Lectures, with Notes*. Printed for C. Nourse, 1785, p. 24.

[26] See Alam, Muzaffar and Sanjay Subrahmanyam. "The Making of a Munshi." *Comparative Studies of South Asia, Africa and the Middle East* 24, no. 2 (2004): 61–72.

[27] Bowen gives an overview of institutional changes to Company education: Bowen, John. "The East India Company's Education of Its Own Servants." *Journal of the Royal Asiatic Society of Great Britain and Ireland* 3–4 (1955): 105–123. On South India in particular, see Raman, Bhavani. *Document Raj: Writing and Scribes in Early Colonial South India*. University of Chicago Press, 2012; Trautmann, T. R., ed. *The Madras School of Orientalism: Producing Knowledge in Colonial South India*. Oxford University Press, 2009.

a matter of trust: "it may be remarked, as an occurrence of no ordinary nature, that the professors of the Brammanical faith should so far renounce their reserve and distrust, as to submit to the direction of a native of Europe, for compiling a digest of their own laws."[28] Hastings would assert that his own government should take credit, attributing the dissipation of the "jealous prejudice" of the "Brahmans" to the "liberal treatment they have of late years experienced from the mildness of our government."[29] However, the root cause of this new alignment of interests lay not with the "mildness" of Company rule but with a totally transformed political economy of education and knowledge production in the region.

This relatively large-scale integration of the local scholarly elite into the Company's administration in the presidencies marked a new shift in the way information was being accumulated by the British. The collections of the generation before the Seven Years' War were made largely through gift exchanges, the occasional plunder of rival French, Spanish or Dutch ships and purchases and exchanges in bazaars. Those of Orme and Dalrymple's time were formed in similar ways, and in addition sometimes drew on the Company's own records. Now, however, there was an increasingly important interpersonal dimension. Through directly hiring local experts, much new work was being produced. The Company writer and future Member of Parliament Nathaniel Brassey Halhed, for example, employed a group of eleven *pandits* and *munshis* (or "most experienced Lawyers" and "Professors of the Ordinances," as he calls them) to gather materials and produce interpretations and translations for one of the first of the British compilations and translation/interpretations of native legal codes, the *Code of Gentoo Laws* (1778).[30] In his orientalist work, William Jones studied closely with Sanskrit tutors such as Ramlochan and *munshis* such as Bahaman, and when on the bench as a judge he depended on the advice of the court *pandits* Goberdhan Kaul and Ramcharan.[31]

Charles Wilkins, the Company's first curator of the India House library-museum, made his name in British India as a printer and authority on Sanskrit after managing to obtain access to renowned tutors in Benares. From a family of printers, Wilkins went to India as a Company

[28] Quoted in Raj. *Relocating Modern Science*, p. 134.
[29] Wilkins. *The Bhăgavăt-Gēētā*, p. 24.
[30] Halhed, Nathaniel Brassey. *A Code of Gentoo Laws* (1776), p. x. See Jain, Nalini. "Colonial Circuits of Power, Indian Raw Materials: An Analysis of Nathaniel Brassey Halhed's A Code of Gentoo Laws (1776)." *South Asian Review* 26, no. 2 (December 1, 2005): 3–20. See also Rocher, Rosane. *Orientalism, Poetry, and the Millennium: The Checkered Life of Nathaniel Brassey Halhed, 1751–1830*. Motilal Banarsidass, 1983.
[31] Raj. *Relocating Modern Science*, p. 100.

writer and quickly became involved in producing the first native-language typefaces for such clients as the Raja of Tanjore.[32] Wilkins also focused on acquiring local languages and was allowed to remain on leave from his usual duties in Calcutta for a year in Benares, an important seat of Hindu scholarship and on the fringe of Company-controlled territory. Here, he was one of the first Europeans to be allowed to study Sanskrit under Indian Brahmin *pandits*.[33] He worked especially closely with Kasinatha Bhattacharya on a set of transcriptions and translations that would establish his reputation as the first English translator of, and leading authority on, Sanskrit. In this collaborative context, collecting often took the form of copying out texts provided by the teachers, transcribing oral lessons, and making vocabularies, word lists and dictionaries.

During this time, under Kasinatha, Wilkins pieced together a first English translation of one of the central texts of Hinduism, the *Bhagavad Gita*. After Wilkins sent Hastings a manuscript copy of the translation, Hastings secured (without Wilkins's knowledge) the patronage of the Court of Directors to publish the work in London. Wilkins's *Gita* was a commercial success, the first Hindu work of literature to be widely read in Europe, and was sold to the public as "one of the greatest curiosities ever presented to the literary world"; becoming something of a sensation, it was published in French, Russian and German within a decade.[34] Prefacing the work was a letter from Hastings to the chairman of the Company, Nathaniel Smith, which argues for the value ("if not *utility*") of such projects to the interests of the Company: cultural exchanges will bond the British and the people of India in mutual understanding. On the one hand, having Company servants read the *Gita* will improve their virtue and trustworthiness.[35] On the other hand, he suggests that in producing a translation of the *Gita*, the British were demonstrating that – in contrast to the Mughal rulers of old – the new rulers of Bengal were respectful of the knowledge of the "Brahmans." Here, Hastings promotes the value of Indian literature for the English along the lines of what Uday Singh Mehta has termed a "cosmopolitanism of *sentiments*," according to which the route to improving imperial rule is increasing understanding and a meeting of minds between the subjects of Britain and the subjects of British India.[36] To Hastings and his

[32] Nair. "'. . . Of Real Use to the People.'" [33] Sivasundaram. "'A Christian Benares.'"

[34] Wilkins's *Bhăgavăt-Gēētā*, from the advertisement on the frontispiece.

[35] It being "on the virtue, not the ability of their servants that the Company must rely for the permanency of their dominion." Wilkins. *Bhăgavăt-Gēētā*, p. 12.

[36] As opposed to a later "cosmopolitanism of *reason*" that would take hold half a century later under the influence of liberal utilitarianism. Mehta, Uday Singh. *Liberalism and Empire: A Study in Nineteenth-Century British Liberal Thought*. University of Chicago Press, 2018.

generation, nothing could be so useful in creating a bond between Britain and India as an exchange of high art and knowledge.

Importantly for the subject of this chapter, Hastings also valorizes the *very process* of collecting and gathering knowledge as part of the meeting of British and Indian minds. In Hastings's presentation, the social interaction around which the "accumulation" of knowledge occurs tends *in itself* to reduce prejudice and ill-feeling among both subjects and rulers, and both the inhabitants of India and of England:

> Every accumulation of knowledge, and especially such as is obtained by social communication with people over whom we exercise a domain founded on the right of conquest, is useful to the state: it is the gain of humanity: in the specific instance which I have stated, it attracts and conciliates distant affections; it lessens the weight of the chain by which the natives are held in subjection; and it imprints on the hearts of our own countrymen the sense and obligation of benevolence.[37]

The native centers of education, such as where Wilkins was based in Benares, are depicted here as a place of exchange, interaction and growing cultural understanding, a place where knowledge generates respect and respect generates further avenues for exchange.

Of particular relevance to the future creation of the Company's library-museum at India House fifteen years later, Hastings argues that a lessening of prejudice among the *inhabitants of England* is also critically needed, hence the importance of accumulating knowledge of India *back in* Britain:

> It is not very long since the inhabitants of India were considered by many, as creatures scarcely elevated above the degree of savage life; nor, I fear, is that prejudice yet wholly eradicated, though surely abated. Every instance which brings their real character home to observation will impress us with a more generous sense of feeling for their natural rights, and teach us to estimate them by the measure of our own.[38]

In direct contrast to some later Company idealogues, most notably James Mill, Hastings is arguing that it is to the mutual benefit of all that the British public and the people of British India to come to know each other through cultural exchange.

By the mid 1780s, the Company was beginning to bring responsibility for the education of its servants under even more formal control. Hastings had established the Calcutta Madrassa. Meanwhile, Kasinatha remained

[37] Wilkins. *Bhăgavăt-Gēētā*, p. 13.

[38] "But such instance can only be obtained in their writings: and these will survive when the British dominion in India shall have long ceased to exist, and when the sources which it once yielded of wealth and power are lost to remembrance." Wilkins. *Bhăgavăt-Gēētā*, p. 13.

in Benares and became personal *pandit* to the British Resident Jonathan Duncan. Kasinatha convinced Duncan, and ultimately the Court of Directors, to establish a Company-funded "Hindoo College or academy" for the collection and transcription of ancient texts; and for the education of new *pandits* who could go on to serve within the colonial government.

Along with the new *madrassas* and colleges, orientalists and naturalists among the Company's servants also organized themselves, under the patronage of Hastings, into a scholarly society. William Jones's idea in 1784 to establish a scholarly society in Calcutta "to enquire into the arts, sciences and literature of Asia" envisioned a new communal organization for what had long been a fragmented and individualistic pursuit.[39] The Asiatic Society of Bengal met at the Grand Jury Rooms of the Supreme Court, the same court of British-trained judges that had been imposed on the Company by the Regulating Act, where Jones himself (and several other founding members) was a junior judge. The Society had eighty-nine members by 1788, many of whom were drawn from the Company administration. The Court of Directors often extended support to the publications of Society fellows, usually subscribing in advance for a significant number (usually fifty) of books.[40] In these formal and informal ways, both native scholars and Company servants were increasingly given material – if irregular – support by the Company.

The Sciences of the Permanent Settlement

By the time the Company's charter was up for renewal again in 1793, France had just invaded Austria, Louis XIV had been executed and the new Republic had declared all old monarchies its enemy. With France a growing presence on the Indian subcontinent as well, the Company's armies were within the penumbra of potential war. Company profits were healthy, as the recent lowering of taxes on tea in 1784 led to a boom in the China trade, and the Company's tea sales at India House had risen from £6.5 million to £15 million within the last two years. In this political climate, the Company's charter was renewed with minimal debate and only minor changes.[41] In India, meanwhile, the Company had initiated a vast reworking of the land-ownership system in Bengal. The Permanent

[39] Jones quoted in Mitra, Rajendralal. *Centenary Review of the Asiatic Society of Bengal, from 1784 to 1883. Part I. History of the* Society. Thacker, Spink and Co., 1885. Mitra was the first Indian president of the Asiatic Society of Bengal: p. 4.

[40] Mitra. *Centenary Review of the Asiatic Society of Bengal*, p. 10.

[41] One change that was important to the Company's outsourcing policies in general was an act allowing the Company to grant licenses to individuals for inter-Asian trade (for example, in the opium trade between India and China).

Settlement Plan, begun in 1786 and finalized in 1793, aimed to increase land productivity by stimulating technical and infrastructural change in agricultural practices. The regional tax collectors (*zamindari*) were compelled to enter into a "permanent" fixed-rate contract with the Company, on the basis that this would secure more reliable revenue while at the same time encouraging the new landed class to reinvest profits in capital development.

Historians have argued that, while Hastings and the early orientalists such as Wilkins and Jones had been focused on understanding and interpreting India, by the 1790s the idiom of the Board of Control, and of British politics in general, was increasingly one of change and improvement.[42] The Mughal instruments of state that had once been seen as the necessary basis for Company policy were increasingly disregarded as degraded relics. If Company rule was not bound to Mughal traditions, the door to bringing British traditions to the subcontinent was opened.[43] In the Permanent Settlement Act, the Company pursued liberal forms of "improvement" on a vast new scale according to a very British model of a landed class. The reworking of property relations would, it was hoped, create in Bengal a new class of improvement-minded landowners.[44] Decades later, similar thinking would also lead Thomas Munro to institute land-ownership reforms (*ryotwari*), giving direct ownership to cultivators, in the Madras and Bombay presidencies.

In other parts of its empire in the final years of the eighteenth century, the Company's pursuit of "improvement" ranged widely, from small interventions such as the introduction of a new plow in St. Helena, to the production of a history of Indian snakes and poison treatments (produced by Madras naturalist Patrick Russell, whose collections would end up split between Joseph Banks and the Company's museum), to much more ambitious attempts to introduce new cash crops to India.[45] Cash crop projects were one of the most contentious issues relating to

[42] The definition and pursuit of "improvement" became a defining feature of British political culture – both domestic and imperial – from the 1780s well into the later nineteenth century. Briggs, Asa. *The Age of Improvement, 1783–1867*. Longmans, 1960. Ranajit Guha has shown how that same discourse came to be even more pronounced within colonial politics in British India. Guha, Ranajit. *Dominance without Hegemony: History and Power in Colonial India*. Harvard University Press, 1997.

[43] Travers, Robert. *Ideology and Empire in Eighteenth-Century India: The British in Bengal*. Cambridge University Press, 2007, pp. 233–234.

[44] The classic work on this is Guha, Ranajit. *A Rule of Property for Bengal: An Essay on the Idea of Permanent Settlement*. Orient BlackSwan and Permanent Black, 2016 [1963].

[45] Each of these and others are described in the introduction to Gupta, P. C., ed. *Fort William-India House Correspondence and Other Contemporary Papers Relating Therto. (Public Series): Vol. XIII: 1796–1800*. National Archives of India, 1959, pp. 147–149.

science under the Company in this period. Especially in the wake of the loss of the North American colonies, manufacturers and politicians pushed for increasing the production of commodities of critical economic interest to Britain. In this period, the directors tended to act on this matter not by managing such projects directly but rather by offering individuals the chance to develop and profit from a new venture. For example, when Joseph Banks and the Board of Trade argued for the expansion of Indian hemp and flax, both being key materials for rope and cordage, and an absolutely critical naval supply, the Company, in response, granted passage to India for one agricultural projector named George Sinclair, "reputed to be well skilled in the culture and management of hemp and flax according to the most approved methods practiced in Europe."[46] Sinclair had submitted a pamphlet to the directors in which he proposed new methods for growing improved hemp in Bengal. Sinclair was then permitted to proceed to Bengal "for the purpose of ascertaining by experiments to be made *on his own private account* how far his ideas, as detailed in this work, shall appear to be well founded." The dispatch makes clear that although the directors viewed research into hemp production as a matter "of important national advantage," and while they offer Sinclair "every degree of protection that may be needful," they stop short of funding the project: "*We have not deemed it expedient that the Company should be subject to any expence on this account.*"[47] Fully flexing its monopoly on access to Asia's nature, the Company offers only the right to travel and experiment in India and the "protection" of his ventures, in exchange for his self-funded pursuit of new agricultural projects. The Company also offers further encouragement to Sinclair in the form of a guarantee of wide freedom of movement and action. Should the hemp and flax project "fail to effect the improvements suggested," Sinclair is given "liberty to engage in any other" so long as his operations remain strictly within the limits of the Company's territories.[48] Sinclair was initially successful and in 1799 the Company agreed to a large experimental station, still funded by Sinclair, for the production of hemp, flax, and *sunn*, contracted with the Marine Board. When Sinclair died later that year, the Court arranged to pass the contract to Banks, who funded the voyage out to India of six flax growers.[49]

Sometimes the directors developed interests in particular projects or regions from their informants or the pages of the Asiatic Society of

[46] Gupta. *Fort William-India House Correspondence*, letter no. 97.
[47] Gupta. *Fort William-India House Correspondence*, letter no. 97. My emphasis.
[48] Gupta. *Fort William-India House Correspondence*, letter no. 97.
[49] Chambers, Neil. *The Indian and Pacific Correspondence of Sir Joseph Banks, 1768–1820, Volume 5*. Routledge, 2021, p. 26, fn. 7. And see *Volume 5*, letter 21, November 25, 1798.

Bengal's journal. In these cases, again, the only direct action taken by the directors would be an expression of interest or a recommendation sent out to the presidencies, with the funding, personnel and details being left to those governments. For example, in 1797 the directors wrote in a public dispatch of their interest in obtaining more information about "the hill people of Tipperah, Garrow and Rajamhal," about whom they had received some "first outlines" from an overland route journal of one "Ensign Blunt" that made its way into the hands of the Court of Directors. Desiring more information ("to be filled up with many particulars") in order to "serve as the basis of opinion or of measures," the directors "recommended this subject" to the attention of the governor-general. Another few paragraphs give some instructions (though the directors say they do not want to "prescribe minutely the mode or the instruments by which enquiry shall be prosecuted") and stress that all should be done at minimum cost, with only "public utility" (not "private emolument") in mind. The letter then goes on to give a list of "the subjects of investigation" – a list very similar to the "instructions for travelers" written by the Royal Society and other bodies seeking in some way to organize and direct the naturalist or surveyor's attention. Also notable is the suggestion (likely at the behest of one of the Company orientalists back in London) of a comparative study of the languages of the hill regions and those of the plains, which "would probably throw much light upon the origin, perhaps also upon the early history, of both races, and upon other points of curious research." But, again, while signaling support for research into questions of current natural philosophical interest, the directors also make it more than clear that the aim of these broad inquiries is "improvement": to answer the question of "how the condition of these wild people may be improved, how they may be civilized." And improvement of this kind, in turn, is meant to aid the Company in bringing these areas "within the boundaries of the Company's government."[50]

Under an intensifying interest in "improving" the agricultural districts and "civilizing" the hill districts, the directors continued some of the old modes of science patronage by way of irregular individual enterprise. But more change was also coming to the organization of Company science in India. One of the earliest of these was the growing administration surrounding medical practice and hospitals. Back in the early 1760s, the three presidencies had each established Medical Service branches. These were staffed by officers under the management of a head surgeon, and civil and military branches of the service were also established. The

[50] Gupta. *Fort William-India House Correspondence*, letter no. 97.

institutional scope of the Medical Service would continue to grow steadily throughout the first half of the nineteenth century.[51] In addition, new institutions of science under the Company stemmed directly from the new land policies being applied to Bengal. These policies would also reorder the Company's relationship to knowledge of India. First and foremost, the Permanent Settlement stimulated a new wave of revenue surveys and maps of property lines (as would the *ryotwari* system in the south).

Geography was now the fastest-growing branch of Company-owned science on the subcontinent. Just as hydrographical knowledge had always been critical to the success of the long-distance trade, geographical knowledge was now essential to the Company's territorial, tax-funded administration. As Colin Mackenzie, the Company's first surveyor-general (and who, as we will see, amassed a vast collection of manuscripts), would put it in 1795, "knowledge of Geography" is merely "a useful preliminary" to the real target of Company interest: "The Revenues, Resources, Populations, Natural Productions and Manufacturers of a Country."[52] But geographical expertise was not (and perhaps could not be) contracted out, as hydrography had been through the ship captains. The Company hired its first in-house surveyor, James Rennell, in 1767, over thirty years before it created (for Dalrymple) the position of in-house hydrographer. Rennell's first assignments were to produced route maps and topographic surveys of the Company's new territories in Bengal and Bihar. After his *Bengal Atlas* of 1780, Rennell next pieced together the first English map of the entire subcontinent, the *Map of Hindustan*. And as surveying and mapmaking grew under the Company, so too did the need for astronomical measurements to determine longitude and keep time. In 1792, the Company formally took over the running of a small private observatory in Madras that had been set up by the administrator William Petrie, thus establishing the first of what would become, by the 1830s, a disparate network of Company-owned observatories across the empire, with Madras the colonial center of that network. Soon the Madras observatory would also be collecting stellar positions from the southern sky, to be exported to India House, which would in turn forward the data to Greenwich Observatory.

Surveying and astronomy were closely connected to the military, and these new institutions show how the Company's army was, around the turn of the century, becoming, as Christopher Bayly puts it, "a most

[51] After the abolition of the Company, this became the unified Indian Medical Service. Crawford, D. G. *A History of the Indian Medical Service: 1600–1913*. W. Thacker, 1914, p. 198.

[52] Quoted in Edney. *Mapping an Empire*, p. 363.

important store of information available to the colonial state, rivalling the civilian service."[53] The army itself was becoming a form of "institutionalized knowledge." As the army moved away from supplying by way of foraging, looting and forced extraction to formally engaging a wide range of supply contractors, procurement officers put together a web of local suppliers who provided to the army everything from maps and route guides to victuals and medicines to horses and armorers.[54]

Meanwhile, a different kind of investment in a new institution of science was underway in Calcutta. In 1788, the Court of Directors approved funding for a new botanical garden. Far from the ornamental pleasure gardens they would become, botanical gardens in the late eighteenth century were experimentation stations for horticultural and agricultural projects such as the hemp and flax investigations of Sinclair. The Company had only tried to establish a botanical garden once before, in 1760, at their small spice-trading factory in Sumatra, Bencoolen. At the time, the capture of Manila had raised hopes of new success for the British spice trade. Earlier attempts to develop spice plantations were revived, and while Dalrymple was plundering the archives of the Spanish Philippines, the Company made a (then) rare effort to formally organize botanical collecting and development. It had been staffed by Philip and Charles Miller, sons of the head gardener at the Chelsea Physic Garden. The Millers were to develop "in the greatest secrecy" nutmeg and clove plantations, and there was hope that tea, ginger, turmeric and mulberries could also be cultivated. Spice plantations run by Chinese migrants already dotted the archipelago, and Bugis traders smuggled in seedlings purchased or gathered from these plantations or areas in which the British were not granted access.[55]

Company support for the idea of a botanical garden at Calcutta was very strong, and its plan, drawn up by a military inspector, Lt Colonel Robert Kyd, was ambitious.[56] Upon receiving Kyd's proposal, the Court of Directors sought the opinion of Joseph Banks, who was very

[53] Bayly. *Information and Empire*, p. 155.

[54] Many large businesses of the colonial era began with successful monopolies on military contracts. The Offices of the Commissariat would be formally established in 1822. Bayly. *Empire and Information*, p. 157.

[55] Marsden, William. *The History of Sumatra: Containing an Account of the Government, Laws, Customs and Manners of the Native Inhabitants, with a Description of the Natural Productions, and a Relation of the Ancient Political State of That Island*. Printed for the author, 1784, p. 307. Also see Kathirithamby-Wells, Jeyamalar. "Peninsular Malaysia in the Context of Natural History and Colonial Science." *New Zealand Journal of Asian Studies* 11, no. 1 (2009): 337–374.

[56] On Kyd and the early Calcutta gardens, see Thomas. "The Establishment of Calcutta Botanic Garden."

enthusiastic about the idea. At this time, the closest the British had to a national botanical garden was George III's Royal Gardens at Kew, where, since the early 1770s, Banks and others had conducted experiments in transplanting exotic flora. In support of a Company garden, Banks echoed Kyd's rhetoric of the joint benefit to the inhabitants of India (food crops alleviating famine), the profit of the Company (developing produce for trade with China) and Britain (establishing export crops of use to British industry such as cotton or hemp). More surprisingly, in a report sent to the Court of Directors via the Royal Society's president, Thomas Morton, Banks also argued that the Company's proposal was *too* ambitious, large and expensive. Banks estimated the proposed garden was of an "immense size ... which cannot be less than 50 or 55 acres" – much too large, he believed, to be managed successfully. Banks compared its proposed size to Kew, which he said was only about two acres, and he noted that even the large nurseries near London that supplied the whole city with "trees, shrubs and plants seldom occupy above 20 [acres]." He cautioned that "if the garden is established on this Extensive plan the seeds of its certain dissolution are sown at the very period of its institution."[57]

Kyd had actually chosen a site that would be 310 acres. And when the directors authorized the proposal, no mention was made of curbing his ambitions. Their usual cautions about overspending were relatively mild: "so sensible are we of the vast importance of the objects in view, that it is by no means our intention to restrict in point of expense in the pursuit of it."[58] Thus, when George Sinclair first set out to make his fortune in hemp and flax (or some other new venture) in 1793, he was not entirely without formal Company support. Ahead of Sinclair, some seeds and other supplies were sent by him to the Calcutta gardens, where Sinclair would first test out some of his methods.[59]

It seems clear that the Court of Directors' enthusiasm for the botanical gardens was tied to both the desperate crisis of the recent famines and new hopes for "improving" the political economy of Bengal via agricultural development. Recent histories of colonial science in India have documented the immense political and scientific significance of the Calcutta gardens.[60] The disastrousness of the first years of Company rule in Bengal

[57] Banks, Joseph. "Banks to Morton on the Calcutta Gardens." In *The Indian and Pacific Correspondence of Sir Joseph Banks*, edited by Thomas Morton. Pickering & Chatto, 1788, pp. 352–354. Banks to Thomas Morton, November 25, 1788, no. 256.

[58] Court of Directors to the governor-general in Council, Calcutta, July 31, 1787. Quoted in Thomas. "Establishment of Calcutta Botanic Garden," p. 170.

[59] Gupta. *Fort William-India House Correspondence*, "Introduction," para. 129.

[60] Axelby, Richard. "Calcutta Botanic Garden and the Colonial Re-Ordering of the Indian Environment." *Archives of Natural History* 35, no. 1 (2008): 150–163; Arnold, David. "Plant Capitalism and Company Science: The Indian Career of Nathaniel Wallich."

had, by the early 1770s, become fully recognized back in Britain. By the mid 1770s, the discourse surrounding how to improve or enrich the colony took on a new urgency.[61] There was, however, little agreement on the cause of Bengal's woes and therefore how to proceed. Some British contemporaries saw the famines as evidence of the inherent defects of the land; others blamed the inhabitants, their character or husbandry.[62] Plenty others blamed the Company – both its policies from London and the apparently rapacious greed of its servants in India. Kyd's proposal framed the key aim of the gardens in terms of alleviating the risk of famine in northern India. And although the directors agreed, the correspondence with Banks was much more focused on resources substitution and the possibility of introducing new cash crops such as tea. Over the next decades, as Zaheer Baber argues, "the possible role of the botanic gardens in alleviating the effects of famine on the population articulated in Kyd's original proposal was forgotten and renewed attention to its contribution to enhanced revenue generation became salient."[63]

It would be many more years before the Calcutta gardens became a distribution hub for the multiplying plantations of new cash crops, including cotton, tea, *chinchona* and eventually rubber. In the meantime, the gardens became a center for the accumulation of botanical and natural historical collections, with ever larger shipments sending many of these materials back to Britain. Kyd died soon after his proposal was accepted, and William Roxburgh, then the Company's naturalist in Madras, took over in 1794. For the next twenty years, Roxburgh experimented with many different kinds of economic plantations (teak was the most successful, though the garden was also propagating coffee from Arabia and tobacco from Virginia and Bengal hemp) and planted thousands of new species in the gardens. He also commissioned hundreds of drawings of native plants, folios of which he had been sending to Joseph Banks since 1790. With Banks as an intermediary, the Company agreed

Modern Asian Studies 42, no. 5 (2008): 899–928; Baber, Zaheer. "The Plants of Empire: Botanic Gardens, Colonial Power and Botanical Knowledge." *Journal of Contemporary Asia* 46, no. 4 (October 2016): 659–679; Desmond. *The European Discovery of the Indian Flora*; Sangwan, Satpal. "Natural History in Colonial Context: Profit or Pursuit? British Botanical Enterprise in India 1778–1820." In *Science and Empires*, edited by Patrick Petitjean, Catherine Jami and Anne Marie Moulin. Boston Studies in the Philosophy of Science. Springer, 1992, pp. 281–298; Thomas, Adrian P. "Calcutta Botanic Garden: Knowledge Formation and the Expectations of Colonial Botany, 1833–1914." Ph.D. diss. King's College London, 2016.

61 Travers. *Ideology and Empire in Eighteenth-Century India*. Travers argues that, beginning in the late 1780s, however, this ideological framework weakened and was eventually replaced by a more strictly imperialistic politics.

62 See Arnold. "Hunger in the Garden of Plenty: The Bengal Famine of 1770."

63 Baber. "The Plants of Empire," p. 668.

to support the publication of the multivolume result of Roxburgh's botanizing around Madras.[64] This would be the beginning of a new pattern, continued first by Colebrooke and then Nathaniel Wallich, of regular exports of botanical collections from the Calcutta gardens to London, where, after 1801, for a time, the deposits would be placed at India House rather than with Banks or Kew.[65]

*

Speaking to the Asiatic Society of Bengal of the challenges and opportunities for scientific investigation in British India, William Jones lamented that what Company servants needed in order to pursue their investigations was *more time*: "'Give me a place to stand on,' said the great mathematician [Archimedes] 'and I will move the whole earth." Give us time, we may say, for our investigators, and we will transfer to Europe all the sciences, arts, and literature of Asia."[66]

Whereas it had once been very difficult and expensive for Europeans to even gain minimal acquaintance with the "science, arts and literature of Asia," now the greatest barrier was, according to Jones, time: time away from the official Company duties to which each and every British orientalist in India was tied. But that barrier, too, was beginning to fall as the Company started to create positions and offices devoted to knowledge management and production. Between the end of the Seven Years' War and the beginning of the Napoleonic wars, the Company was beginning to engage much more directly in the organization and management of the sciences upon which its trade and governance depended. Kapil Raj calls this period "the first step in the transformation of the study of exotic peoples from an individual activity – mainly European missionaries – into a massive and institutionalized activity ... [and] the first step in the transformation of the emerging British empire from one held by force of arms to one held – at least in theory – by information."[67] As we have seen, those changes began with the wartime transformations of the political economy of knowledge in Bengal. While foreigners in India had always

[64] It was published in multiple volumes between 1805 and 1820: Roxburgh, William. *Plants of the Coast of Coromandel: Selected from Drawings and Descriptions Presented to the Hon. Court of Directors of the East India Company*. Vol. 1. Printed by W. Bulmer and Co. for G. Nicol, Bookseller, 1795.

[65] Rocher, Rosane and Ludo Rocher. *The Making of Western Indology: Henry Thomas Colebrooke and the East India Company*. Routledge, 2012, pp. 126–127.

[66] Jones, William. "The Design of a Treatise on the Plants of India." In *The Works of Sir William Jones*. J. Stockdale and J. Walker, 1807, p. 1. On Jones's work between botany and philology, see Menon, Minakshi. "What's in a Name? William Jones, 'Philological Empiricism' and Botanical Knowledge Making in Eighteenth-Century India." *South Asian History and Culture* 13, no. 1 (January 2, 2022): 87–111.

[67] Raj. *Relocating Modern Science*, p. 109.

collected, both wartime plundering and the Company's new position relative to the Mughal Empire would open up many new avenues of access for Britons intent on acquiring manuscripts, curiosities and other knowledge resources. The Company's financial support for the publication of the *Gita* is one example of the kind of patronage that was extended to naturalists and orientalists. But the large collections that were beginning to be brought back to London would remain, for now, part of the private trade, destined for personal collections or sale by individuals. And, as we will see in Chapter 3, as more and more servants returned to London with their collections and skills in tow, a new form of Company public–private science would begin to form by the 1790s, eventually reshaping the institutional structures of science in London.

In this earlier period, as we have seen, virtually all of the new developments in Company science were happening in the colonies: the ambitious new terrestrial surveys, which deployed state-of-the-art techniques; the generously funded botanical gardens at Calcutta; and the wider, more formalized employment of *pandits*, *munshis* and other native educators and scholars. These new spaces for knowledge production and management within the Company represent some of the many changes that accompanied the Company's structural transformation during the late eighteenth century from a relatively marginal militarized maritime trading company to the subcontinent's dominant territorial imperial power. As we will see, however, within a few short years, pressures from both the subcontinent and the home country would lead the Company to sharply increase investment in institutions of science and education back in Britain.

3 The Pull of Company Science to London

The Tigers of Leadenhall Street

In December 1803, while digging for new sewer lines at the Leadenhall Street entrance to the East India Company's headquarters, workers hit upon something unexpected. Charles Wilkins, the orientalist, was called outside to investigate. Workers slowly dug out and around the object. It turned out to be a piece of mosaic Roman pavement – one of the first to be found and preserved in London – roughly 9 feet square and decorated at its center with a well-known scene from Greco-Roman mythology: Bacchus, dressed in purple and green robes, holding his wine cup and fennel frond, and reclining on the back of a growling tiger.[1] (See Figure 3.3.)

The sewer work was part of a major reconstruction of the Company's headquarters at India House. This plot in the middle of the City of London had been the Company's administrative and commercial base since the mid seventeenth century (see Figure 3.1). In 1798, after nearly five decades of steady territorial and commercial expansion in Asia, the directors had found they also needed to expand their office space. At that time, the Company employed around 55,000 individuals, with 30,000 of those based in London.[2] The new India House, completed in 1801, had consumed some of its old neighbors and now occupied nearly a full city block. What is now the site of the Lloyds of London skyscraper was then a sprawling five-story set of interconnected structures, gathered together behind a grand neoclassical façade. Just to the east of India House, across Whittington Lane, was the skin market, home to slaughterhouses, candle-makers, tanneries and leatherworks. The southern end of India House abutted the huge Leadenhall Market, one of the City's oldest and largest centers for the sale of meat, vegetables and herbs. A few streets away in

[1] Brayley, Edward Westlake and John Britton. *The Beauties of England and Wales, Or, Delineations, Topographical, Historical, and Descriptive, of Each County* .... Vol. 10. T. Maiden, 1810, p. 96.

[2] Bowen. *Business of Empire*, pp. 265–267.

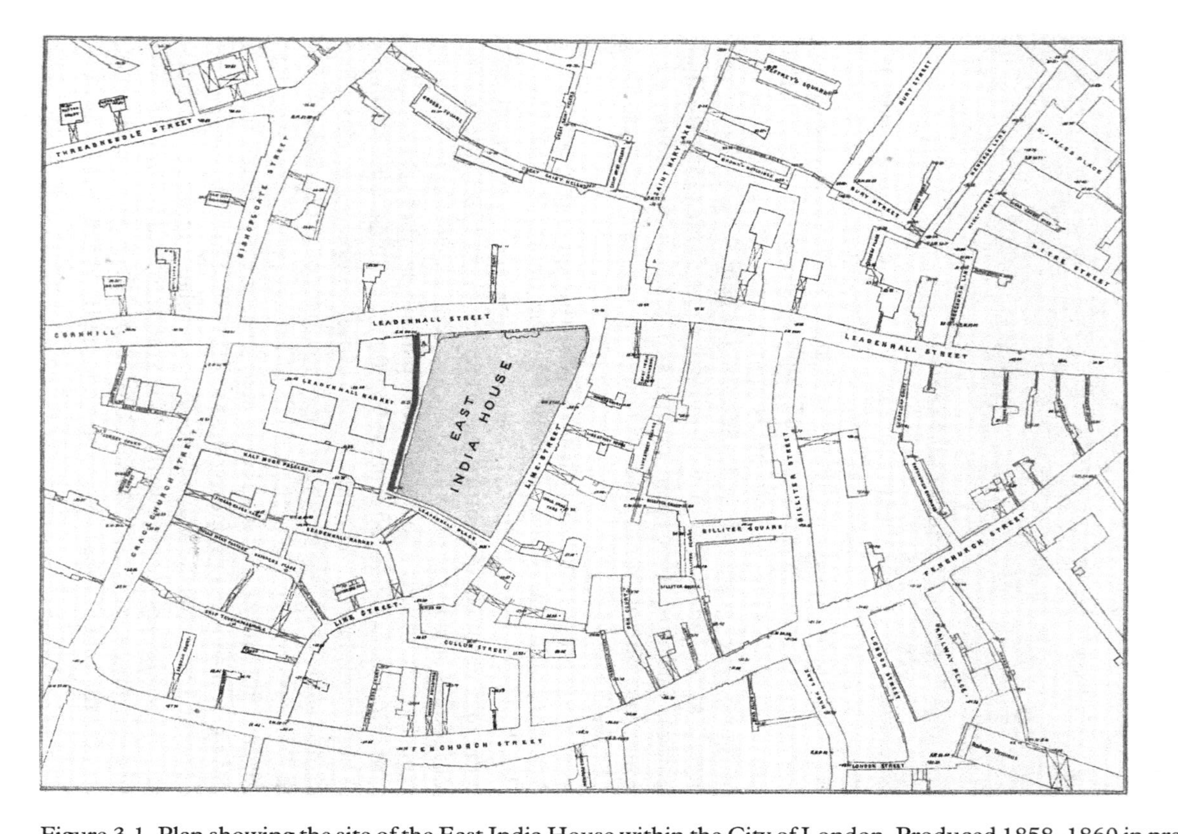

Figure 3.1 Plan showing the site of the East India House within the City of London. Produced 1858–1860 in preparation for the demolition of India House. Reprinted in Birdwood, George C. M. *Relics of the Honourable East India Company:* [illegible]

Figure 3.2 View of India House looking south down Leadenhall Street. From *Ackermann's Repository of Arts*, June 1, 1817. Copyright British Library Board (asset no. Maps K.Top.24.10.c).

Figure 3.3 The remaining central design of the Leadenhall Street mosaic. © The Trustees of the British Museum (asset # 103463001).

New Street and Cutler Street a massive new complex of Company warehouses was going up, covering five acres and twenty-five warehouses, each six stories high, and containing nearly 150 rooms. The warehouses were so impressive that foreign dignitaries often asked for tours.[3]

The old India House had a simple pilastered façade and a naval scene painted over the doorway. Now, drovers on the way to the market would herd cattle past a huge ionic portico.[4] The new building was much statelier, and it fully embraced the popular discursive parallels between the ancient Roman and new British Empire. Some observers were, in fact, disappointed with the lack of any "Asiatic" design for the home of not only the center of Britain's Asian trade but also the seat of government of British India: "there is nothing relative to the eastern world that *presents itself* to observation," one architecture critic complained. But others argued that "it would be too shocking to the London eye were the

[3] Makepeace. *EIC's London Workers*, p. 22.
[4] Foster, William. *The East India House, Its History and Associations*. John Lane, 1924.

building totally 'oriental.'"[5] How serendipitous, then, for a part of Roman Britain to be unearthed at the entrance, and for this particular scene to neatly tie ancient Rome and Asia together. Bacchus, or Dionysus, was closely connected in Greek mythology to Eastern conquest. Various stories (most famously the poem "Dionysiaca") tell of Bacchus traveling to India. The god of fertility, wine, and reproduction, Bacchus romped around the subcontinent, battled local gods and armies, made alliances, and introduced his favorite food and drink. Eventually he made a triumphant return with a great procession of captured treasure, including a long train of exotic animals (hence the tiger in the mosaic).

This ancient, if not entirely reputable, representation of an imperial collector now became part of the world's latest imperial collection. The Bacchus mosaic was carefully excavated and carried into India House. Behind the architectural unity of the grand classical façade, India House was a sprawling and top-heavy set of offices through which any decisions passed painfully slowly. The Court of Proprietors still elected the chairman, deputy-chairman and twenty-four directors. But the Board of Control, appointed by the Crown, now oversaw, on political matters, the Court of Directors. Crucially, the Court of Directors still controlled the vast majority of the Company's patronage; that is, the issuing of new army and civil service positions in British India. Company writerships were highly sought after, and control of the patronage gave the directors a significant amount of political capital in Britain. A wide range of sometimes clashing political and economic interests were gathered together here. For one thing, the Company's commercial functions encompassed both the interests of those whose profits depended upon the sale of goods and those whose profits depended upon the shipping of goods. In addition, the close ties between the Company and Parliament meant that the Company's administration also reflected or imported traditional political divisions at play in British politics. The East India Company in 1800 was, in form, with its formal monopoly over the Eastern trade, a deeply conservative (trade protectionist, anti-reform, Tory) institution. At the same time, the growing strands of liberalism (free trade, pro-reform, utilitarian, Whig) of the early nineteenth century were also increasingly represented within the Company.[6]

British India was by now divided into governorships of three geographically distinct presidencies at Madras (which also included all British

[5] Brayley, Edward Wedlake and Joseph Nightingale. *A Topographical and Historical Description of London and Middlesex* .... Vol. 2. Sherwood, Neely and Jones and G. Cowie, 1814, pp. 764–765.

[6] Mehta. *Liberalism and Empire*; Ince, Onur Ulas. *Colonial Capitalism and the Dilemmas of Liberalism*. Oxford University Press, 2018.

Indian regions east of the subcontinent such as the Straits Settlements and the Company's factories in Chinese treaty ports), Bengal and Bombay. The Crown, with the approval of the Company, appointed and, in theory, presided over the governors in charge of each presidency. The governor-general of Bengal was the highest-ranking official in British India and in many ways the supreme authority on the ground. But the Court of Directors could (and often did) criticize, censure, revise and revoke the decisions of the governor-general. At the same time, however, with communication between Britain and British India taking six months at least, the governor-general was also able to subvert, ignore or otherwise disrupt the instructions from India House.

Having passed under the portico, the Baccus mosaic was then carried into a large central atrium, from which extended a maze of hallways and rooms cobbled together around a central open yard. Most of the Company business was conducted in the great rooms off the central hallway downstairs from the library and museum. Within these offices, hundreds of clerks kept the paper machinery of the Company's empire running. To the right was the Grand Court Room, where the proprietors (stockholders with large enough holdings to be able to vote on Company matters) met to debate and vote. The room was richly decorated, with "an uncommonly fine Turkey carpet covering the whole flooring" and a vast marble chimneypiece. There was a bas relief of Britannia sitting on a globe being attended by figures representing Asia and Africa, who offered to Britannia various gifts and commodities. Clocks, mirrors and mathematical instruments associated with navigation – a signal of how fundamental the science of navigation and surveying was to the Company's interests – were hung on the walls. On the panels of the "uncommonly handsome doors" were six large paintings giving a panorama of views of the Company's key ports in the late eighteenth century: Fort St. George (Madras), Bombay, St. Helena, the Cape of Good Hope, Fort William (Calcutta) and Tellicherry. All of the India House spaces, or the public rooms anyway, were a spectacle of the Company's geographical reach.[7]

Another centerpiece of the new India House was the "New Sale Room." Because of their monopoly on trade east of the Cape of Good Hope, all goods brought from Asia to Britain were sold by auction in India House. The New Sale Room was theater-like, with stepped seating and a large staging area at the front. Like some of the new shopping arcades cropping up at the turn of the century, the room was equipped with natural lighting from a glass and iron ceiling, a technique also now

[7] Brayley. *A Topographical and Historical Description of London and Middlesex*, pp. 159–164.

used in the Company's most important warehouses. The New Sale Room was also comfortably warm, heated "without any visible fire, the result of a subterranean conveyance of heat." The pilastered walls displayed more scenes of Asia's "commercial attributes."[8]

Within the new India House were many spectacular spaces, and displays of oriental curiosities and works of art were scattered around the hallways and committee rooms. But the new India House also had within it a new space intended in part to differentiate mere plunder from a more public-minded (so it was argued) kind of collecting: the "Oriental Repository." It was here that Wilkins directed the mosaic to be moved. Up a set of stairs off to the left of the vestibule was the newly added library and museum space. Here, as one early visitor guide put it, "every book known to have been published in any language whatsoever is to be found here, relative to the history, laws or the jurisprudence of Asia," as well as "an unparalleled collection of oriental manuscripts in all the Oriental languages," including the only printed Chinese-language books in England.[9] The library was not large, about 60 feet long and 20 feet wide, and was well lit with large circular skylights and tall windows facing Leadenhall Street and Lime Street. Above an ornate mantlepiece hung a painting of "the Emperor of Persia a young man with a long black beard in magnificent jeweled dress."[10] Recesses in the wall displayed busts of Robert Orme and Warren Hastings. But the main attraction were the walls covered in bookcases and shelving designed specially to house a great material variety of written forms: from "the smooth silky paper of India" to "the Malayan manuscripts ... etched with a sharp tool upon the leaves of the palm tree, joined at the ends and made to open like a fan." Still others "folded up in the ancient manner [and] extend several yards in length when opened."[11] Two very personal items of Tipu Sultan's – his personal copy of the Quran and a journal in which he wrote down his dreams – were on prominent display.

Adjoining the library was the museum, where the visitor guide reports seeing a stone covered in "Babylonian inscriptions," a 2-foot-long fragment of jasper covered in carvings, antiquities from India, Chinese works of art including jade carvings, paintings and a massive silk lantern, and the Bacchus mosaic. A few years after the mosaic had been found, Bacchus would be joined by many more tigers, the loot from the storming of Seringapatam in 1799 finally having made its way back to

[8] Brayley. *A Topographical and Historical Description of London and Middlesex*, p. 764.
[9] Brayley. *A Topographical and Historical Description of London and Middlesex*, p. 766.
[10] Brayley. *A Topographical and Historical Description of London and Middlesex*, p. 766.
[11] Brayley. *A Topographical and Historical Description of London and Middlesex*, p. 766.

India House. The solid-gold tiger-themed throne of Tipu Sultan, Britain's great rival in southern India, had been broken up to be divided into customary prize payouts for army officers, but one of its solid-gold tiger heads did make it back to the museum. The most famous item from Tipu's palace, however, was the celebrated "Musical Tiger" (see Figure 3.5).[12] Now in the Victoria and Albert Museum, Tipu's tiger was like a mechanical Enlightenment version of the Bacchus mosaic, but with the political imagery inverted. It was a life-sized painted wooden automaton of a tiger (representing Mysore) atop a pale-skinned soldier in redcoat, which, when wound up, would growl and claw at the squealing, squirming soldier.

This chapter follows the creation and early growth of Company science in London. As we have seen, in the late eighteenth century, the Company's new investments in education, knowledge management and institutions of science were largely focused on British India. But around the turn of the century, the foundation of the new library-museum and colleges in Britain would sharply redirect the growth of new Company-run initiatives for science and education back to Britain and crystalize that shift into a new set of institutions and priorities related to knowledge management. It was a shift that took full advantage of the Company's legal monopoly on access to Asia's knowledge resources. And it would begin with the stepwise incorporation into the administration at India House of the work of the orientalists, naturalists and collectors covered in the previous chapters. The London careers of a set of nabob-scholars – Robert Orme, Alexander Dalrymple and Charles Wilkins from Chapter 2, as well as William Marsden – illustrate how the early beginnings of Company science in London flourished at the porous boundary between individual and corporate ownership.

Incorporating the Nabobs

At a meeting of the Antiquarian Society in London in 1772, Matthew Mite, a wealthy former servant of the East India Company, offered a procession of new presents to the Society's museum. The gifts included a piece of lava from Vesuvius and a box of natural history specimens, which contained "for the use of my country ... a large catalogue of petrifications, bones, beetles and butterflies." The piece of lava, given special attention by Mite, was collected as a sample of foreign natural

[12] The arrival of the tiger is recorded in BL Mss Eur F303/1, July 29, 1808. See, for example, Wilkes, John, ed. *Encyclopaedia Londinensis*, 1815, p. 452; Britton, John. *Illustrations of the Public Buildings of London* (vol. 2). London, 1828, p. 88.

production worthy of study for the "useful" aim of introducing and propagating volcanoes within the English landscape:[13]

MITE: By a chymical analysis, it will be easy to discover the constituent parts of this mass, which by properly preparing it, will make it no difficult task to propagate burning mountains in England, if encouraged by premiums.
FELLOWS: Which it will, no doubt!

To this and Mite's other contributions to "national knowledge," the fellows responded enthusiastically ("What a fund of learning!" "Amazing acuteness of erudition!" "Let this discovery be made public directly!").[14]

Matthew Mite is the titular character in playwright Samuel Foote's satire *The Nabob*, which brings together and skewers late Georgian fashionable culture (including natural philosophy) and imperial politics. First staged in 1772, *The Nabob* – then a derogatory for the *nouveau riche* among returned Company servants – follows the schemes of Mr. Matthew Mite, who has recently returned from India and is now exercising his new wealth to advance his social position.[15] Mite's riches have enabled him to buy his way into the elite world of learning embodied by such new institutions as the Society of Antiquaries (*f.* 1751), parodied here.[16] This is just one part of Mite's larger scheme to secure for himself a place among the aristocracy, who, for their part, regard him as no more than a thief and plunderer. After being accused by one such family of impoverishing India to acquire his ill-got wealth, Mite retorts: "I am sorry . . . to see one of your fashion concur in the common cry of the times; but such is the gratitude of this country to those who have given it dominion and wealth." To which the patriarch replies: "I wish even that fact was well founded, Sir Matthew. Your riches (which perhaps too are only ideal) by introducing a general spirit of dissipation, have extinguished [here in Britain] labor and industry, the slow, but sure source of national wealth."[17]

Mite claims his *individual* riches are but part of a larger contribution to the *nation*'s wealth and strength; the old aristocratic family, however, claims the opposite: colonial exploits are disrupting England's traditional

[13] Foote, Samuel. *The Nabob; a Comedy, in Three Acts. As It Is Performed at the Theatre-Royal in the Haymarket.* London, 1778, III (I), p. 54.
[14] Foote. *The Nabob*, III (I).
[15] See Nechtman, Tillman W. "Nabobs Revisited: A Cultural History of British Imperialism and the Indian Question in Late-Eighteenth-Century Britain." *History Compass* 4, no. 4 (July 2006): 645–667.
[16] For a classic study of scientific institutions and social mobility, see Thackray, Arnold. "Natural Knowledge in Cultural Context: The Manchester Mode." *The American Historical Review* 79, no. 3 (1974): 672–709.
[17] *The Nabob*, III (I), p. 63.

and reliable patterns of commerce and political economy. Act III, at the Antiquarian Society, brings those issues to bear on Enlightenment learned culture, suggesting that the colonial "spirit of adventure" and the dubious collections brought home offer only an illusion of progress in knowledge. Mite's visit to the Antiquarian Society mocks the idea that nabob-scholars and their curious collections are contributing useful knowledge to the nation. Picking out for ridicule the growing discourse of "improvement" among the learned societies at the time, Foote presents Mite and his Antiquarian Society as deluded with the self-image of the Society as an important resource for national utility and publicly useful knowledge.[18] But, in fact, Mite's contributions are trivial, misdirected and distinctly useless. Mite's ideas for economic "improvement" by way of foreign resource substitution (i.e. propagating volcanoes) are downright destructive.

The character Matthew Mite could plausibly have been based on any number of returning Company servants whose wealth and status was, in part, based on an engagement with learned societies and cultures of collecting.[19] As in Foote's *Nabob*, such collecting was of a piece with the wider debate about just how valuable to the *nation* was the mass of private wealth captured in the Company's recent wars. In the next section, we will see how, with the establishment of the new oriental repository at India House, the Company would step in and attempt to gain control of – or at least a stake in – this thriving world of private collecting of Asia in Britain. But the first steps toward instituting new spaces for science at the Company would involve the returned nabob-scholars we met in the last chapter. These figures played a crucial role in establishing both a new London-based orientalism and new London-based institutions of science at India House Robert Orme would become the Company's first historiographer; Alexander Dalrymple, the Company's first hydrographer; and Charles Wilkins, the first curator of the Company's library and museum. Several would also become part of the circle of Joseph Banks (who, at the time *The Nabob* was staged, had

[18] For a recent review of the discourse of utility and improvement in the period, see Stewart, Larry and Kelly J. Whitmer. "Expectations and Utility in Eighteenth-Century Knowledge Economies Notes and Records Special Issue Introduction." *Notes and Records of the Royal Society of London* 72, no. 2 (2018): 111–117. Also see Ehrlich, Joshua. "Empire and Enlightenment in Three Letters from Sir William Jones to Governor-General John Macpherson." *The Historical Journal* 62, no. 2 (2019): 541–555.

[19] However, Wimsatt deals with the question of who, if anyone, Matthew Mite may have been based upon and concludes it is a generic satire. Wimsatt, William K. Jr. "Foote and a Friend of Boswell's: A Note on the Nabob." *Modern Language Notes* 57, no. 5 (Johns Hopkins University Press, 1942): 325–335.

just returned from the expedition of the *Endeavour* to the Southern Ocean with Captain James Cook. And Banks, in particular, would, through resource substitution schemes such as those satirized in *The Nabob*, bring orientalism and natural history to bear on schemes to improve Britain's trade balance).[20]

Increasingly, returning orientalists were able to find not only comfortable social networks but also, in the best cases, lucrative new positions in the home government at India House. To be sure, the majority of Company servants pursued more directly financially interested projects during their time in Asia. But personal collections generated significant financial as well as cultural capital and opened doors to new economic opportunities. One such nabob-scholar clearly explained his worldly interest in pursuing "disinterested" scholarship while stationed in Asia. William Marsden was born in Ireland to an Anglo-Irish family of bankers. He (like Orme) also went to Harrow, and he joined his brother as a Company writer at Fort Marlborough, the Company factory near Bencoolen (Bengkulu, Indonesia), at the age of sixteen. He spent nearly a decade in Sumatra, from 1771 to 1779. As he recalls this time in an autobiography, he spent much of those years devoting himself to "the Muses": "what I had acquired of classical learning at school was not neglected, as after my arrival in Sumatra, I made translations of the Greek odes of Anacreon and Sappho." And in fact, in the same year that *The Nabob* was staged in London, Marsden and his brother were staging Greek tragedies in a playhouse they had built in Bencoolen. But, as he continues, "my curiosity being ever awake to the objects around me, [t]he objects, indeed, of my literary pursuits were by no means of a confined nature. I had an ardent thirst for knowledge, both for its own sake and from the flattering, however distant, hope, of it *enabling me to distinguish myself in the event of my future return to London*."[21] And so, to that end, he continues, "I seriously directed my attention to collecting materials for giving an account of the island."[22]

Marsden decided (or was forced by bad health) to return to England before having achieved the customary level of financial success, which, as he explains, was "until the annual savings from the emoluments of offices

[20] "Transplantation," or resource substitution, schemes were popular targets of mercantilist policies to do just that. They had been popularized by Carl Linnaeus a few decades earlier. See Koerner, Lisbet. *Linnaeus: Nature and Nation*. Harvard University Press, 2009. On Banks as a neo-mercantilist, see Gascoigne. *Science in the Service of Empire*.

[21] Marsden, Elizabeth, ed. William Marsden. *A Brief Memoir of the Life and Writings of the Late William Marsden, Written by Himself*. Cox, 1838, p. 11. My italics.

[22] Marsden. *A Brief Memoir*, p. 15.

would accumulate to what is termed a fortune – that is, such a sum as, when invested English securities, would permit the owner to enjoy the conveniences of life, without further exertions on his part."[23] Marsden's understanding of the cultural capital and future financial possibilities of developing an expertise in Sumatra would have been unremarkable at the time. In both the colonial and the home-country context, many of this new class of well-off merchants, colonialists and industrialists participated in the growing number of scholarly clubs and societies all around Britain.

By the 1770s, returning employees of the Company were bringing huge quantities of goods from Asia, filling their homes and estates to such an extent that the material culture of empire became a defining feature of the English country home.[24] Often having made a very comfortable sum abroad, sometimes having captured vast riches, returning nabobs as well as families connected with shipping, banking and Company administration filled their homes with materials from the Asian trade and, increasingly, wartime plunder. This included weapons, cloth, jewels, utensils and paintings, but also manuscripts and cabinets filled with *naturalia*, as well as exotic plants and live animals. The ultra-wealthy Child family, for example, which had been involved in Company shipping and administration since the seventeenth century, maintained at their lavish Osterly Park estate extensive gardens as well as a "menagerie full of birds that comes from a thousand islands which Mr. Banks has not yet discovered," as Horace Walpole put it.[25] The family of Edward Clive (son of Robert Clive) amassed back in England a vast "treasure," as Lady Clive put it, of natural specimens, works of art and craft, and stuffed and live animals, many collected (some by the Clive daughters themselves) and others purchased or given as gifts to Lady Clive (Company officers were, by this time, barred from receiving personal gifts, but family members were a different story) while the family was on tour to the recently plundered kingdom of Mysore.[26] And even beyond those who had served the Company or been to Asia, cabinets of natural history and foreign

[23] Marsden. *A Brief Memoir*, p. 16.

[24] Finn, Margot and Kate Smith. *The East India Company at Home, 1757–1857*. UCL Press, 2018.

[25] Horace Walpole to Lady Ossory, June 21, 1773. In *Yale Edition of Horace Walpole's Correspondence*, edited by W. S. Lewis. Yale University Press, 1937, p. 125, quoted in "Osterley Park and House Case Study – East India Company at Home, 1757–1857," n.d., http://blogs.ucl.ac.uk/eicah/osterley-park-middlesex.

[26] Jasanoff. *Edge of Empire*, p. 187. Also see Archer. *Treasures from India*. On the character of the nabob, the role of material culture in the creation of nabob identity and the vitriol surrounding it, see Nechtman, Tillman W. "A Jewel in the Crown? Indian Wealth in Domestic Britain in the Late Eighteenth Century." *Eighteenth-Century Studies* 41, no. 1 (2007): 71–86.

works of art, especially textiles and tableware, were common in upper-class households.[27] The ultra-wealthy had, for example, their own glass-houses for exotic flora, "China rooms" to display porcelain collections, museum-like natural history displays and even menageries. For example, Margaret Cavendish Bentinck, duchess of Portland (who was an investor in the Company), put together the largest natural history collection in mid eighteenth-century Britain. The "Portland Museum" at Bulstrode was also home to a menagerie, aviary and large botanical garden (see Figure 3.4). The vast, well-curated collection would be auctioned off, after her death, in 1787.[28]

Back in London, in the 1760s and 1770s, in the wake of the Company's expansion after the Seven Years' War, some of those individual collectors would parlay their private material gains into key positions within both India House and London's wealthy philosophical circles. These collections would often initially occupy a "semiprivate" space in which they were owned by individuals but hired out in a newly formal way by the Company.[29] One such collection is that of Robert Orme, who returned permanently (some said fleeing – with Clive) to London in 1760 during the war. He was now well off by English standards but not nearly as rich as those of the great nabobs such as Clive himself. More importantly for the future trajectory of his career, Orme had returned to London with records and archives that allowed him to produce the first detailed account of the Company's recent wars, together with a study of Bengal. He bought a house in Harley Street, where he installed his personal library and settled into the life of a nabob-scholar. He began work on what would become the *History of the Military Transactions of the British Nation in Hindoostan*. It was in its time regarded as the most authoritative and complete English work on "India" (a relatively new term, which Orme explains is distinct from the "East Indies") and it was also the first account of the recent wars that had so radically extended the Company's territorial reach.[30]

[27] See, for example, Retford, Kate and Susanna Avery-Quash, eds. *The Georgian London Town House: Building, Collecting and Display*. Bloomsbury Visual Arts, 2021.

[28] Pelling, Madeleine. "Collecting the World: Female Friendship and Domestic Craft at Bulstrode Park." *Journal for Eighteenth-Century Studies* 41, no. 1 (2018): 101–120.

[29] Semiprivate here follows Holger Hook and his study of the making of the British Museum collections: Hoock, Holger. "The British State and the Anglo-French Wars over Antiquities, 1798–1858." *The Historical Journal* 50, no. 1 (March 2007): 49–72.

[30] The Member of Parliament Edmund Burke, a great critic of the Company and later of Hastings, regarded Orme and his book very highly, praising its impartiality, and remarked of Orme: "no historian seems to have been more perfectly informed of the subject on which he has undertaken to write." Anon., *The Emerald*, 1.48 (1808), p. 567.

Figure 3.4 Frontispiece to the auction catalog for the duchess of Portland's museum, which then contained the largest natural history collection in Britain. Skinner and Co. (London, England) and John Lightfoot. *A Catalogue of the Portland Museum, Lately the Property of the Duchess Dowager of Portland, Deceased: Which Will Be Sold by Auction by Mr. Skinner and Co. on Monday the 24th of April, 1786, and the Thirty-Seven Following Days . . . at Her Late Dwelling-House, in Privy-Garden, Whitehall: By Order of the Acting Executrix*. [London], [Mr. Skinner and Co.], 1786. From the Biodiversity Heritage Library. www.biodiversitylibrary.org/item/243075.

Orme became a fellow of the Society of Antiquaries in 1769. That year, he was also appointed to the newly created position of historiographer to the Company, with a salary of £400. Although in some ways "the world's first in-house corporate historian," Orme's own private library was most critical to his work.[31] Orme had a desk within the Examiner's Office, but generally worked from home, surrounded by his massive collection. A few years later, in 1771, the directors, responding to complaints about the "confused and disorderly state" of Company records, established a new office devoted to document management.[32] The new Registrar's Office was to take custody and arrange into a numbered catalog all books, papers and records. The office was also to oversee registering incoming and outgoing materials. With these two moves, and just in time for the coming debate over the Company's charter renewal in 1773, the Company formally took over management of the production of its own history. The political importance of the Company historiographer was such that, after the charter renewal of 1773, part of the new Regulating Act, which increased Crown control of Company policy via the new Board of Control, was to allow the Board to also appoint another historiographer. Orme would thus soon be joined by the historian and philosopher John Bruce, a close ally of Henry Dundas, then president of the Board of Control.

The first volume of Orme's *History* dealt with the years between 1751 and 1755, and a key duty of the new historiographer was to produce subsequent editions. It would not be until almost fifteen years later, in 1778, that Orme finally produced the next volume of his history. Some historians argue that his slow progress was largely due to his growing unease with the political events unfolding in India. As Orme wrote to a friend in 1767: "it is these cursed presents [i.e. gifts and bribes] that stop my history. Why should I be doomed to commemorate the ignominy of my countrymen, and without giving the money story [i.e. the question of corruption] that has accompanied every event since the first of 1757, I shall not relate all the springs of the action."[33] In later years, Orme switched to less fraught territory, working on a history of the Mughal Empire, publishing *Historical Fragments of the Mughal Empire* in 1782. The "fragments" in the title points to the fact that, as Orme explains, for

[31] Smith, Andrew and Daniel Simeone. "Learning to Use the Past: The Development of a Rhetorical History Strategy by the London Headquarters of the Hudson's Bay Company." *Management & Organizational History* 12, no. 4 (October 2, 2017): 334–356, p. 334.

[32] Bowen. *The Business of Empire*, p. 173.

[33] Delgoda. "'Nabob, Historian and Orientalist,'" p. 370.

this project, back in England, Orme had very little manuscript Persian or Sanskrit material, yet the number of manuscripts and printed works about India now circulating in England still made such a study possible.[34]

Alexander Dalrymple's career followed a similar pattern as that of Orme, and was equally dependent upon his private library. After returning to London in 1764, Dalrymple established himself as an authority on the South Seas and the Southern Indian Ocean, joined the Royal Society and began advocating for further exploration of the Southern Ocean along with others who believed there was an as-yet-undiscovered major continent. The British government decided to support such exploration, to which the Company also gave £2,000. But when he lost out leadership of the expedition to James Cook, Dalrymple instead turned to the world of print. Dalrymple became a prolific editor and publisher of works related to seafaring, exploration and navigation in Asia. He began, as had Richard Hakluyt, with collected histories of voyages based on sources gathered from a wide chronological and geographical range. His first of these, *An Historical Collection of the Several Voyages and Discoveries in the South Pacific Ocean* (1770), brought together materials to make the case for there being an undiscovered continent to the south of Borneo and the Philippines. The subtitle declares the work's value "Being chiefly a LITERAL TRANSLATION from the SPANISH WRITERS" and indicates, as Dalrymple elaborates in the introduction, that the work is not intended for popular enjoyment but for the information of the serious student of Eastern navigation. Distancing himself from the thriving travel writing genre, he presents the decision to produce a "LITERAL TRANSLATION" as a matter of unsparing scientific utility (and personal financial sacrifice): "the undress and uncouth sound of a literal translation is enough to frighten all readers except the *very few* who take up a book *merely* for information; but it was to these *few* I have devoted my labors."[35] And he signals his authority on the subject by way of his own history as a ship captain and in the East India Company, which, crucially, gave him access to this very rare information and enabled him to form "a collection of all the discoveries in the South-Sea" during the Seven Years' War.[36]

[34] He also mentions using materials from his collections of copies of Company records from Bengal, Madras, Bombay, Surat, Rajapore and Carwar. See list of "Authorities" cited or mentioned in *Historical Fragments of the Mogul Empire. Of the Morattoes, and of the English Concerns, in Indostan*. London, 1782.

[35] Dalrymple. *An Historical Collection*, p. xi.

[36] Dalrymple. *An Historical Collection*, p. xi.

Dalrymple also continued collecting after he returned to London in 1764. One contemporary describes him as a constant presence in certain bookseller shops and auction houses:

> His yellow antiquarian chariot seemed to be immovably fixed in the street just opposite the entrance door of the long passage leading to the sale room of Messrs King and Lochee in King Street Covent Garden, and, towards the bottom of the table in the sale room, Mr. Dalrymple used to sit, a cane in his hand, his hat always upon his head, a thin slightly twisted queue and silver hairs that hardly shaded his temple. His biddings were usually silent, accompanied by the elevation and fall of his cane, or by an abrupt nod of the head.[37]

Of his London shopping, Dalrymple notes in particular being able to acquire a "curious collection of Spanish memorials; these greatly elucidate the printed relations, which without having this assistance, must have remained unintelligible."[38]

Dalrymple's *History* marked the start of a period of about three years in which he mined his own library to produce nautical and hydrographical charts and pilot guides about the China routes. His focus was on the area around Borneo, the Sulu archipelago (where for years he had advocated establishing a Company settlement at Balambangan) and the Philippines, based first on his own travels and later on the manuscripts and rare prints in his collection, becoming more of an editor than an author. In this period, the standard means of transferring nautical information was via manuscript copying. The Court of Directors was not entirely happy to have such routes and port descriptions published, and during a period when a Balambangan station was under active consideration, for example, they discouraged him from making the material publicly available.[39] Dalrymple was also in close contact with the French hydrographer D'Aprés de Mannevillette, and in 1772 sent him the six publications he had so far produced, sanctioning him to "make what use you please of the Charts I have sent you," which D'Aprés did in later editions of his widely read *Neptune Oriental.*[40] Here Dalrymple found his focus: gathering, organizing, editing and publishing nautical works. His first set of charts comprised six drawn from a Dutch collection (of Van Keulen) and the rest from his own collection, producing in 1774 *A Collection of Plans and Ports in the East Indies*. This work also marked a new relationship between Dalrymple and the East India Company. Earlier tensions now

[37] Roberts, William. *The Book-Hunter in London: Historical and Other Studies of Collectors and Collecting*. E. Stock, 1895, p. 56.
[38] Dalrymple. *An Historical Collection*, p. xxiv.
[39] Cook. "Alexander Dalrymple," p. 51.
[40] Cook. "Alexander Dalrymple," p. 61.

dissipated, and Dalrymple now worked with an eye to gaining significant Company subscriptions for forthcoming works. The Company agreed, for example, to purchase fifty copies of *A Collection of Plans and Ports in the East Indies*.

In 1772, Dalrymple was formally re-hired by the Company as the Company's in-house hydrographer. In 1777, his projects turned from his own collections to those at India House, examining ships' logs and publishing charts on behalf of the Company. The Company hadn't produced its own nautical charts since the early seventeenth century. By 1779, Dalrymple was receiving an annual salary of £500 for his hydrographic work. He still did most of his work at his home (in the household of his patron the former governor George Pigot). He also hired out engravers and printers and, when necessary, supplemented the Company's annual £500 stipend for publication costs with his own funds.[41] But, beyond providing the Company with a required number of copies, he was allowed to do what he wished with the material and any further profit would be his.

By the mid 1780s, Dalrymple had produced hundreds of charts for the Company, based increasingly on the logbook collections at India House. He was also now selling collections of these to the public, packaging and repackaging the Company materials in an ever-changing number of forms. In 1791 he began his annual series *The Oriental Repertory*, which became a very widely used resource, a compendium of all kinds of travel and shipping information, updated route guides, charts, tables and so on. He had also been successful in getting the logs in more regular forms and in convincing captains to produce information of the kind he was seeking, including making use of new marine chronometers for longitude. In 1795, as tensions escalated between France and Britain yet again, the Admiralty followed the Company's lead and began a program of organizing and processing its own vast archive of ships logs and miscellaneous charts. Dalrymple also became hydrographer to the Royal Navy, receiving the same salary from the state as from the Company. He would retain both positions until his death in 1809.

Not all Asian collections in London were being managed in semiprivate collaboration with the Company. There were plenty of much more independent collections being formed as well. Williams Jones, the puisne judge and orientalist in Calcutta, sent a significant collection of his manuscripts to the Royal Society in 1792 (the rest were retained by Lady Jones and eventually auctioned off with her library after her death

[41] Cook. "Alexander Dalrymple," p. 27.

in 1832).[42] In addition, the orientalist and administrator Henry Thomas Colebrooke, who was also Jones's close associate and successor in the project to create a digest of Hindu laws, would also remain independent from the Company upon his return to London in 1815. Colebrooke instead founded in 1823 the Royal Asiatic Society (modeled on the Asiatic Society of Bengal). He had amassed an even larger collection than Jones, and though he left much of his library back in India with the missionary and printer William Carey (who had agreed to copy the lot and eventually ship the originals back to London), he also shipped about seven tons of books and manuscripts, over 2,500 items, then said to be the largest collection of material from Asia in Europe.[43]

William Marsden, who also formed a great private library at this time, presents a different trajectory. After nearly a decade in Bencoolen, he resigned and returned to London, where he helped to run an agency house. He also spent a great deal of time and money collecting Asian coins and manuscripts within London, from booksellers and private library auctions. Marsden was, above all, a collector of information about languages, particularly the languages of present-day Austronesia. He published the first English history of Sumatra and the first English dictionary of Malayan. The study of languages encompassed both the extremely practical, with word lists and dictionaries that circulated constantly among ship captains and seamen, and the philosophical, in which language was considered a window onto the history of civilizations.[44] As the war with France took shape, Marsden also, like Dalrymple, joined the Admiralty as an undersecretary in 1795, becoming first secretary with a massive salary of £4,000 in 1807.

Along with Dalrymple and Rennell, Marsden also became part of the Company contingent of the naturalist Joseph Banks's circle, joining his supper and breakfast clubs as well as being elected a fellow of the Royal Society.[45] Banks, for his part, facilitated Marsden's philological research by passing on material related to languages that came into his

[42] Lawrence. "Building a Library."
[43] Rocher. *The Making of Western Indology*. On Colebrooke's collections and their dispersal and shipping, see p. 129.
[44] For example, in his *History of Sumatra*, Marsden gathers linguistic evidence to argue for the central Asian ("tartaric") origin of all the tribes of the region. He also contests Blumenbach's history of humankind, which uses skin color as the primary feature with which to separate out four (later five) distinct groups, and argues against classifying literate Sumatrans as "savages and barbarians." See Carroll, Diana J. "William Marsden, the Scholar behind the History of Sumatra." *Indonesia and the Malay World* 47, no. 137 (2019): 66–89, p. 70.
[45] See, for example, Marsden, Elizabeth, ed. *William Marsden: A Brief Memoir of the Life and Writings of the Late William Marsden, Written by Himself*. Cox, 1838, p. 72.

hands.[46] Although his research interests never left South and Southeast Asia, Marsden did not court the Company for any official position – perhaps running an agency house provided him with sufficient financial independence. And perhaps this is why, as with Banks himself, the British Museum was much more his institutional base than was India House.[47] He even purchased a home in Bedford Square to be close to, as he put it, "my philosophical friends in Soho Square."[48] In the last quarter of the eighteenth century, this area of London, about 2.5 miles west of India House, was the cultural center of the sciences, housing two of the most significant institutions: the British Museum and the Royal Society. Banks's circle of influence also stretched further west to the botanical collections of the Chelsea Physic Garden, run by the Society of Apothecaries, the main suppliers of drugs to Company surgeons, and to King George III's gardens at Kew. At his death, Marsden bequeathed his coin collection to the British Museum, and his library went to the newly founded King's College (the Marsden library is now shared between King's and the School of Oriental and African Studies [SOAS]).[49]

All of these institutions were connected in various ways to science under the Company in London in this period. Materials from Asia would make their way via Company servants into Banks's vast herbarium, the king's gardens, the Royal Society's repository and the cases and stores of the British Museum. Social and intellectual ties were strong: surgeons and naturalists returning from Company service were frequently elected to the Royal Society, sometimes becoming prominent members. As we will see, Banks and other members of the Royal Society were also at times part of Company deliberations on administrative matters, and the Company subscribed funds to some voyages of exploration that the Society promoted. It should also be noted that many wealthy members of the Royal Society and other naturalists and collectors also had financial ties to the Company as stockholders. The towering Glaswegian collector William Hunter, for example, sometimes funded his massive purchases through the sale of Company stock.[50]

[46] Carroll. "William Marsden." Contra John Gascoigne, who describes Marsden as a protege or acolyte of Banks, Carroll makes the case that Marsden was then considered a scholar in his own right.

[47] Chambers, Neil. *Joseph Banks and the British Museum: The World of Collecting, 1770–1830.* Taylor & Francis Group, 2007.

[48] Marsden. *A Brief Memoir*, p. 49.

[49] See Marsden, William. *Bibliotheca Marsdeniana Philologica et Orientalis.* London, 1827.

[50] Brock, C. Helen. "The Happiness of Riches." In *William Hunter and the Eighteenth-Century Medical World*, edited by William F. Bynum and Roy Porter. Cambridge University Press, 1985, pp. 35–56.

The New Library-Museum at India House

In 1798, the same year the Company first broke ground on the expansion of the new India House in the City of London, the directors announced their plan (see earlier) for a new "public repository."[51]

105. You will have observed by our dispatches from time to time that we have invariably manifested, as the occasion required, our disposition for the encouragement of Indian literature. We understand it has been of late years a frequent practice among our servants, especially in Bengal, to make collections of oriental manuscripts, many of which have afterwards been brought into this country. These remaining in private hands, and being likely in a course of time to pass into others, in which probably no use can be made of them, they are in danger of being neglected, and at length in a great measure lost to Europe as well as to India. We think this issue a matter of great regret, because we apprehend that, since the decline of the Mogul Empire, the encouragement formerly given in it to Persian literature has ceased, that hardly any new works of celebrity appear, and that few copies of books of established character are now made; *so that there being by the accidents of time, and the exportation of many of the best manuscripts, a progressive diminution of the original stock, Hindostan may at length be much thinned of its literary stores, without greatly enriching Europe. To prevent in part this injury to letters, we have thought that the institution of a public repository in this country for oriental writings would be useful*, and that a thing professedly of this kind is still a bibliothecal desideratum here. It is not our meaning that the Company should go into any considerable expence in forming a collection of Eastern books, but we think the India House might with particular propriety be the centre of an ample accumulation of that nature; and conceiving also that gentlemen might chuse to lodge valuable compositions where they could be safely preserved and become useful to the public, we, therefore, desire it to be made known that we are willing to allot a suitable apartment for the purpose of an oriental repository, in the additional buildings now erecting in Leadenhall Street; and that all Eastern manuscripts transmitted to that repository will be carefully preserved and registered there.

106. By such a collection, the literature of Persia and Mahomedan India may be preserved in this country, after, perhaps, it shall, from further changes and the further declension of taste for it, be partly lost in its original seats.

107. Nor would we confine this collection to Persian and Arabian manuscripts. The Shanscrit writing from the long subjection of the Hindoos to a foreign Government, from the discouragement their literature in consequence experienced, and from the ravages of time, must have suffered greatly. We should be glad, therefore, that copies of all the valuable books which remain in that language, or in any ancient dialects of the Hindoos, might through the industry of individuals at length be placed in safety in this island, and form a part of the proposed collection.[52]

[51] "Repository" was then a common term for a sale-house or storehouse.

[52] BL IOR E/4/467: India and Bengal Dispatches, May 25, 1798, ff. 430–439, my emphasis. Quoted in Desmond. *India Museum*, p. 5.

As first presented in the 1798 Dispatch, the primary aim of a Company repository was the "preservation" of an endangered "oriental literature." The threat to India's "literary stores" was described as threefold: first, according to the directors, very few new literary works of note were now being produced (the golden age of Persian and Arabic literature lay in the past); second (and echoing the argument in Foote's *Nabob* that individual nabob wealth did not contribute to national wealth), the voracious private collecting and exportation of oriental manuscripts by Company servants was "thinning" the stock of original works in India (while at the same time this collecting was "not greatly enriching Britain"); finally, Sanskrit and other languages of pre- (and non-)Mughal India had "suffered greatly" under "foreign" (i.e. Mughal) rule. For all these reasons, so it was proposed, a safe harbor was needed *in Britain* for the literary and scientific material of the subcontinent.

Robert Orme had been making the case for a Company library at India House since at least 1792.[53] It is unclear how directly Orme was involved in the directors' final decision to allocate space in the new India House for a repository. The dispatch's depiction of an endangered Persian and Arabic and a long-oppressed "Shanscrit" is much the same as arguments made by Charles Wilkins in his introduction to the *Bhagavad Gita*. In its first expression the repository was conceived as a collection designed to intervene in and incorporate the private manuscript trade; there is no discourse of "improvement" here, or even of "usefulness" beyond the narrowly presented aim of "preventing further injury to letters." The directors' first vision of the repository was *not*, like that of the botanical gardens or the surveyor's offices on the subcontinent, as a new arm of an improving mission. And it was also, in expecting to grow at "no great expence" and by way of donations, not much of a departure from the old outsourcing model of knowledge management. It was merely a signal of a willingness to take "public" but, in effect, corporate ownership of these materials. But in that little shift was the making of what would become an important new institutional space for Company science in Britain.

London newspapers immediately reported on the plans and developments for a repository at India House. In the same year that the directors announced their plans for the library and museum, *The Oracle* reported on some "presents" including "two chests containing some very valuable jewels" for the "Oriental Museum at the India House" as well as alerting readers that "a magnificent and extensive Library is to make a part of the

[53] See Arberry, Arthur J. *The Library of the India Office: A Historical Sketch*. India Office, 1938.

additions to the India-House."[54] Years before Tipu's tiger arrived in London, the *Morning Post* had already learned that "a most curious piece of mechanism," which was "proof (if any were yet wanting) of the deep hatred and extreme loathing of Tippoo Saib [sic] towards the English nation," had been captured at Seringapatam and was to be shipped back to London (so the *Post* thought) to the Tower to go on display with other national war trophies.[55] (Tipu's tiger also appears in the *Times*'s earliest reference to the "Company's Museum" in an account of the visit of a Mamluk envoy, Elfin Bey, in 1803, when the group played "Rule Britannia" on the organ of the tiger.)[56]

A year later, this modest idea had been spun into a plan for a small British Museum-like library and museum at India House. This new proposal was the work of Warren Hastings and Charles Wilkins, both of whom had returned to London in the mid 1780s. Wilkins had, as we now would expect, returned to London with a trove of Indological material he had collected while in India. He continued his translation work, publishing a book of Sanskrit fables, the *Hitopadesha*, in 1787, was elected to the Royal Society in 1788 and issued more sections of the *Mahabharata* in 1794 and 1795. A fire at Wilkins's home in 1796 damaged much of his collections and destroyed all of his Bengali and Sanskrit typeface. Hastings had originally been called back to face accusations of corruption and mismanagement, and a subsequent trial of impeachment had dominated nearly a decade of intra-London Company politics. During the trial, which lasted from 1786 to 1795, Wilkins often attended court to speak in support of him. In addition, Kasinatha and the "Bengali Pandits of Benares" sent letters of support for Hastings, one with 112 signatures.[57]

After Hastings was acquitted, he passed to the Court of Directors Wilkins's proposal for "A Sketch of a Plan for an Oriental Museum proposed to be established at the India House."[58] Wilkins's proposal described a repository-like archive of "maps, charts, plans, views, manuscripts, printed books, coins, medals, statues and inscriptions." But it also

[54] *Oracle*. "[Gift from Peshwa of Jewels to Museum]." October 23, 1798; Oracle. "[Announcing the Museum]." November 28, 1798.
[55] *Morning Post*. "A Musical Tyger." April 19, 1800.
[56] *The Times*. "Elfi Bey," December 9, 1803. The Times Digital Archive. Reprinted in MacGregor, p. 176.
[57] Davis, Richard H. "Wilkins, Kasinatha, Hastings, and the First English Bhagavad Gītā." *International Journal of Hindu Studies* 19, no. 1–2 (2015): 39–57, p. 49.
[58] Proposal reprinted in Forbes Watson, John. *On the Measures Required for the Efficient Working of the India Museum and Library with Suggestions for the Foundation, in Connection with Them, of an Indian Institute for Enquiry, Lecture, and Teaching*. Her Majesty's Stationery Office, 1874, appendix B, pp. 55–56.

included three "cabinets": "natural productions," "artificial productions" and "miscellaneous articles." Among the "natural productions" were the three categories of "Animal, Vegetable, and Mineral." The general organization mirrored that of the British Museum, but the proposed items for collection were distinctly drawn from the world of the Company and its empire. Under "Animal" Wilkins included many highly sought-after items of trade and commerce, for example "animals, parts of animals, or produce of animals as are objects of commerce, and all in its natural state: the tusk of the elephant, the wool of the shawl goat, the musk in its bag, the cocoons of the different silk worms, lack [lac] with its colouring substance intact, the cochineal, and the edible birds' nest."

Wilkins does also suggest that the more typical specimens of natural history – stuffed or preserved animal trophies or large and merely "curious" animals offered as gifts – should also be found a place if they are "accompanied by an Abstract of its Natural history."

In the curating of the "vegetable productions," Wilkins suggests that the focus should be on the plants "whose produce is an article of commerce." For example: "timber ... for ship-building, [plants with] medicinal virtues or fragrant scents ... sugar canes and tea trees, cotton plants ... indigo and other plants used in staining and dyeing ... oils, gums and resins which are the natural produce of the plants of Asia." And all of these specimens should be well documented, accompanied by a "memorandum of its peculiar qualities, place of growth &c."

Wilkins's description of the "mineral productions" to be collected focuses, yet again, almost exclusively on known and economically significant materials:

> that species of steel which is known at Bombay by the name of bat or coots [wootz] ... bitumens and petroleums ... it would be a curiosity to our chymists to see the saltpetre and fossil alkali in the native earth, unmanufactured, as well as the borax ... precious stones, marble and alabaster ... particular attention should be given to those stones, earths and clays as might be useful in our manufacture.

Similarly, it was British manufacturing and trade interests, and the intense importance of the textile trade to Company profits, that dominated the collection of "artificial productions" proposed by Wilkins. He suggests ambitiously to procure "generally samples of all the manufactures of Asia, and, particularly, every article in silk and cotton, in every stage from the cocoon and pod to the cloth ready for the market; of the different sorts of colouring substances prepared in India; of sugar and sugar-candy; of saltpetre and borax &c &c." Importantly, Wilkins also proposed to form a collection of Asian technology: "models of the various machines and tools used in the manufactures of Asia should form a part of the Collection; and also of the implements of husbandry, and instruments used in their sciences, mathematical, astronomical, musical &c. &c." Finally, in acknowledgment of the centrality of gifts and the place of treasure within India House, he also proposed a space for "miscellaneous articles," which he describes as "Curiosities, chiefly presents, such as cannot conveniently be classed under any of the former heads."

Wilkins's description of an imagined Company museum was entirely unique to the nature of British interest in, and the state of Company knowledge relative to, Asia. His is an ideal collection produced by someone with detailed knowledge of materials of commercial interest to the Company and its customers. Unlike the directors' initial description of a "repository," intended merely for "preservation," and unlike the curiosity cabinets of the wealthy amateurs, Wilkins repeats again and again the aim of collecting well-studied objects of interest to *commerce and manufacturing*; this is a plan for an industrial or economic museum forty years before the boom in economic museums would begin (see Chapter 7). Although similar in some ways to the British Museum model, the Company's model of a museum was also clearly *not* an Enlightenment-style encyclopedic or universal collection. James Delbourgo has stressed how the growth of Hans Sloane's collection (which would become the founding collection of the British Museum) illustrates how global trade "enabled the pursuit of universal natural history, aimed at gathering as much of the world's variety as possible."[59] The initial plan for the Company's museum was, in contrast, specific to Asia, and with the apparent intention of gathering together not an ark of all of God's creations, but a great warehouse of all of humankind's material desires.

Hastings wrote enthusiastically to the Court of Directors, urging "the formation of a new and untried system for ingrafting the knowledge of India on the commercial persuits of the Company."[60] Echoing the early orientalism of Jones and Wilkins, but somewhat at odds with the directly commercial character of Wilkins's proposal, Hastings suggested that an "oriental museum" at India House would distinguish the Company from other trading companies ("men associated for the purposes of pecuniary gain") and demonstrate a more enlightened corporate character. As Hastings puts it (and does so in the language of trade), the Company has "joined a desire to add the acquisition of knowledge (and wonderful will be the stores which the projected institution under such

[59] Delbourgo. *Collecting the World*, p. xxx.
[60] Forbes Watson. *On the Measures Required*, appendix B.

auspices will lay open to them) to the power, the riches, and the glory which its acts have already so largely contributed to the British Empire and Name."[61]

Wilkins's original proposal for the repository had also included a request to fund a scholarly society devoted to "Eastern learning … and the cause of science in general."[62] Based on the model of the Asiatic Society of Bengal ("a Society similar to that now flourishing in Calcutta"), Wilkins proposed that the Society's meetings could be held in the India House repository, members would be allowed free use of the collection for their research and the Society would also run a printing office, with the ability to print "in the Oriental Characters" for use by both the Company and Society business, with the Society payments in effect funding the press office itself.[63] Apparently moved by at least part of Wilkins's vision, the directors voted in favor of the repository portion of the proposal, although they declined to fund the scholarly society, and also voted to offer the new position of curator to Wilkins with a salary of £200 per annum.

Wilkins's plan for the repository was well ordered, focused and comprehensive. But, for many years, the actual collection would be very different. In the first few years of the library and museum's existence, its collections "grew" largely through the disgorging of materials – often curiosities and presents – from other rooms and shelves in India House. A new policy had, in November 1801, required all books and "articles of curiosity" dispersed throughout India House and Company warehouses to be deposited in the new repository.[64] Thus, for example, one director, Hugh Inglis, transferred to the museum a collection of rare books "of the Maharatta character" and a "silver image of the God Buddha or as he is called at Ava [Myanmar], Gowdona … with a curious Japanned [lacquered] Box."[65] John Roberts, another director, deposited typeface for the Telegu language cut by Vincent Figgins. Wilkins himself presented a set of newspaper clippings. Then came a "Persian manuscript," "a silver ring with a black stone bearing an inscription in Nagari Characters," "A Crystal with an Arabick Inscription," "A Manuscript dictionary and grammar in French and Tamul [sic] by Father Dominique Pondicherry from 1743" and an "Egyptian Idol" from the Chairman's Office. From the Treasurer's

[61] Quoted in Desmond. *India Museum*, p. 13 (IOR: E/1/101 Misc. letters received November 15, 1799).

[62] Quoted in Forbes Watson. *On the Measures Required for the Efficient Working of the India Museum*, appendix B.

[63] Quoted in Forbes Watson. *On the Measures Required for the Efficient Working of the India Museum*, appendix B.

[64] BL Mss Eur F303/1 frontispiece.

[65] BL Mss Eur F303/1, December 20, 1801.

Office arrived "the Horn, or rather Tusk, of a fish said to have been found many years ago sticking in the bottom of an Indiaman."[66] The same day, from the Secretary's Office came a "curious ring, an opal set in gold" and an "Arabic manuscript on the small-pox printed at Cairo by the French," as well as a set of books heavy on history and travel (i.e. "Dr Halde's" *Description of China* [1738]), dictionaries (i.e. Richardson's *Persian Dictionary*), and natural history and particularly botany (i.e. Hill's *British Herbal* [1756], Dillenius's *Hortus Elthamensis* [1732], Gerard's *Herbal* [1633] and Grew's *Anatomy of Plants* [1682]). That same day, seventy-five printed books were transferred from the Registrar's Office to be cataloged by the library. The list included standard English reference works such as the journals of the House of Commons, volumes of Treatises and Charters, and Chamber's *Cyclopaedia*. What might be called Company reference works were also well represented: Ben Marsden's *History of Sumatra*, Nathaniel Halhed's *Code of Gentoo Laws*, the *East India Acts*, Dalrymple's *South Sea Directory*, Rennell's *Bengal Atlas*. Recent works that the Company had subscribed for would have been deposited in the Registrar's Office as well; thus the list also contained Symes's *Embassy to the Kingdom of Ava* (1800), Vincent's *Periplus of the Erythrean Sea* (1800) and a dozen other histories, travels and geographies printed in the previous decade.[67] And so it went for the rest of 1802 and into 1803. The whole building seemed to turn out its pockets, and into the library was swept everything that fell out.

On December 30, 1802, amid a pile of recently published books sent to the library from the Director's Office, was one curious manuscript: "The Original Manuscript Record of Tippu [Tipu] Sultan's Dreams." This was among the first shipments of plunder from the siege of Seringapatam in 1799 to arrive at India House. It is also an exceedingly intimate relic of one of the greatest challengers to Company rule on the subcontinent, first cherished in the room where the Court of Directors' met and, as we have seen, later put on display in a glass case in the library.[68] A month later, the Examiner's Office deposited "Proceedings of a Jacobin Club at Seringapatam" and "a gold medal commemorating

[66] BL Mss Eur F303/1, December 23, 1801.

[67] BL Mss Eur F303/1, December 23, 1801.

[68] Brittlebank, Kate. "Accessing the Unseen Realm: The Historical and Textual Contexts of Tipu Sultan's Dream Register." *Journal of the Royal Asiatic Society* 21, no. 2 (2011): 159–175. Also see Stewart, Charles. *A Descriptive Catalogue of the Oriental Library of the Late Tippoo Sultan of Mysore to which are prefixed Memoirs of Hyder Aly Khan and his son Tippoo Sultan*. Cambridge, 1809. p. 94. For an early English translation of six of the dreams, see Beatson, Alexander. *A View of the Origin and Conduct of the War with Tippoo Sultaun Comprising a Narrative of the Operations of the Army under the Command of Lieutenant-General Harris, and of the Siege of Seringapatam*. London, 1800, pp. cix–cxiii.

the fall of Seringapatam."[69] Soon these items were joined by a trickle then a flood of materials plundered from Mysore, although what arrived at India House was only a fraction of what had been taken when the last of the Anglo-Mysore wars ended with the storming of Tipu Sultan's palace. The day-book records for later 1802 to 1806 are missing, but when the entries pick up again, another large deposit is recorded from India.[70] The first waves of Mysore plunder enriched the India House collection in the form of manuscripts, rare books, the famous mechanical tiger (described earlier, Figure 3.5), war trophies and gold and jewels. In addition to what was considered Company booty, many valuable jewels, precious metals, household wares, works of art and military souvenirs would make their way in a semiorganized manner through the Company's prize agents into private hands and hence back to family homes in Britain.[71] But, at least initially, the Company's army

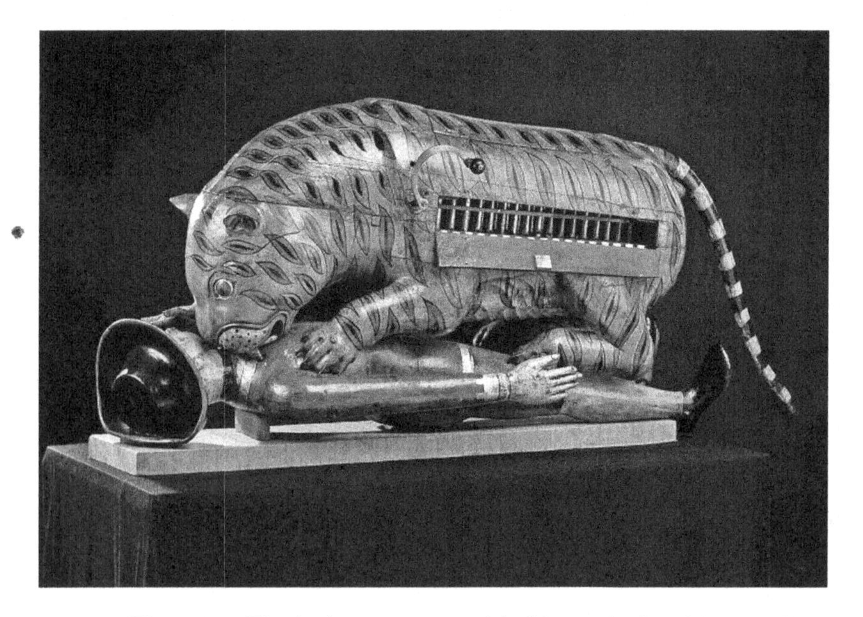

Figure 3.5 Tipu's tiger, constructed in Mysore in the 1780s or 1790s. Now at the Victoria and Albert Museum (no. 2545 IS). Copyright Victoria and Albert Museum, London. Pictures From History/ Universal Images Group via Getty Images.

[69] BL Mss Eur F303/1, January 8, 1803. [70] This arrived on July 18, 1806.
[71] A recent Sotheby's auction, "Mughals and Maharajas" in 2019, featured material from the capture of Seringapatam.

retained in Calcutta the renowned royal library of Seringapatam. Far from a planned accumulation of examples of economically important materials as Wilkins had proposed, the early patterns of object accumulation in the museum followed the serendipitous patterns of wartime acquisition.

Wilkins was, however, somewhat intentionally shaping the printed book collection through regular book-buying trips. For the first decade or so, Wilkins preferred to shop at the Strand bookseller Francis Wingrave, who also counted among his devoted customers the Astronomer Royal Nevil Maskelyne. The first purchases for the Company's library come in late 1801, just a few weeks after the library begins to register acquisitions. Again, the subjects are botanical; Wilkins purchased two of the most authoritative Indian herbals from a sale at the auctioneers Leigh and Sotheby's: the twelve-volume *Hortus Malabaricus* and the *Thesaurus Zeylandicus*. In early 1802, Wilkins was stocking the new library with dictionaries (Johnson's *Dictionary*, Ainsworth's *Latin Dictionary* and Boyer's *French Dictionary*, in addition to a Portuguese dictionary). A week later, Wilkins was back in the Strand, this time on a larger shopping spree, and this time focused on Dutch, French, Italian and Prussian works of science, including, for example, Foucquet's *Observations Astronomiques de Chinois*, Manucci's *Storia do Mogor*, Du Bec's *Histoire du Grand Tamerlanes*, Palafox's *History of China*, Kolbe's *Description of the Cape of Good*, several accounts of voyages, Bayer's *Museum Sinicum* and Rumphus's *Herbarium*.[72]

And so it went for the next three decades. The growing library purchased books in a wide range of subjects, but invested especially heavily in Asian languages, history, geography and natural history (especially botany).[73] The librarian generally had at his disposal a budget of £200–£300 per year (about the same as his base salary) and seems to have done his best to collect for the Company as many new publications on Asia or the Middle East that would be sold in Britain, as well as many from continental Europe, that he could get his hands on through his favorite booksellers. In the first decade or so, much of what Wilkins purchased were French or Dutch works, but from the 1810s onwards, as more and more British publications relating to Asia appeared, the buying trips became largely a matter of trying to keeping up with all that was being published about Asia or the Company itself in (primarily) London. Starting in the 1830s, Prussian works would also be purchased in

[72] BL Mss Eur F303/1, December to January 1801–1802.

[73] The first published catalog shows a similar subject distribution. East India Company. *A Catalogue of the Library of the Hon. East-India Company*. J. & H. Cox, 1845.

increasing numbers.[74] In the 1820s, Wilkins shopped most often at Black and Company, and in the early 1830s also frequented Parbury and Company. But soon after William H. Allen and Co. opened a few doors down from India House on Leadenhall Street, it became the library's main source for British and European printed books. W. H. Allen were also the booksellers through which the Company sold, in increasing numbers, their own publications (some printed in London, others in India). It also became a leading publisher of colonial material in general, including the *Asiatic Journal and Monthly Register*. In addition to growing through purchases, many more books were "deposited" in the library by other branches of the Company or by shipments from presidency offices. Hence, for example, when the Court of Directors agreed to support a new publication by subscribing for (i.e. pre-purchasing) multiple copies of the work, a few copies would usually be deposited in the library. And from the 1830s onwards, many more books began to arrive from abroad. Most importantly, the Committee of Public Instruction of the Bengal Government and the Calcutta School Book Society, both of which regularly published works in native language, would (when prodded by the Court of Directors) send copies of these in bulk to be deposited in the library, where they often would be reserved for use by another pair of new institutions closely connected to the library: the Company's civil and military colleges.[75]

Sample of extracts of library day-book entries, January 1810. BL MSS EUR F/303/2

January 2

Sent to the Persian Ambassador by order of the Chairman […] History of Shah Jahan in 23 vols (Returned)

9

Taken hence by Mr Simons the Book of Drawings containing … of the silk worm (Tea Plant) &c for the Committee of Warehouses (Returned)

10

Received from the [opening] room 40 Copies of Gladwin's Persian & Hindoostani Dictionary 2 Vols Calcutta 1809 unbound

11

Sent to Professor Hamilton, Haileybury 1 Copy of … […] Sanskrita Grammar … which had been left out might be …

[74] In the 1860s German works were bought from Williams & Norgate.

[75] See, for example, BL MSS EUR F303/36, Copy India Public Department, September 14, no. 15 of 1836.

Sent to Hennington's 40 Copies of Sanskrita Grammar to be half bound with 40 copies each of the five plates (Returned)

The Bengal Journal and Ledger received by Mr Kendall on the 10th instant appears to have been transmitted by the resident at [Mundh]ihabad for the use of the students at Hertford and Streatham – Rec'd copy of a letter from the secretary to the Council of the College dated 19th April 1809

12

Sent to his Excellency the Persian Ambassador the [Hagbbi s Dirna] three books (Returned)

15

Received from [Asperne]'s – Asiatic annual Register for 1807 Vol 9th

Rec'd from Dr Roxburgh Calcutta – resin of Naturnia Indicia and Produce of the Tree [Bosseillia seriata] in two small parcels

Rec'd from Committee of Correspondence room, a couple of framed Paintings of Trichnopoly and of a romantic rock in [To . . .] Wood

18

Rec'd in last December, 4 large Framed China Paintings descriptive of

[a procession], likewise 3 carved landscapes from China intended as a present to the Consul of the French Republic or rather to the Empress Josephine the whole of which have lain in the Baggage Warehouse several months. Rec'd Indian Register for 1810

22

Received from Mr Hudson together with six plans &c relating to the River Hooghley the following plans:

Dr Barnes Plan of the defensive post at Ladder Hill. 1806MS

Tract of the Cuddalore along the East Coast of ... 1761

Chart of Part of the Eastern Sea from about [...] North

Chart of the [Cuyos] and Parts of [Panay] 1775

Brought by Mr Malth [us]

27

Sent to be bound in Marble Covers 20 Copies of 3 first sheets of the [Histopadesi]

30

Sent to Professor Hamilton 29 copies of the first three sheets of the Library edition of the Sanskrita Histopadesi taken hence by Mr [Hin ...]

The Company's British Colleges

At the same time as Wilkins's proposal for a library-museum was being considered by the Court of Directors, developments in India were forming that would eventually propel the Company's plans for London-based institutions of science in new directions. The new governor-general of India, the Anglo-Irish aristocrat Richard Wellesley was setting in motion an ambitious set of plans to make Calcutta the center of Company education and training. After his successful defeat of the kingdom of Mysore, Wellesley's next move was to reform the system of civil service education. Wellesley, himself a collector, especially of natural history drawings and paintings, set up a natural history museum and menagerie at Barrackpore in 1800 (with Francis Buchanan the first director). In the same year, much more ambitiously, he announced plans for a sprawling new college at Fort William in Calcutta.[76]

Wellesley's new college was symbolically and materially tied to the British defeat of Mysore. Wellesley wanted the college itself to stand as a perpetual monument to that victory, so he set a symbolic founding date of the same day as the fall of Seringapatam, May 4, 1800, even though the college was actually founded about four months later.[77] In a more direct material way, the vast and valuable library of Tipu Sultan, which had been captured along with the rest of the state treasury, was transferred to Fort William with the intention of making it the founding collection of the college's new library. Wellesley envisioned the college as an extension of the "benevolent intent" of the British Empire, which rules "in the mild and benign spirit of the British constitution." Its aims were to promote "the prosperity and happiness of the people" of British India by way of better training for the (at this point exclusively British) civil service.[78]

Wellesley's college plan was much more ambitious than what was then offered at the existing Calcutta Madrassa or other regional schools such as at Varanasi. At the time, the majority of writers, many of whom went to British India as young as sixteen, would be sent out with no special training. Company officers would start out as copying clerks, learning the business of Company administration on the job. They were given no special training in languages but were allowed a "Munshi allowance," an extra sum of money that could be used to hire a native teacher. In 1798 Frank Gilchrist, author of *A Hindustani Grammar and Dictionary*, proposed to the Company directors that he be allowed to take on Company officers as students in exchange for the Munshi allowance. Out of Gilchrist's plan came the first formal examinations; students of Gilchrist

[76] Sisir Kumar Das. *Sahibs and Munshis: An Account of the College of Fort William*. Orion Publications, 1960. Raj. *Relocating Modern Science*, chapter 2. For a broader analysis of the British-imposed education system from the 1830s onwards, see Seth, Sanjay. *Subject Lessons: The Western Education of Colonial India*. Duke University Press, 2007.

[77] Bowen. "The East India Company's Education," p. 108. Also see Pramod Nayar's five-volume resource for the history of education in British India: Nayar, Pramod K., ed. *Colonial Education and India, 1781–1945*. Routledge, 2019.

[78] "Notes on the Regulations of the College," July 10, 1800, reprinted in Pearce, Robert Rouiere. *Memoirs of the Most Noble Richard Marquess Wellesley*. Richard Bentley, 1847, p. 195.

would from 1801 be given a standard examination, testing their knowledge of the laws and regulations of the Company as well as their proficiency in oriental languages.[79]

Wellesley proposed to expand this slight and loosely structured system of education into a three-year residential college, where all newly arrived Company officers would be taught both "European" and "Asiatic" subjects. The list of subjects to be taught was extensive, including nine of the major languages of the subcontinent. It also included Greek, Latin, "Mahomedan law, Hindoo law . . . English law," political economy, classics, history, geography, mathematics, botany, natural history, chemistry and astronomy.[80] Clearly many of these subjects fell under the then-expansive category of "useful knowledge" but Wellesley, who himself was educated at both Eton and Harrow and then Oxford, imagined the real utility of such an education was in the construction of habits and the shaping of the minds of Company servants. With chauvenistic venom, Wellesley claimed this was especially critical for the case of civil servants working in British India because the general environment of India was, so he argued, dangerously degraded and depraved:

> [the students'] early habits should be so formed, as to establish in their minds such solid foundations of industry, prudence, integrity, and religion as should effectually guard them against those temptations and corruptions with which the nature of this climate and the peculiar depravity of the people of India will surround and assail them in every station, especially upon their first arrival in India. The only discipline of the service should be calculated to counteract the defects of the climate, and the vices of the people, and to form a natural barrier against habitual indolence, dissipation and licentious indulgence.[81]

Wellesley had brazenly put this grand plan into motion without informing or receiving approval from the Court of Directors. And, upon finally receiving the communications of Wellesley's plans in April 1801, the directors ordered the college – already up and running – to be immediately abolished and replaced by a much less ambitious school for language instruction only. A Wellesley biographer and admirer, writing in 1847, put the harsh decision down to a clash of sensibilities between the old corporation and its new role as state: "the men of mere facts, figures and money bags were not reasoned out of their predilection for the old routine of mere mercantile utility."[82] But, in fact, the Court was not at all against such an ambitious education plan; it was only against basing it in Calcutta

[79] Bowen. "East India Company's Education," p. 108.
[80] Wellesley. *Notes*, excerpted in Pearce. *Memoirs*, p. 197.
[81] Wellesley. *Notes*, excerpted in Pearce. *Memoirs*, pp. 188–189.
[82] Pearce. *Memoirs*, p. 201.

rather than in England. Wellesley's vision for the education of the Company's servants would indirectly form the foundation for a new education regime back in the home country. The Company's stated reasons for closing Wellesley's college were that the expense was far too great, there were too many professorships, the emphasis on European subjects was unnecessary and a better plan would in any case place separate colleges in each presidency. Unstated was the fact that, if civil service appointments were made by the governor-general after graduation from a college in Calcutta, this would greatly reduce the highly valued patronage power held by the Court of Directors. Furthermore, Wellesley had taken individual initiative one step too far.[83]

The Court's decision to close Wellesley's college was widely criticized in London by both critics and some supporters of the Company. The Board of Control and the Court debated the future of the college throughout 1802 and finally in September 1803 an agreement was reached. The college at Fort William was to focus on language training, for writers within its presidency only, and its funding was reduced by half.[84] Then, in 1805, in a public dispatch the Court of Directors announced the opening of a new college, to be located 30 miles north of London in Hertford, for the training of all civil servants headed to India.[85] The powerful Director and Clapham Sect evangelical Charles Grant was one of the driving forces behind the scope and scale of Haileybury, which he often referred to as his "child."[86] It was to be temporarily located in Hertford Castle while the architect William Wilkins, who had previously worked on several Cambridge colleges, was commissioned to build a grand neoclassical college, complete with a vast interior quadrangle and an elegant dining hall.[87]

Company professorships paid well, and they attracted and retained eminent scholars, virtually all of whom were clergymen and fellows of Cambridge colleges (the connection between Cambridge and Haileybury was tight, although after Haileybury closed in 1855, its successor in

[83] Not only in the college itself but also in a simultaneously communicated assertion that Indian-built (and owned) ships would (mostly because of wartime conditions) be used for the upcoming shipping season. For heated debate regarding Wellesley's college and shipping plans at a Court of Proprietors, see *London Chronicle*, December 17, 1800.

[84] See Bowen. "East India Company's Education," pp. 120–122.

[85] Bowen. "East India Company's Education," p. 112.

[86] Carson, Penelope. "Grant, Charles (1746–1823), Director of the East India Company and Philanthropist." *Oxford Dictionary of National Biography*. Oxford University Press, 2004.

[87] Liscombe, R. Windsor. "Wilkins, William (1778–1839), Architect and Antiquary." *Oxford Dictionary of National Biography*. Oxford University Press, 2004.

Figure 3.6 Sketch of Haileybury College, with cows and students in the foreground, by Thomas Medland, 1810. Copyright British Library Board (asset maps_k_top_15_74).

Indian civil service training would be the Oxford India Institute [*f.* 1875]). The first principal was Reverend Samuel Henley, a religious writer, Shakespeare scholar and fellow of Queen's College, Cambridge.[88] Wilkins accepted the additional position (and pay) of examiner and visitor to the college. The curriculum was very similar to the one Wellesley had designed, with professors of mathematics; natural philosophy; classical and general literature; political economy and history; the general laws and policy of England; Persian, Arabic, Hindustani, Bengali and Sanskrit (and later Telugu); Hindu literature and the history of Asia; and drawing and penmanship. Of the first generation of Company professors, two of the most prominent were the political philosopher the Reverend Thomas Robert Malthus (whom the students took to calling "Pop" or "Old Pop," i.e. Population Malthus), who taught political economy, and the Scottish orientalist and philosopher Alexander Hamilton, who taught Bengali and Sanskrit.[89] Charles Stewart, who had initially been appointed a professor at Wellesley's college, returned to Britain in 1806 to teach Arabic, Persian and Hindustani. The Reverend William Dealtry, a Cambridge wrangler, friend of Grant and fellow Church Missionary Society co-founder, was appointed the first mathematics professor, along with another wrangler and future vicar Bewick Bridge. Yet another clergyman and Cambridge wrangler, Henry Walter, was hired to teach natural philosophy. The artist and engraver Thomas Medland was hired to teach drawing and oriental penmanship.[90]

On July 16, 1806, a large section of Tipu Sultan's famed library finally arrived in Britain: 197 boxes of highly valuable Arabic and Persian manuscripts, plus a dozen especially rare volumes that Wellesley had suggested should be presented "to the King and Universities of the United Kingdom."[91] This effectively represented the final dismantling and relocation of Wellesley's vision for Company science from Calcutta to Britain. The supporters of Wellesley's Calcutta plan had once hoped that it would "be rendered brilliant and dazzling" and "attract the notice of surrounding nations and attract the various literati."[92] Now, to the

[88] Moriarty, G. P. and John D. Haigh. "Henley, Samuel (1740–1815), Church of England Clergyman and Writer." *Oxford Dictionary of National Biography*. Oxford University Press, 2004.

[89] On Malthus at Haileybury, see Tribe, Keith. "Professors Malthus and Jones: Political Economy at the East India College 1806–1858." *The European Journal of the History of Economic Thought* 2, no. 2 (September 1, 1995): 327–354.

[90] According to Kapil Raj, the college also brought over instructors from South Asia. See Raj. *Relocating Modern Science*, p. 154.

[91] BL Mss Eur F303/1.

[92] William Wilberforce in December 1807, quoted in Pearce. *Memoirs*, p. 226.

satisfaction of India House, the centrifugal force of Haileybury College, together with the India House repository, would draw attention to London, to the British orientalists and the growing network of institutions of knowledge connected to the Company in Britain.

Despite the failure of Wellesley's grand plan, however, later in the nineteenth century Calcutta would become a major center of education and intellectual output in the region.[93] But the growth of Calcutta's scientific and educational capacity was also undoubtedly slowed by the pull of these Company resources back to Britain. Wellesley's college continued in a different form, without dedicated buildings or housing for students, into the 1830s. Still, under the umbrella of language studies, the Calcutta College eventually employed over 100 native scholars. Its library would also grow immensely, eventually to become a founding collection for the first Calcutta Public Library. But the native teaching positions were subordinated to the heads of language departments, which were invariably held by Europeans. These positions were often held by collectors and orientalists such as William Carey, Frank Gilchrist and Henry Thomas Colebrooke. The college also fit awkwardly within the machinery of the Calcutta administration, and students attached to the college often fell into debt. Even the presidency government itself suggested abolishing it and returning to the Munshi system. In the Bombay presidency, no college was ever established, although the government occasionally floated the idea. The Court of Directors had asked the Madras government to propose a plan for a language college in 1802, and in May 1812 the College of Fort St. George opened. It had no professors (the examinations were conducted by Company writers and translators) but a large staff of Munshis.[94]

With the establishment of the East India College at Haileybury, the Company had taken full control of the training of its servants.[95] It had done so without reducing the Court of Directors' patronage power. And Haileybury, together with the library and museum at India House, would form a critical new institutional infrastructure for the growth of orientalist

[93] See Raj. *Relocating Modern Science*, chapter 2.
[94] Raj. *Relocating Modern Science*, chapter 2.
[95] See Alborn, Timothy L. "Boys to Men: Moral Restraint at Haileybury College." *Malthus, Medicine, & Morality* (January 1, 2000): 33–55; Tribe, Keith. "Professors Malthus and Jones: Political Economy at the East India College 1806–1858." *The European Journal of the History of Economic Thought* 2, no. 2 (September 1, 1995): 327–354; Moore, R. J. "The Abolition of Patronage in the Indian Civil Service and the Closure of Hailybury College." *The Historical Journal* 7, no. 2 (1964): 246–257; McCartor, Robert Lynn. "The John Company's College: Haileybury and the British Government's Attempt to Control the Indian Civil Service." Ph.D. Texas Tech University, 1981, p. 24; Milford, Lionel Sumner. *Haileybury College, Past and Present*. T. F. Unwin, 1909.

studies and all that entailed, including Asian-language printing capacity, translators and tutors, and manuscript and print libraries in Britain. In 1809 the Company expanded its educational landscape even further, establishing a separate training college for cadets joining its large private army. Previously, those entering the engineering or artillery branches of the Company's army would have been trained either privately or at the Royal Military Academy Woolwich (*f.* 1741). Now they would spend two years at the Company's Military Seminary in Addiscombe. The curriculum included mathematics, natural philosophy, chemistry, surveying, gunnery and fortification, as well as French, Latin and Hindustani.

All of this represented a vastly expanded role for India House in science and education.[96] Before the foundation of Haileybury, although outgoing servants were required to demonstrate proof of certain skills (legible writing and account or bookkeeping), the Company did not involve itself in providing those skills. The only Britain-based education provided by the Company was a period of internship required for junior servants starting out in the China tea trade: after 1789, writers elected for Canton were required to spend a year in London in order to observe the quarterly tea sales and, under a tea broker and the head warehouse keeper, be taught "knowledge of the different qualities of the Teas ... [and] the nicer distinctions necessary to guide the buyers in their purchases."[97]

The directors had firmly asserted the Company's interest in pulling both patronage power and the growing domain of British orientalism and Company science back into the orbit of London. But this new geography of Company science remained controversial for several more decades. In British India there was, initially, little interest in voluntarily feeding the growth of the institutions back in London. At a 1798 meeting of the Asiatic Society of Bengal, for example, the issue was raised whether the Society should endeavor to directly support the new India House repository by collecting materials for it. The motion was ridiculed and dismissed. In fact, in general, while wartime collecting and surveying was steadily yielding results, *voluntary* donations from Company servants in Asia were so slow to materialize that, in 1805, the directors sent out another public letter, once again soliciting donations: "As our Original views in establishing this library have by no means been abandoned, and we still entertain hopes, that the invitation held out to individuals in India ... would be successful, if properly

[96] Raj interprets the effective relocation of Wellesley's college back to Britain in terms of the Anglo-French rivalry of the Napoleonic wars. See Raj. *Relocating Modern Science*, ch. 4.

[97] Quoted in *Haileybury Memorials*, p. 9.

[98] June 25, 1805. Excerpted in *The Asiatic Annual Register or a View of the History of Hindustan and of the Politics, Commerce and Literature of Asia*, vol. 9 (1806 [1809]), p. 95.

seconded by our supreme government, we again refer you to them, and desire that the subject may be entered into with alacrity and zeal."[98]

There were, to be sure, several very significant donations in the first few decades, but these would come from individuals who had already returned to Britain. The vast manuscript collection of Robert Orme finally came under Company control after his death in 1801 when his will left the collection to one of the directors. And, in 1819, Henry Thomas Colebrooke donated his well-known collection to India House.[99] The largest private individual donation to India House up to that time, Colebrooke's collection was instrumental in establishing the India House library-museum as the most important repository of Asian materials in Britain, if not Europe. As Roseane Rocher and Ludo Rocher have argued, it was Colebrooke's collection, which "surpassed all in size and scope," that catalyzed a shift in the European center of orientalism from the National Library of France (and its largely missionary-sourced collections) across the Channel to Leadenhall Street.[100] Colebrooke's donation cemented the authoritative status of the Company's collections, but it also indicates how far the reputation of Company science had come by this time. Colebrooke was very active in London's scientific networks and was deeply invested in promoting orientalist scholarship in Britain, but his relationship with the Company was always uneasy to say the least.[101] It is thus especially striking for such a prominent figure in British science to choose to support the fledgling India House library-museum in this way.

Still, by no means did all major Company collectors based in London follow suit. William Marsden's collection was auctioned off at his death in 1836. When administrator and naturalist Sir Stamford Raffles died in 1826, his wife donated his large southeast asian collection to the Linnean Society and the new Zoological Society of London, which he and the Company naturalist Thomas Horsfield had been instrumental in launching. William Farquhar, while the British Resident in Melaka from 1803 to 1818, sent many specimens and drawings to the Asiatic Society of Bengal. After returning to Britain, he donated a valuable collection of natural history drawings to the Royal Asiatic Society in the 1820s.[102] Even the vast private collection acquired by Colin Mackenzie while surveyor-general (see Chapter 2) was not donated; instead, the Company had to

[99] Rocher. *The Making of Western Indology*, pp. 139–140.
[100] Rocher. *The Making of Western Indology*, p. 145.
[101] During his time in India he had been in conflict with the directors multiple times and was skipped over for advancement.
[102] Farquhar, William, John Sturgus Bastin, and Chong Guan Kwa. *Natural History Drawings: The Complete William Farquhar Collection – Malay Peninsula, 1803–1818*. Editions Didier Millet, 2010, pp. 31–33.

purchase the materials – much of which controversially remained in India – from his widow for the eyebrow-raising sum of £10,000.

Meanwhile, Haileybury was off to a rocky start. The new college buildings boasted the largest quadrangle in England, but the structures were ridiculed in the press as quite possibly *the* ugliest in Britain. Pupil misbehavior was regularly reported in the press, and student "riots" in 1809 and 1811, during which students marauded around the school breaking windows, blowing horns, and otherwise terrorizing the staff, led to much handwringing and worry over the moral and ethical qualities of the future India administrators.[103] Yet the directors were seen to be lax in their punishment of disruptive students. Most expelled for misbehavior were readmitted a year later; evidence, according to critics, of corrupt patronage and the iron influence of the old "Indian families." In other cases, students expelled for taking part in the rioting were still given a writership and sent off to India, and a parliamentary committee found in 1809 that a few writerships were still being sold by directors (in one case, a seat in Parliament was offered in exchange for a writership, though the exchange never seems to have happened). In all of these ways, as the Board of Control complained, the Company was not following its own new rules for education and the distribution of patronage.[104] If, as has sometimes been argued, Haileybury in the first decades was "the Directors' public relations gimmick to protect its patronage," in many ways the public relations part of the gimmick went terribly awry.[105] At the same time, however, with the high salaries for professors and unique resources including its connections to the Company's library and museum, Haileybury was attracting and retaining prominent and influential scholars.

*

The Company's investment in a new "oriental repository" came just as new threats to the Company's monopoly were gaining traction. However, what seems even more consequential to the Company's decision to establish the library, museum and colleges were two dimensions of a changing geography of knowledge resource accumulation within the empire: first, the now-flourishing domestic trade in manuscripts, antiquities and specimens, and second, Wellesley's plan to make Calcutta a major new center of learning. These were both threats to the perceived authoritative status of India House and also represented new opportunities for the expansion of

[103] Alborn. "Boys to Men."

[104] McCartor, Robert Lynn. "The John Company's College: Haileybury and the British Government's Attempt to Control the Indian Civil Service." Ph.D. thesis submitted to Texas Tech University, 1981.

[105] McCartor. "The John Company's College." Texas Tech University, 1981.

the domain of India House authority. Together, they drove the directors to intervene in and take more direct control over the political economy of knowledge between Britain and colonial India.

The growth of the Company's repository may not have been off to a very strong start. Those generous donations that the directors had hoped would be tucked in among their tea and textile imports did not at first materialize. But grow it eventually did, and the rate of accumulation at India House would, as we will see in the next chapter, continue to accelerate over the next half-century. The discovery of Bacchus on his tiger would also turn out to be a very fitting prelude to the wartime plunder that would flow into India House from the Company's successful campaigns during the Napoleonic wars. For the next half-century, Bacchus's tiger and Tipu's tiger were witness to the continual procession of material from Asia, some purchased or received as gifts, much of it plundered or collected in the context of wartime territorial expansion. In France, in the 1790s, Napoleon's army went so far as to explicitly copy the Bacchic parades of wartime plunder, with crowds gathering to cheer the arrival at the Musée Napoleon of precious works of art, cabinets of rare natural curiosities, manuscripts and books, jewels and gems, and even live animals.[106] In the case of the East India Company, however, the wartime loot and plunder that would go on display at India House (and later at Britain's other national museums) arrived with little fanfare. More and more, the unremarkable carts moving between the warehouses and India House were transporting crates of manuscripts, specimens, works of art and antiquities along with the textiles, tea and spices on their way to auction. Instead of crowds cheering, the steadily increasing flow of acquisitions was marked only by a carefully kept logbook of objects incoming and outgoing.

At some point in the next several decades, even the Bacchus mosaic would be taken down from the library and museum to make room for something else. The soot-filled basements, the dusty attics and the damp, open courtyards all became places where deposits for the library and museum would end up being stored – sometimes unopened and uncatalogued – over the next fifty years. The courtyards were where sculptures and other large metal or stone objects were deposited – left to the polluted air and fluctuating temperatures. The Bacchus mosaic appears to have been left largely forgotten in one of these yards for several decades. On some accounts this is when the pavement disintegrated, leaving only the

[106] Lipkowitz, Elise S. "Seized Natural-History Collections and the Redefinition of Scientific Cosmopolitanism in the Era of the French Revolution." *The British Journal for the History of Science* 47, no. 1 (March 2014): 15–41. On French collecting and the state in this period, see Engberg-Pedersen, Anders. *Empire of Chance: The Napoleonic Wars and the Disorder of Things*. Harvard University Press, 2015.

central image of Bacchus and the tiger.[107] At some point one of Wilkins's successors rescued it from the elements, had it fastened to a slab of slate and hung it in the old Tea Sale Room, now (in 1856) refashioned (much to the tastes of those who wanted India House to have a more "oriental" appearance) as a Mughal-style sculpture gallery. In its last installation at India House before eventually being donated to the British Museum (where it now resides), Bacchus and his tiger shared space with ancient Indian sculptures from the Amaravati temple complex in central India, fossils from the Siwalik Hills below the Himalayas and ghostly white plaster-casts of the faces of Tibetan and Nepali villagers.

The establishment of the Company's library, museum and colleges in London represented an important turning point in the Company's knowledge management. But as the fate of the Bacchus mosaic, and many other manuscripts and objects that passed into India House, makes clear, this is not a simple story of growing power-knowledge. Scholars once took the growth of information collection by nineteenth-century states as evidence of, or even synonymous with, the extension of modern forms of state authority. However, recent studies have emphasized a more complex relationship between state power and systems of knowledge. The growth of these London collections and colleges did not unproblematically establish new forms of authority or power. Rather, the new library-museum and colleges would become yet another site for ongoing contests among competing interests of the sciences, the Company and other commercial and colonial stakeholders. In consequence, as the century progressed, the growing tension between the advance of liberalism and the Company's defense of its monopoly privileges would begin to be felt within Britain's emerging new cultures of science.

[107] Smith, Charles Roach. *Illustrations of Roman London*. London, 1859, pp. 57–59.

Part II

From Company Science to Public Science, 1813–1858

4 Patterns of Accumulation

Alexander von Humboldt's Rejection

In 1812, the Prussian naturalist Alexander von Humboldt, recently returned from his celebrated expedition to South America, wrote to Emperor Napoleon describing his next goal: to explore the Himalayas, that "high mountain chain that stretches from the source of the Indus to the source of the Ganges."[1] Two years later, in June 1814, he was in London accompanying the prince of Prussia on a diplomatic tour. Humboldt visited the library at India House and presented to the Court of Directors his desire to travel to India. Humboldt's case was also pressed with the prince regent. Nothing came of these requests, and Humboldt was not granted access to India. Three years after that, this time accompanied by his friend the French astronomer François Arago, Humboldt once again visited London and once again sought approval to travel to India, and, once again, he was unsuccessful. Over the next several years, Humboldt continued to press his case with the directors while also gathering financial support. But he never made it to India, most likely because he was never granted permission by the Company.[2]

Instead of Humboldt, a series of Company servants would be the first Europeans to explore the ecologically unique and politically significant string of high mountain ranges. At exactly the time Humboldt was enquiring about access, the Company was employing the Scottish surveyor Alexander Gerard to explore various roads and routes through the Himalayas. His expedition eventually reached nearly 20,000-feet altitude and the border asserted by China. Gerard sent his notebooks with barometric, trigonometrical and meteorological readings, as well as maps and accounts of villages (in regions previously assumed to have been

[1] Théodoridès, Jean. "Humboldt and England." *The British Journal for the History of Science* 3, no. 1 (June 1966): 39–55, p. 43.

[2] According to Théodoridès, there is no surviving archival record of the directors' decisions regarding Humboldt's requests. "Humboldt and England," p. 44.

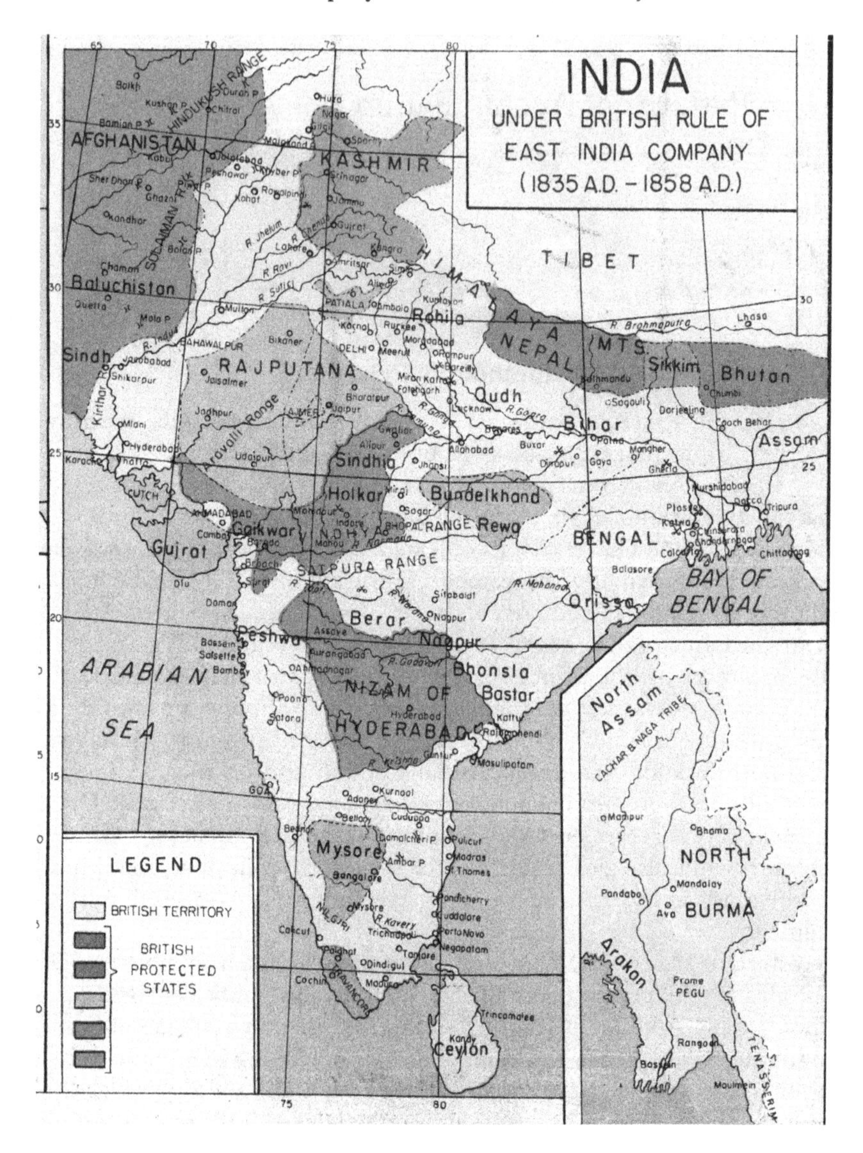

Figure 4.1 Map of India under British rule, 1833–1858. Image courtesy of Vidya Chitr Prakashan, New Delhi.

uninhabited and uninhabitable), together with a geological collection gathered at 19,000 feet, back to India House. They would sit in storage for several years before the Company administrator and orientalist Henry

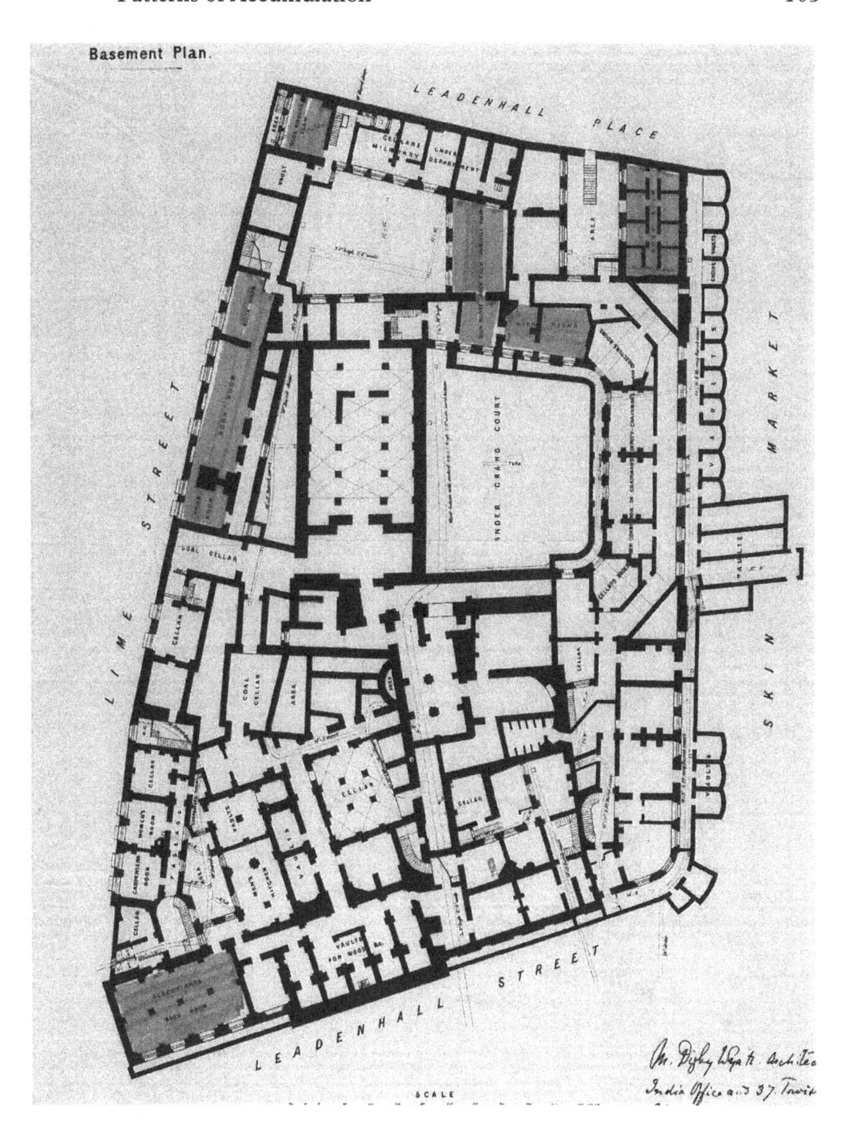

Figure 4.2 Basement floor with "Book Rooms" highlighted. In the early 1800s much of the basement of India House was used to store Madeira wine, wood, coal and other necessities. Between 1800 and 1858, more and more rooms in the basement were given over to book and record storage, accountants' rooms and a bookbinder. At some point a "Women's Room" was added in the southeast corner. Based on plans of the East India House produced by W. Digby Wyatt in 1860, just before its demolition. From a reproduction in Birdwood, George C. M. *Relics of the Honourable East India Company: A Series of Fifty Plates*. Quaritch, 1909. From the Collection of the Cornell University Library.

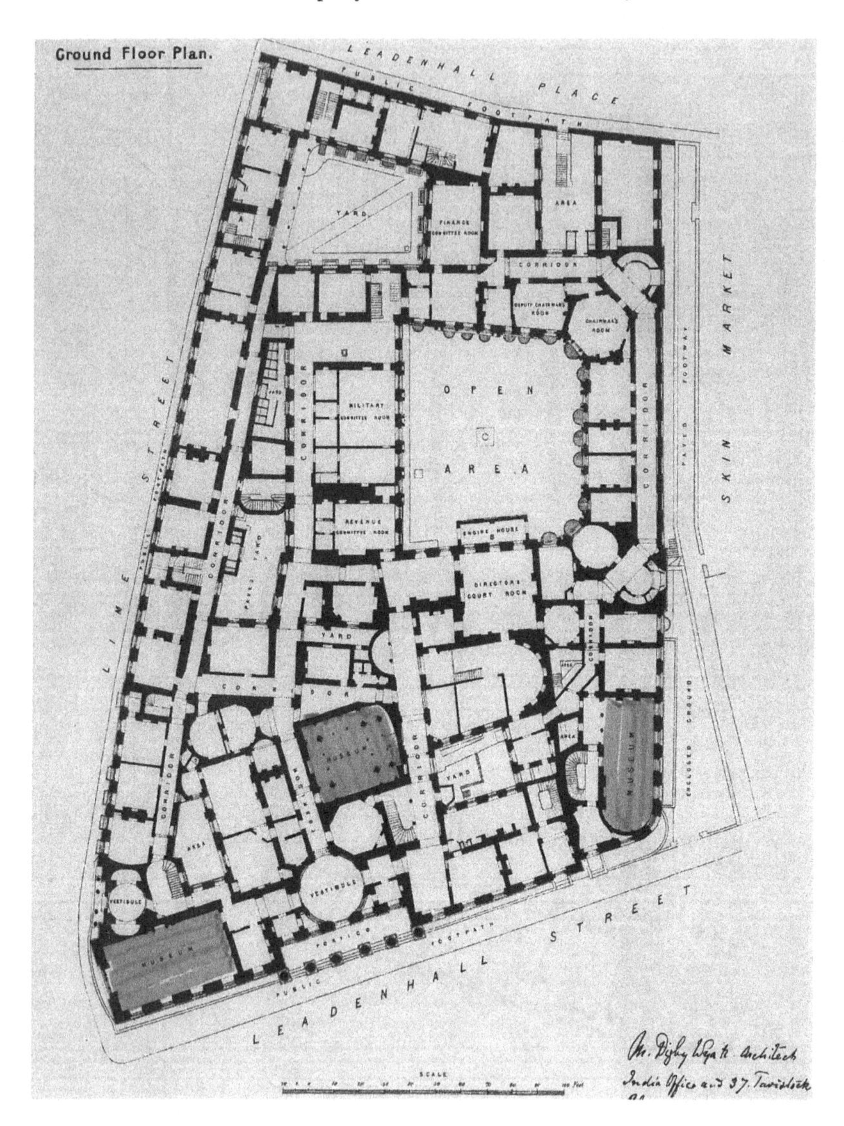

Figure 4.3 Plan of the ground floor of East India House as it was in 1860, with museum spaces highlighted. The museum has expanded to fill the large rooms in each corner along Leadenhall Street as well as the old Tea Sale Room just past the main vestibule on the right. Based on plans of the East India House produced by W. Digby Wyatt in 1860, just before its demolition. From a reproduction in Birdwood, George C. M. *Relics of the Honourable East India Company: A Series of Fifty Plates*. Quaritch, 1909. From the Collection of the Cornell University Library.

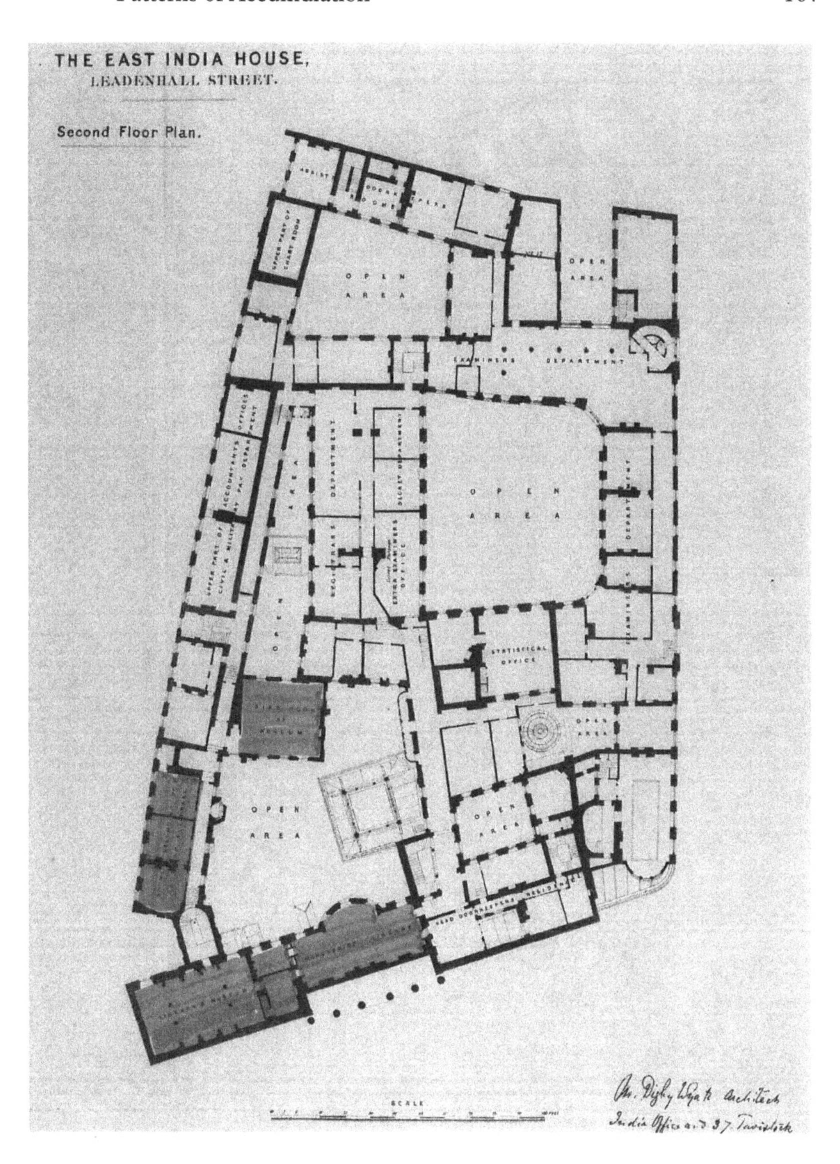

Figure 4.4 Plan of the second floor. The original library and museum space from 1801 is in the bottom-left corner. By 1860, the museum and library space had expanded to both adjacent rooms as well as the old surveyor's office down the hall and to the right, here labeled "bird room of museum." Based on plans of the East India House produced by W. Digby Wyatt in 1860, just before its demolition. From a reproduction in Birdwood, George C. M. *Relics of the Honourable East India Company: A Series of Fifty Plates*. Quaritch, 1909. From the Collection of the Cornell University Library.

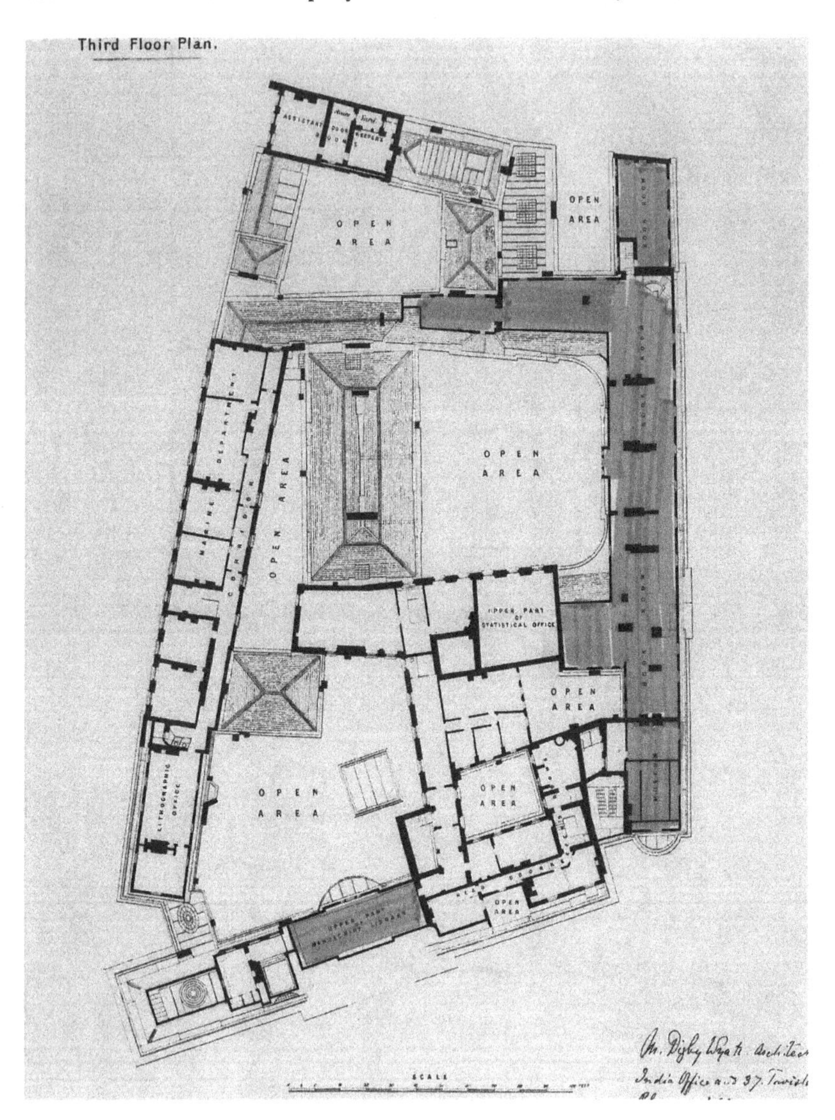

Figure 4.5 By 1860, the third floor of India House now contained more museum galleries as well as the lithographic office, the upper part of the statistical office and more "Book Rooms." Based on plans of the East India House produced by W. Digby Wyatt in 1860, just before its demolition. From a reproduction in Birdwood, George C. M. *Relics of the Honourable East India Company: A Series of Fifty Plates*. Quaritch, 1909. From the Collection of the Cornell University Library.

Thomas Colebrooke requested access to the materials.[3] Colebrooke then arranged for the duplicates in the collection to be separated out and donated to the newly established Geological Society of London (founded in part by Company servants). He also edited Gerard's journal into a publication sponsored by the Royal Asiatic Society.[4] Less than a decade later, under the patronage of the Company, the Company surgeon, naturalist and curator John Forbes Royle would begin to publish his extensive and influential biogeographical study of the Himalayas (more on Royle in the next chapters).[5]

Napoleon had believed that Humboldt was a Prussian spy, and it is possible the directors were uncomfortable with Humboldt's relationship with both Napoleon and the Prussian court, both of which were potential imperial rivals. But, as David Arnold notes, it was generally routine for the directors to deny access to naturalists and explorers who weren't connected to the Company.[6] Humboldt's great legacy makes his inability to access the Himalayas seem all the more historically significant, but his rejection is one especially striking example or illustration of how the Company's powerful monopoly on access to India directly shaped the political economy of science in the period. In exercising their control of access to India, the directors may have been trying to, as Arnold argues, "preserve [their] commercial privileges and prevent outsiders from undermining [their] authority."[7] But perhaps it was not so much a worry about their authority being undermined by outside explorers as it was a more basic calculation of a lost opportunity to expand their own circle of authority that drove such policies. And, given the high value, both in direct economic and indirect social and intellectual terms, of the rights to explore, collect and publish on those results, it should not be surprising that access to those resources was routinely restricted by the directors at this time.

This chapter is about the patterns of accumulation at India House after the foundation of the library-museum and during the period in which, despite changes to its charter, the Company retained control over British

[3] BL Mss Eur F303/345, July 8, 1825, Library Committee Minutes.

[4] Colebrooke requested access to the unopened chests containing Gerard's [Garrard's] collections, to which the directors agreed and requested Wilkins superintend the opening and dividing up of the material. The resulting publication is: *On the Valley of the Setlej River in the Himalaya Mountains: from the journal of Captain A. Gerard / with remarks by Henry Thomas Colebrooke*. Alexander Gerard, 1792–1839. Cox and Baylis, 1825.

[5] Royle, J. Forbes. *Illustrations of the Botany and Other Branches of the Natural History of the Himalayan Mountains and of the Flora of Cashmere*. Wm. H. Allen, 1839.

[6] Arnold, David. *New Cambridge History of India: Science Technology and Medicine in India*, Vol. III.5, Cambridge, 2000, p. 20.

[7] Arnold. *New Cambridge History of India*, p. 20.

access to the natural and knowledge resources of Asia. The chapter begins by describing how the Company came to play a more direct role in the acquisition and management of knowledge resources for repositories in Britain. Between the opening of the library and museum and the Great Exhibition of 1851, survey collecting for the Company and private collecting by Company surveyors were primary means by which the Company's new institutions of knowledge management were enriched. Company surveys during this period became closely tied to both military plundering and biogeographical collecting. Embedded in a series of ongoing conflicts over territory and trade, the making of these collections served as a means of further weakening rival states. Once back in London, these collections would also be crucial to the early development of the Company's library-museum.

During the same period, Crown support for the old monopoly was beginning to wobble. The last section of this chapter considers the place of knowledge accumulation and management in the tumultuous period around the charter debate of 1813, when many of the Company's monopoly privileges would be annulled. During these debates, a key defense of the monopoly was for the directors to present the administration at India House as the most trustworthy, authoritative source of knowledge regarding Asia in Britain, and thus the institution most suited to controlling trade and exercising governance. Within the Company, however, confidence in the Company's grasp of knowledge about Asia was far less absolute, and after the Company's losses in the 1813 charter, new worries about the Company's knowledge management practices would lead to even further efforts to centralize and better organize the stores of information accumulating at India House.

Territorial Expansion and Postwar Surveys

> June 2, 1802: "Three Chests containing a collection of insect shells, minerals and other objects of natural history made at Ceylon by Mr. Jonville accompanied by a memoir in French and sundry drawings. Received from the Baggage warehouse."[8]

These chests were the first substantial natural history collections to arrive at the Company's new museum from abroad. They were also among the first acquisitions made in the context of the Napoleonic wars in Asia. "Mr. Jonville" was Joseph Marie Eudelin Mervé de Jonville, a Corsican hired by the first British governor of Ceylon, Frederick

[8] BL IOR MSS Eur F303/2, June 2, 1802.

North, when the British occupation began in 1798.[9] Jonville was first employed to study the lucrative cinnamon plantations. Meanwhile, North had also instructed Jonville to "inquire into, and Collect, whatever regards the Natural philosophy, the natural history, and the meteorology of this island ... likewise ... the Customs, usages, history, and even languages of the Country."[10] Jonville reported in 1800 that while he had managed to acquire "a Considerable Collection of Natural Curiosities," including roughly 500 specimens and 800 draft drawings, his collecting had been severely limited by his not having been allowed to survey beyond the cinnamon country. Within a few months of writing to North of his desire to collect more broadly across the island, Jonville had been sent with a diplomatic expedition beyond the Company's territory to the Court of Kandy, where he was to act as official collector, illustrator and interpreter. On returning to the British base at Colombo, he was appointed surveyor-general of the British possessions and was instructed to begin a new survey of the entire island. He began producing a steady stream of maps and reports on subjects such as rice cultivation, elephant hunting, the pearl fisheries and the different regions of the island. Along with two locals, whom Jonville identifies as Andrisaratchi and Adrian Rajapakse, he collected accounts of the religions of the island, and attempted to produce translations of Sinhalese into English.[11]

Among the most prominent of the new generation of collectors were now the Company's surveyors such as Jonville, formally employed to chart and report upon the vast new territories, simultaneously accumulating observations, measurements, manuscripts and specimens. In the first decades of the nineteenth century, the Company would initiate three different surveys of the vast central territory in India over which it had gained control after the Mysore wars, in 1798. William Lambton would begin a "general survey" using trigonometrical methods, measuring two arcs of the meridian through the Carnatic and one parallel running west from Madras. When transferred to Bengal in 1817, this project would become the root of the Great

[9] On Jonville, see the introductory biography by Marie-Hélène Estève in Jonville, Eudelin de. *Quelques Notions Sur l'Isle de Ceylon*. Ginkgo Editeur, 2012. On geography and empire in Sri Lanka more broadly, see Sivasundaram, Sujit. *Islanded: Britain, Sri Lanka, and the Bounds of an Indian Ocean Colony*. University of Chicago Press, 2013; Sivasundaram, Sujit. "Tales of the Land: British Geography and Kandyan Resistance in Sri Lanka, c. 1803–1850." *Modern Asian Studies* 41, no. 5 (2007): 925–965.

[10] Jonville to Governor North, February 1, 1800. Quoted in Archer, Mildred. *Natural History Drawings in the India Office Library*. Published for the Commonwealth Relations Office by HM Stationery Office, 1962, pp. 35–36.

[11] Estève. *Quelques Notions*, p. 37.

Trigonometrical Survey of India. A second survey was the geographically detailed topographical survey led by the future first surveyor-general of India, Colin Mackenzie. Colonial surveying and collecting intersect vividly in the life of Mackenzie.[12] Over the course of a thirty-eight-year career in India, Mackenzie would supervise the construction of dozens of charts and maps and lead the new All-India Survey. Again and again, Mackenzie was dispatched to follow in the wake of Company expansion of formal territorial or informal political control. He was sent to Hyderabad in the 1790s, Mysore between 1799 and 1810 and Java from 1811 to 1813. He became the surveyor-general of Madras in 1810 and then of British India in 1814, a post he held until 1821. On all of his survey assignments, he collected voraciously, usually with the help of local scholars and guides. At his death, Mackenzie had amassed arguably the most significant European collection of information on South and Southeast Asia produced before 1830. Nowhere else was there a government surveyor who was simultaneously one of the most active collectors of literature and manuscripts.[13]

For the Mysore survey, Mackenzie would spend 1800–1807 in the field, during which time he was running two parallel surveys. First, there was the Company's topographical survey, for which the Company provided assistants and sub-assistants For the Company, he regularly produced large regional maps at 4 miles to the inch, together with district maps at 2 miles to the inch plus an assortment of more detailed maps of areas of special interest.[14] He also directed some of these Company-funded assistants to pay attention to a wide range of subjects: Mackenzie's "Hints or Heads of inquiry for Facilitating our Knowledge of the More Southerly Parts of the Deckan [Deccan], 1800" arranged sets of enquiries in different categories ranging from the geographical (modern and ancient names of towns, districts, features of landscape, rivers; local and British distances

[12] On Mackenzie as a collector, and on the Mackenzie collections, see Mantena, Rama Sundari. *The Origins of Modern Historiography in India: Antiquarianism and Philology, 1780–1880*. Palgrave Macmillan, 2012; Robb, Peter. "Completing 'Our Stock of Geography', or an Object 'Still More Sublime': Colin Mackenzie's Survey of Mysore, 1799–1810." *Journal of the Royal Asiatic Society* 8, no. 2 (1998): 181–206; Dirks, Nicholas B. "Colonial Histories and Native Informants: Biography of an Archive." In *Orientalism and the Postcolonial Predicament: Perspectives on South Asia*, edited by Carol A. Breckenridge and Peter van der Veer. University of Pennsylvania Press, 1993, pp. 279–313; Blake, David M. "Colin Mackenzie: Collector Extraordinary." *The British Library Journal* 17, no. 2 (1991): 128–150; Srinivasachariar. "Robert Orme and Colin Mackenzie."

[13] Ratcliff, Jessica. "Hand-in-Hand with the Survey: Surveying and the Accumulation of Knowledge Capital at India House during the Napoleonic Wars." *Notes and Records: The Royal Society Journal of the History of Science* 73, no. 2 (June 20, 2019): 149–166.

[14] Edney. *Mapping an Empire*, p. 179.

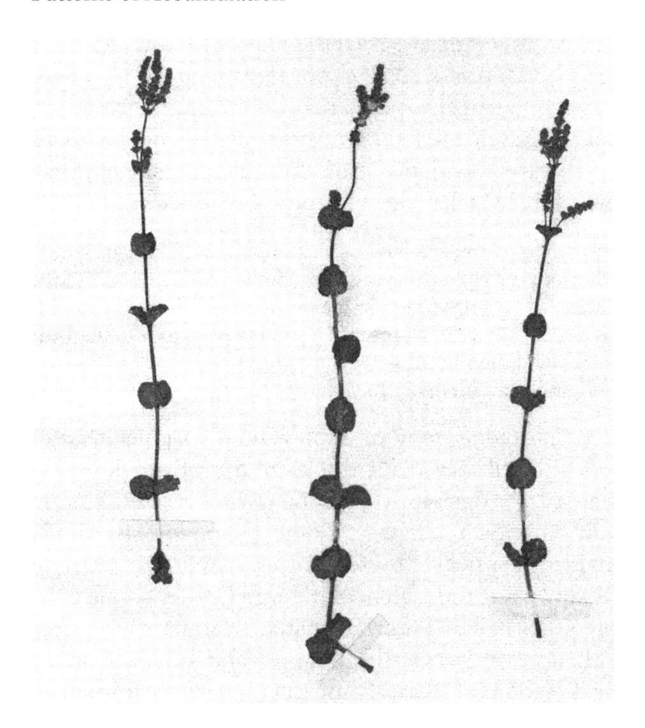

Figure 4.6 Type specimen of *Rotala Rotundifolia* collected by Francis Buchanan-Hamilton during his survey of Mysore in the early 1800s. Now at the Natural History Museum, London. By permission of the Trustees of the Natural History Museum, London.

between locations) to the natural and social (population, languages, ancient and modern history, legal and land revenue systems, local diseases and medicines, "Productions of the country," plants and animals, "minerals, fossils, ores etc.," meteorology, arts and sciences, commerce, customs, and [perhaps the category he was most interested in himself] "books and depositories of native learning").[15] Some assistants specialized in different areas. Benjamin Heyne, for example, focused in part on natural history. In addition to his notes and publications or "statistical tracts," Heyne's collection of birds arrived at the museum in 1813.[16]

[15] BL IOR Mss Eur F/128/213. Summarized in Edney. *Mapping an Empire*, p. 44.

[16] Horsfield, Thomas. *A Catalogue of the Mammalia in the Museum of the Hon. East-India Company*. J. & H. Cox, 1851. Also see Moore, Frederic. *A Catalogue of the Birds in the Museum of the Hon. East-India Company*. W. H. Allen and Co., 1854.

In tandem with this official surveying and collecting, Mackenzie was also building up a vast private collection. For much of this work, he paid for his own assistants and sub-assistants, who were trained largely by the Kavali brothers, Borayya and later Lakshmayya.[17] In practice, the surveys ran as one. As Horace Hayman Wilson, future Company librarian and an early cataloger of the Mackenzie collection, described it:

> The collection of books, papers, and inscriptions went hand in hand with the survey . . .. In the course of his surveying operations [Mackenzie visited] . . . all the remarkable places . . .. Accompanied in his journeys by his native assistants, who were employed to take copies of all inscriptions, and obtain from the Bhramans of the temples, or learned men in the towns or villages, copies of all records in their possession or original statements of local traditions.[18]

Meanwhile, yet another survey of Tipu Sultan's former territories was commissioned by Wellesley, with the intention of providing, relatively quickly, a "statistical" (descriptive, for the purposes of the state) account of the virtually unknown territory. Some Company shareholders and directors remained skeptical of these territorial expansions. On the advice of William Roxburgh of the Calcutta Botanic Gardens, this commission went to Francis Buchanan (later Buchanan-Hamilton), a surgeon and botanist who had recently acted as botanist and surveyor on a political mission to the Kingdom of Ava (Myanmar) (and whose journal had made its way into Tipu Sultan's library.)[19] According to Buchanan's instructions, the "first and great object" of his survey was to describe the agriculture of the region, but the full list of the information he was tasked with collecting included not only animal and vegetable productions and modes of farming but also climate, mineralogy, manufactures and "the condition of the inhabitants."[20] Under this last category, Buchanan was asked to collect whatever information he could about the societies he

[17] On the role of the Kavali brothers and Mackenzie's native assistants, see Mantena. *Origins of Modern Historiography*, chapter 3.

[18] Wilson, Horace H. *Mackenzie Collection: A Descriptive Catalogue of the Oriental Manuscripts and Other Articles Illustrative of the Literature, History, Statistics and Antiquities of the South of India, Collected by the Late Lieut.-Col. Colin Mackenzie*. Asiatic Society Press, 1828.

[19] On Buchanan and survey collecting, see Menon, Minakshi. "Transferrable Surveys: Natural History from the Hebrides to South India." *Journal of Scottish Historical Studies* 38, no. 1 (May 1, 2018): 143–159; Vicziany, Marika. "Imperialism, Botany and Statistics in Early Nineteenth-Century India: The Surveys of Francis Buchanan (1762–1829)." *Modern Asian Studies* 20, no. 4 (1986): 625–660. For Buchanan's wider collecting, see Watson, Mark F. and Henry J. Noltie. "Career, Collections, Reports and Publications of Dr Francis Buchanan (Later Hamilton), 1762–1829: Natural History Studies in Nepal, Burma (Myanmar), Bangladesh and India. Part 1." *Annals of Science* 73, no. 4 (2016): 392–424.

[20] Buchanan, Francis. *A Journey from Madras through the Countries of Mysore, Canara and Malabar . . . Vol. 1*. Black, Parry and Kingsbury, 1807, pp. viii–xii.

encountered: food, clothing, religion, history, law, police, custom, commerce. For virtually all categories, Buchanan was asked to pay particular attention to the opportunities for "improvement" of methods, materials and so on.[21]

British occupation of Dutch Java began in 1811, after Napoleon invaded the Netherlands and the Stadtholder sought British protection of its colonies in Asia. Mackenzie was sent from Madras to lead initial reconnaissance and surveying. During the first two years of the British occupation, he would play a key role in reorganizing the administration of Dutch Java under the British lieutenant-governor Stamford Raffles. As head of a committee tasked with comprehending the state of the country, this included conducting a massive survey of the Dutch colonial archives, from which (similar to Orme's and Dalrymple's collecting during the British invasion of the Manila in the Seven Years' War) copies were made and originals taken, resulting in a huge transfer of information about administration, land tenure, revenue and trade, history, natural history, and arts. This archival survey was then complemented with a series of military and topographical surveys. Mackenzie also participated in (and made valuable personal collections during) Raffles's surprise invasion of the wealthy royal city of Yogyakarta in October 1812.[22]

Mackenzie's summary of his surveying and collecting work in Java makes clear that large sections of the Dutch archives, which included information on not only Java but also all of the Dutch possessions in the East, had been either copied or removed by the British.[23] But, as in Mysore, Mackenzie also conducted in parallel a "private" collecting expedition that targeted all kinds of Dutch and Javanese collections. Again, Mackenzie's work was made possible by the hiring of local informants, one of whom – "an ingenious native of Java" – would return with Mackenzie to India to continue translating the Javanese materials.[24]

[21] David Arnold has focused on Buchanan's instructions in particular as a way of exploring the "prehistory" of development: Arnold, David. "Agriculture and 'Improvement' in Early Colonial India: A Pre-History of Development." *Journal of Agrarian Change* 5, no. 4 (2005): 505–525.

[22] Some were "saved from the wreck of the Sultan's library at the storm of the Craten [Castle] of Djocjacarta, by permission of the prize agents," in Blagden. *Mackenzie Collection*, p. vii.

[23] The Java collections arrived separately from Mackenzie's other materials and were cataloged separately. See Blagden, Charles Otto et al. *Catalogue of Manuscripts in European Languages Belonging to the Library of the India Office ...: The Mackenzie Collections. Pt. I. The 1822 Collection & the Private Collection*. Oxford University Press, 1916, pp. xxiv–v.

[24] Blagden. *Catalogue of Manuscripts*, p. xxvii.

A certain amount of coercion in obtaining access to some of the privately held collections seems to have been required:

> The colonists were found willing to assist and produce their stores, and the natives were soon reconciled, even the class whose interests might be presumed to travers[e], if not oppose these enquiries. The regents and their dependents were, though at first shy, ultimately cordially assisting to the objects of investigation: and . . . to the last moment of stay at Batavia (18 July 1833) materials, MSS [manuscripts] and memoirs, in copy or original, with letters in reply to the questions circulated, were transmitted from the most distant parts.[25]

Meanwhile, Raffles was conducting his own surveys of Java, employing the naturalist and surgeon Thomas Horsfield. Horsfield was an American doctor trained at the University of Pennsylvania who had been working for the Dutch East India Company in Java since 1801.[26] He had been employed as a surveyor and naturalist since 1804, traveling and collecting more or less continuously. In the early decades of his time in Java, he was focused on *materia medica* and published – mostly in the Batavian Society's transactions – descriptions of over sixty plants with medical uses.[27] Together with the Batavian Society, he had also produced a plan for a great *Hortus Medicus* for the region, to be paid for by sale of the medicines. When he received funding from the government for an expedition along the northwest coast of the island, his attention was also drawn to mineralogy and geology, and to natural history more broadly, especially entomology. By the time the British took over, Horsfield had amassed a large botanical, mineralogical and zoological collection. Now under the enthusiastic patronage of Raffles, Horsfield continued his travels and collecting. A first shipment of Horsfield's specimens arrived at India House in 1813. Raffles, meanwhile, was leading his own military expeditions to the interior, including the attack on Yogyakarta, to assert British rule. Topographical surveys and collections of archaeological and ethnographic material were gathered along the military route.[28]

[25] Blagden. *Catalogue of Manuscripts*, p. xxviii.

[26] On Horsfield, see Bastin, John Sturgus. *The Natural History Researches of Dr. Thomas Horsfield (1773–1859): First American Naturalist of Indonesia*. Oxford University Press, 1990; Horsfield, Thomas and John Sturgus Bastin. *Zoological Researches in Java, and the Neighbouring Islands*. Oxford University Press, 1990; Bastin, John Sturgus "The Geological Researches of Dr. Thomas Horsfield in Indonesia 1801–1819." *Bulletin of the British Museum* 10 (1982): 75–115.

[27] McNair, James B. "Thomas Horsfield: American Naturalist and Explorer." *Torreya* 42, no. 1 (1942): 1–9, p. 4.

[28] On Raffles's recording of and interest in Javanese "ruins" that came out of these campaigns, see Tiffin, Sarah. *Southeast Asia in Ruins: Art and Empire in the Early 19th Century*. NUS Press, 2016.

Although the French Empire was formally defeated in 1815, the Company remained aggressively entangled in related territorial conflicts for a decade or more. Very soon began some of the so-called little wars of the British Empire in Asia during Europe's "long peace" of the nineteenth century. On the Indian subcontinent, British defeat of Mysore precipitated a vicious series of conflicts with the powerful Maratha Confederacy, which ended in Britain's favor in 1818 and was followed by a survey of the Company's new territories on the Deccan plateau. One of the largest single collections form this period was the massive set of papers, maps and specimens from William Sykes's survey of this vast central Indian territory. After a furlough back in Britain, in 1824 Sykes returned to India and was hired as the "statistical reporter to the government of Bombay," a new position that coincided with his survey work but which was cut from the budget in 1829.[29] Over 4,000 specimens in addition to several hundred drawings and other papers arrived at India House in 1831.[30] On the northeastern frontier, the wealthy and extensive Burmese Empire attempted to push back the Company's expansion into northeastern India. The first Anglo-Burmese war from 1824 to 1825 was hugely costly to the British and the Burmese, in terms of both money and lives. Again, following Britain's success, surveyors would follow and the library and museum would receive a wave of materials.

Material bought, plundered, collected and otherwise acquired in the context of Company surveys made its way back to India House in a variety of ways. Some, like Jonville's collection from Ceylon, arrived as "gifts," which the administration had actively encouraged its officers to produce, and for which the administration dangled the possibility of future preference or reward. This was the first material from the island to arrive in the Company collections, possibly in all of Britain. Their status as a "gift" suggests the London administration had no prior knowledge of, or claim to, the materials collected by Jonville, despite the collections having been made while North was surveying land for the Company. Jonville's salary was likely paid by North rather than by the Company directly, and what he collected seems to thus have been treated as his own (or possibly North's) property.

As with Orme and Dalrymple decades earlier, Mackenzie, Raffles, Horsfield and Buchanan benefited from the wartime upheaval that created favorable conditions for acquiring manuscripts and other materials. And, as in earlier years, Company servants in this period were still

[29] Talbot, Philip A. "Colonel William Henry Sykes: His Contribution to Statistical Accounting." *Accounting History* 15, no. 2 (May 1, 2010): 253–276.
[30] UK NHM, Documents of the India Museum, Z MSS Horsfield no. 252–268.

permitted to amass private collections. But, unlike in the days of Dalrymple and Orme, the *primary* purpose of these surveys was to amass information *for the Company*, and specifically for the collections being developed at India House. The directors now exercised much greater control over the acquisition of knowledge resources in these newly occupied territories, and encouraged it in formal and direct ways. Thus, Buchanan's collections and drawings (and also, in the tradition of ship captains' journals, all of his journals) were considered Company property from the moment the hired officer had acquired or produced them. By 1802, therefore, Buchanan had forwarded a seed collection to Roxburgh in Calcutta and the rest of his materials to the Company library and museum. He also produced a large collection of zoological and botanical drawings.[31] A portion of these were first claimed by Calcutta; after a tussle, the India House library-museum managed to acquire those in several shipments between 1817 and 1819.[32] But the vast majority of information collected by him sat unexamined at Madras, and by the early 1830s the Buchanan survey had developed into a scandal, as it emerged during the charter debates that the Company had spent over £30,000 on the survey to little effect. In response, in 1835, the Company took the unusual measure of sending an officer, Montgomery Martin, to India for the sole purpose of reviewing and reporting on the materials.

This would be the first of multiple cases in which the directors were accused by shareholders of egregiously neglecting expensive and valuable information resources. The cost of the collections of Colin Mackenzie also came under scrutiny. The status of a catalog and access to the collection was asked about at the Court of Proprietors on March 19, 1834, to which the directors responded that one Captain Harkness was currently employed at India House cataloging the collections.[33] Around the same time, a liberal periodical was probing the "actual situation of the Mackenzie collection, with reference to it having been rendered accessible to the public," and suggested that the recently established Public Records Commission should investigate the status of the collection.[34] Around the same time, in 1836, the Company's new surveyor-general of India, Thomas Best Jervis, used part of a period of leave back in London

31 Archer. *Natural History Drawings*, pp. 38–39.

32 Other materials arrived in different shipments; for example, some were unloaded from the *Sovereign* on June 9, 1813: "Buchanan's survey and account ... and 32 statistical tables, drawings, maps and plans, 8 books, a folio containing vocabulary." BL MSS EUR F303/1, no. 1, 1813.

33 *Alexander's East India and Colonial Magazine*, 1836, 1/5, p. 400.

34 *Alexander's East India and Colonial Magazine*, 1834, 11/12, p. 476; *Alexander's East India and Colonial Magazine*, 1836, 1/2, p. 221; *Alexander's East India and Colonial Magazine*, 1836, 1/5, p. 124.

to launch a public campaign against what he saw as outdated cartographic and geographic practices at the Company. Jervis was especially critical of the Company's lack of support back in India for printing surveys, maps and memoirs.[35] Since 1821, the Company had employed London lithographic engravers to print the Company's principal topographical series, the *Atlas of India*. But the Court continued to resist establishing lithographic presses in British India, arguing that the climate was not conducive to quality printing (to which Jervis replied that the best prints in the world are produced in Italy, which has "as bright a sky and as high a temperature" as the subcontinent).[36] Thus many maps and plans produced in India remained in manuscript form, and, according to Jervis, much time and money was being wasted on making copies of these by hand.

Like Jonville, Horsfield had been employed by Raffles as a subcontracted collector and explorer, which meant the Company did not claim direct ownership of the material he collected on his surveys. The first set of Horsfield's collections from Java thus also arrived as a "gift" (when he also sent specimens to Joseph Banks).[37] In 1813 a collection of zoological specimens, medical and botanical reports and mineralogical studies by Horsfield were gathered in Jakarta and sent on to the Company museum. Charles Wilkins described the first set of Horsfield's specimens as "a very curious collection of stuffed Birds and Quadrupeds with a great many beautiful and rare insects ... in the highest state of preservation." Wilkins, on behalf of the Library Committee, then requested funds to purchase glazed cases so that the collections "be mounted or set up in the way usually practiced for Cabinets of Natural History."[38] Meanwhile, news of the end of the Napoleonic wars, and the orders to return Java to Dutch control, caught Raffles and Horsfield by surprise. Raffles's first plan was to pack everything up immediately in 1815 and return, with Horsfield in tow, to London. This didn't happen. Horsfield and the bulk of the collections stayed, while Raffles, now removed from lieutenant-governorship amid

[35] See Jervis, Thomas Best. "Address Delivered at the Geographical Section of the British Association, Friday August 26th 1839. Descriptive of the State, Progress and Prospects of the Various Surveys, and Other Scientific Inquiries, Instituted by the Honorable East India Company Throughout Asia." *The Transactions of the Bombay Geographical Society* 4 (August 1840): 157–189. On Jervis's campaign for a "scientific survey," see Edney. *Mapping an Empire*, pp. 268–280.

[36] Jervis. "Address Delivered at the Geographical Section," p. 187. On the *Atlas of India*, see Edney, Matthew H. "The Atlas of India 1823–1947/The Natural History of a Topographic Map Series." *Cartographica: The International Journal for Geographic Information and Geovisualization* 28, no. 4 (December 1991): 59–91.

[37] Around this time, Horsfield also (through Raffles) began a correspondence with Joseph Banks, and prepared several collections of some 300 dried plants for Banks's herbarium (for which Banks's keeper and librarian Robert Brown prepared a catalog).

[38] BL IOR Mss Eur F303/35. Library Committee Minutes, November 24, 1813.

accusations of financial impropriety in Java, was called back to England. And while it seems Horsfield did consider either remaining in Java or moving on with his collections to Holland, in the end it was the Company that gained possession of them. Horsfield was hired to be an assistant in the Company museum, then "sold" his entire collection to the East India Company by exchanging the material for a significant salary advance.[39] Some hurried plans for shipments thus commenced. He would move to London with the last of this material in 1819 in order to join the museum and begin work on publications. Raffles was relieved, writing to Banks in a letter accompanying Horsfield back to London that although "the Dutch have offered to him every possible inducement as far as money and fame would go, to join their party and send his Collections to Holland," all such offers were refused and "his collections are securely manifested for the Port of London."[40] Horsfield would remain working at the museum for the rest of his life, publishing its first catalogs and going on to succeed Wilkins as curator in 1836. For his own part, Raffles would soon return to Java, conduct more surveys and expeditions, and, in 1819, in his most consequential "purchase" of all, acquiring for the Company, under dubious circumstances, the island of Singapore. In 1824, Raffles was returning to London with his family and yet another even bigger haul, over 120 cases of manuscripts and other items, when, just off Bencoolen (Kota Bengkulu, Indonesia), a fire broke out on the ship. All persons were safely evacuated before the fire reached the magazines and the ship exploded.

The Charter of 1813: Company Science in Defense of the Monopoly

When the next charter renewal season began in 1812, the Company's overall reputation was on a much more positive footing than it had been during the previous charter renewal of 1793. The generally more positive perception of the Company is all the more remarkable given how little had changed in the Company's structure between the 1790s and 1813. Although the Board of Control now gave the government a more direct role in Company affairs, British India was still being ruled by a corporation, which was led by a board of directors and beholden to shareholders.[41] Even with the Company's improved reputation,

[39] Bastin. *Natural History*, p. 61. Also see Desmond. *India Museum* and MacGregor. *Company Curiosities*.
[40] Quoted in Bastin. *Zoological Researches*, p. 68.
[41] Bowen's *Business of Empire* follows Chaudhuri and others in revising older arguments that the Company lost its monopoly largely because it was inefficient.

provincial port representatives and anti-monopoly petitioners organized a lively campaign against the dominance of the Company and of the London ports. In the years leading up to 1813, Britain's economic order was under multiple stresses: the loss of overseas markets in the wake of the wars, high unemployment, sharply rising food prices and associated riots in towns such as Bristol and Sheffield. All of this put old trade arrangements under new pressure, since they were widely seen as maintaining artificially high import prices. Continuing to allow the Company's control of the Asian import trade (under which, for example, auctions at India House set floor prices under which sales would not proceed) now carried heavier liabilities for the government. The expanding textile industries were particularly influential, and the government was eager to support them.

When Parliament collated a raft of complaints against the monopoly and presented them to the Company, the rhetorical force of the Company's reply relied heavily upon two points: first, that the Company's monopoly was a unique thing, and a *political* arrangement, with control over trade merely part of that arrangement; and second, that the Company's institutional expertise regarding Asia was unrivalled, therefore the Company's ability to govern in and trade with British India was *also* unrivalled.[42] It was the Committee of Correspondence, by now the committee from which many policy statements and decisions issued, that drew up the initial defense in reply to Parliament's questions.[43] The Committee projected the Company's authority as not only a state in terms of its current function but also a state in its epistemological position with respect to other British interests involved in the debate. The provincial port owners and merchants, as well as the would-be missionaries pushing to end the ban on evangelizing, were painted by the Committee as dangerously ignorant.[44] The critics, wrote the Committee, display "so many proofs of want of knowledge on Indian subjects" that their complaints against the monopoly and their proposals for India could not be taken seriously.[45]

[42] Report of the Committee of Correspondences, February 9, 1813, in East India Company. *Papers Respecting the Negociation with His Majesty's Ministers for a Renewal of the East-India Company's Exclusive Privileges . . . for the Use of the Court of Proprietors*. E. Cox and Son, 1813.

[43] On the Committee of Correspondence in the early nineteenth century, see Bowen. *Business of Empire*, p. 160.

[44] Webster also notes that the ignorance of the provincial merchants regarding India was mocked more widely as well. See Webster, Anthony. "The Political Economy of Trade Liberalization: The East India Company Charter Act of 1813." *The Economic History Review* 43, no. 3 (1990): 404–419, p. 407.

[45] East India Company. *Papers Respecting the Negociation . . .*, p. 236.

Figure 4.7 “The Storming of Monopoly Fort,” an 1813 satirical cartoon by Charles Williams, showing the Court of Directors defending “Monopoly Fort” with “long speeches,” “solipsism” and dissertations on the “utility of the EIC.” c. British Library Board (asset P1009).

In support of the monopoly, the Committee also presented its own views of India and the India trade, drawing on a particular worldview, a combination of orientalism, political economy, conjectural history and the history of the Company itself. Parts of the argument could have been pulled straight from the classrooms of Haileybury. At the Company's new college, the students were introduced, via Malthus in particular, to Adam Smith's *The Wealth of Nations* as a foundational work. But Smith's economic arguments against trade monopolies, and his understanding of the Asia trade, were a point of controversy.[46] True understanding of the India trade was represented by the Committee as a matter of both knowledge of the culture of India and of the landscape of global trade. Such a combination, it was asserted, came only out of the institutional expertise that the Company had amassed. Scrutiny of the Company and its greatly expanded possessions now entailed ever more detailed renderings of regional history and political structures across a variety of Asian polities. Here, the Committee relied upon the survey reports of Buchanan and Mackenzie for evidence of the social and economic status of the vast new territories now under Company rule. The work of the Company orientalists was required to translate, quite literally, many of the descriptions and features of India that were now taken up as evidence for and against the monopoly. For example, for the largest and most consequential parliamentary report on the monopoly question, the so-called Fifth Report, Wilkins was tasked with producing a glossary of terms. The glossary runs to fifty-eight pages and gives definitions and etymology for, says Wilkins, "all Oriental Terms" that appear in the Report, coming not only from Persian and Arabic but also from "Sanskrit, Hindustany, Bengaly, Telinga, Tamul, Canara, Malabar . . . Turkish and Malay."[47]

Many critics of the Company's monopoly were especially focused on the longstanding trade imbalance between Britain and India. British merchants were still struggling to build markets in British India. Thus, a central, and controversial, issue in the debates was to do with whether it was even possible to increase the Asian market for British exports. And at the core of *this* question were broad historical-philosophical speculations about the nature of "civilization" in British India, and the degree to which the region could be materially "improved." The Company had, since the charter of 1793, been required to take on a certain minimum quantity of

[46] On the curriculum at Haileybury, see Tribe, Keith. "Professors Malthus and Jones: Political Economy at the East India College 1806–1858." *The European Journal of the History of Economic Thought* 2, no. 2 (September 1, 1995): 327–354.

[47] Wilkins's glossary for the Fifth Report can be found in *The Fifth Report from the Select Committee of the House of Commons on the Affairs of the East India Company, Dated 28th July, 1812*. Volume III.

British exports every year, and had therefore, the Committee claimed, "in a long course of years, made numerous, persevering, costly experiments, in attempting to push the vent [i.e. sale] of British commodities."[48] From this experience, as well as from the long history of Asian trade in general, they argued that the India trade – and the failures of British exports to succeed – was as much a fact of nature as the Indian climate itself. Quoting Montesquieu, the Committee agreed that such fixed differences as soil and climate

> shall for ever fix the character of commerce .... Every nation which has traded with India has uniformly brought precious metals thither, and brought back precious goods in return. *Nature herself produces this effect* ... India always has been, and always will be what it is now; and those who trade to India will carry money thither and bring none back.[49]

That Indian culture, in the abstract, was ancient, heathen, unchanging and static was a belief shared by all sides of the monopoly debate.[50] The essence of the debate came down to whether this perceived stagnant civilization *could* be improved (i.e. made more like Britain), and if so, how. The Committee asserted repeatedly (and in some ways contradicting its own claims of having also effected "improvement" on the subcontinent) that Indian culture, and thus its markets, is largely unchanged and *unchangeable.*[51] But to the anti-monopolists, and to the aligned causes of utilitarianism and Christian evangelism, it did not follow from recent history that the Eastern trade would forever remain fossilized. This group argued that the true experiment – that of free trade – had yet to begin. Free trade would be "a substitute and a cure for all commercial evils; would open an unbounded field to British manufacturers, British capital, skill enterprise and knowledge, which would not only supply the wants of the vast population of the East but create wants where they do not exist."[52] The anchor of these arguments was (again) Adam Smith, who had claimed in *The Wealth of Nations* that "the East-Indies offer a market for the manufacturers of Europe, greater and more extensive than both Europe and America put together."[53]

[48] Report, February 9, 1813, p. 214. [49] Report, February 9, 1813, p. 233. My emphasis.

[50] Margaret Schabas, Fredrik Jonsson and others have recently explored what Jonsson calls the "political ambivalence" of natural history in the eighteenth- and nineteenth-century debates over political economy and empire. In this moment, the Smithian stadial views of civilization, and the historical view of an unchanging India, served both the liberals and the conservatives. See Schabas, Margaret. "John Stuart Mill: Evolutionary Economics and Liberalism." *Journal of Bioeconomics; New York* 17, no. 1 (2015): 97–111; Jonsson, Fredrik Albritton. "Rival Ecologies of Global Commerce: Adam Smith and the Natural Historians." *American Historical Review* 115, no. 5 (2010): 1342–1363.

[51] Report, February 9, 1813, p. 212. [52] Report, February 9, 1813, p. 232.

[53] Quoted in Report, February 9, 1813, p. 232.

In response, the Committee asked, "But who should be trusted in judging what would happen ... if the monopoly were ended?" Even the theory of "Dr Adam Smith" did not "anticipate any sudden burst of commerce," and in any case Smith had very little reliable information on India: "His information respecting India was very defective, and erroneous; his prejudices against the East-India Company extreme, and his prognostics concerning the Indian government wholly mistaken."[54] Perhaps the greatest strike against the application of Smith's theories to the Eastern trade, continued the Committee, was the history of that trade *since* the publication of *The Wealth of Nations* nearly forty years earlier. Although "all Europe and America" had searched for "that immense market for European manufactures" that Smith said should be found in the East, none had yet been found. Furthermore, they pointed out (without any reference to the slave trade) that Britain's trade with Africa was nearly as weak as with India and China, yet no similar monopoly existed.

But the real weight of their argument, the Committee asserted repeatedly, came from the Company's long accumulation of experience and knowledge of India, China and the Eastern trade. The Company derived its opinion "not from any single authority [i.e. Smith] but from the broad page of history and practice." "On the side of the merchants there is, in truth, nothing but a sanguine theory. On the side of the Company there is the experience of all the nations of Europe for three centuries; there is the testimony of ancient history; there are the climate, the nature, the usages, tastes, prejudices, religious and political institutions of the Eastern people."[55] In concluding, the Committee painted the decision of whether to extend the Company's monopoly as not a matter of policy but a matter of expertise and risk assessment. It asks the Crown, will it risk "such a mighty convulsion" as to put the India trade into the hands of the ignorant? Or will the free trade lobby's "rage of theory, speculation and innovation" excite instead what it should: a "salutary fear" and a move to "stop short of the precipice" and "rest at some place, so far safe, as not to expose the whole of the empire, Indian and European, to the terrible alternative here brought to view"?[56]

In the end, the Committee's epistemological tactics for defending the monopoly were largely unsuccessful. The new charter, which would take effect in 1814, maintained the Company's China trade monopoly and left the government of the Asian colonies in the Company's hands. But the India and Southeast Asia trade was opened to any private ships of 350 tons and larger, and those traders were also free to enter and leave most

[54] Report, February 9, 1813, p. 232. [55] Report, February 9, 1813, pp. 232–234.
[56] Report, February 9, 1813, p. 246.

British ports rather than being restricted to London. Within British India, though, trade and movement were generally still limited to the main Company settlements. Taken together, these provisions, although opening up trade competition, still left the Company with a significant degree of control over access to India's natural and knowledge resources; for example, naturalists such as Humboldt would still need permission from the Court of Directors to proceed.

The fate of Haileybury College under the new charter was also initially unclear, but Malthus and others came strongly to the defense of their employer and reiterated (somewhat unconvincingly given the scandals involving student behavior) Haileybury's goals of educating moral improvement for better governance.[57] The college was increasingly under pressure from the growing political importance of education in both domestic and colonial circles.[58] The college, mixing both elements of classical liberal education and specialized training in oriental languages and literature, sat uneasily within the growing controversy between, on the one hand, political conservatives, who broadly favored maintaining the Company's monopoly as well as India's traditional legal and cultural systems, along with civil servant training in Indian languages, and, on the other hand, political liberals who broadly favored abolishing the monopoly and introducing European systems of law, languages and even religion to India, as well as civil service training in European subjects. In the end, Haileybury College survived, and the Company retained control over it. However, the Board of Control gained new powers over the Company colleges in India, and the Company was now required to spend at least 1,000,000 rupees (roughly £10,000 at the time) per year on supporting education in India. In addition, the restrictions on Christian missionary activity were lifted and evangelicals were free to establish new schools and societies. This change would be immensely consequential for the future of education in British India.

The Company's unsuccessful defense of the monopoly also had an immediate material impact on the Company's knowledge management practices. By the end of the charter debate period, the library-museum was both growing steadily and continuously being used by different

[57] Malthus, Thomas R. *Statements Respecting the East-India College: With an Appeal to Facts, in Refutation of the Charges Lately Brought against It, in the Court of Proprietors*. John Murray, 1817. Malthus also discusses college issues, including misbehaving students and the precarious position of the college, in his letters to his friend the political economist David Ricardo (who Malthus also tried to recruit to Haileybury). See Ricardo, David. *Works and Correspondence*. Edited by Piero Sraffa, with the collaboration of M. H. Dobb, Cambridge University Press, 1951.

[58] See Zastoupil, Lynn and Moir, Martin, eds. *The Great Indian Education Debate: Documents Relating to the Orientalist-Anglicist controversy, 1781–1843*. Curzon, 1999.

Committees within India House. The day books of the library record steady movement of maps, plans, books and other kinds of records to and from different committees within India House. Briefly reviewing the day books from April to December 1815, for example, reveals the following: the Examiner's Office sending "maps" to the library and receiving, on June 27, the *Materia Medica of Hindustan* by W. Arnold (Madras, 1813); letters from the court of Persia from 1790 on August 30; and various manuscript reports, such as, on October 6, "a copy of Captain Canning's Report of his proceedings at Acheen." In the same period, the Secretary's Office returned to the library plans of districts (Prince of Wales Island and George Town), bills related to the regulation of shipping, printed copies of the minutes of the Court of Directors' meetings, and other proceedings, minutes and letters sent to the Company. In November and December 1815, the Chairman's Office frequently requested the use of certain books about China (e.g. "Staunton's Chinese Embassy, Barrow's China, Milburn's Oriental Commerce").

But the loss of the India monopoly also further spurred a drive toward internal information organization. In the immediate aftermath of 1813, the directors and the Library Committee seemed intent on reforming how the Company managed and made use of its own archives and collections. During the negotiations over the charter renewal, Committees and the Court reported being frustrated with the state of the Company's records.[59] Soon after, the Library Committee proposed a variety of measures that would apply throughout India House and would further centralize archiving and record-keeping. The Registrar's historical and record-keeping duties were to be folded under the wing of the library. The "particular duties ... and object" of the office of the current historiographer, John Bruce, were to be investigated.[60] The organization of the Secretary's Office also came under scrutiny. This seems to have been part of a broader conversation over how the Company's official history was to be generated going forward, and what role the library and the librarian would have in this. New archiving regulations for debates in the Court of Proprietors were proposed, and, most importantly, a regular publication of the debates was to be commenced.[61] The Library Committee also wanted to take control of the distribution of new publications subscribed for by the Company. All publications coming into

[59] "Considerable inconvenience having been felt during the negotiation for the present charter ...." BL IOR F303/35, November 30, 1814.

[60] BL IOR F303/35, November 30, 1814.

[61] "For want of an accurate and easy reference to the various opinions which were exchanged win the debate in the General Court of Proprietors on the renewal of the charter in 1793, and also upon other occasions." BL IOR F303/35, November 30, 1814.

India House were now to go through the library and be recorded in the day books. It is unclear when, or if, all the orders were formally put in place, but by 1817, in one way or another, Wilkins had taken over the role of historiographer and the library had absorbed many of the duties of the former Register Office. Along with these new roles came more clerks (including three former assistants to the historiographer) for the library staff and a doubling of the budget for the librarian's book purchases and other costs.[62] In these ways, out of the crisis of 1813 came an even stronger institutional commitment to, and pride in, the accumulation and management of knowledge resources at India House.

Missions and Subterfuge

After 1813, with the loss of the Company's monopoly on the India trade, the remaining monopoly on the China trade gained new importance. The directors had always had a particular focus on gathering material from China, although British access to Chinese territories was relatively limited and constrained. Since its opening, the library and museum had welcomed a small but steady stream of gifts of curiosities for the museum from China, usually opportunistic gifts, such as four paintings and three jade or stone carved landscapes, intended for Empress Josephine but found aboard a ship captured by the British.[63] Around 1800, there were very few Chinese-language books in Britain; by far the greatest collections of Chinese books and manuscripts were in France.[64] Jesuit missionaries had been able to supply France and the Vatican with materials from China, but the Company had no similar means of access. There was no chance of obtaining permission to conduct surveys or purchase books or maps for export. The directors had to rely on the initiative and subterfuge of their writers and factors. One set of documents records the interest and means by which the directors attempted to obtain material from China for India House. After having received instructions to obtain books and drawings for the Company's library, George Staunton, the resident at Canton, replied on January 29, 1804 that

> [a] Botanical Painter has been employed in capturing the plants, fruits and flowers of this Country, as they come successively in Season, and we shall continue him till all that is curious in vegetable nature shall be designed. Mr. Kerr His Majesty's Botanical Gardener directs his employment and sends us descriptions of those

[62] Typically £300–500 per year. BL IOR F303/35, October 3, 1817.
[63] BL Mss Eur F303/1 18 January 1810.
[64] Standaert, Nicolas. "Jean-François Foucquet's Contribution to the Establishment of Chinese Book Collections in European Libraries: Circulation of Chinese Books." *Monumenta Serica* 63, no. 2 (3 July 2015): 361–424.

already painted which go in the Earl Camden's Packet together with Drawings of the Malacca Fruits by the same Artist.[65]

These drawings, along with the first shipment of Chinese books, arrived in 1805 on the return of the *Earl Camden*'s first voyage. With the help of one of the resident's Chinese interpreters, Staunton had gathered what he said were some of "the most valuable [books] in Chinese literature." Staunton promised to continue hunting down books, noting his interpreter had provided him with a second list of valuable books to be searched out as well.[66] In reply, the directors noted one title – described as "Whu-Frou" – appeared to be missing on arrival. They also asked for books specifically on "History, Art and Manufactures" and for materials – "some Elementary Books and implements of writing" – that would be useful for teaching Chinese languages at Haileybury.[67] In reply, on February 26, 1806, the resident promised that more botanical drawings were to be sent along, and a title that Staunton thought matched the missing title had also been procured, as well as a small collection of writing instruments and "books employed by the Chinese youths." However, Staunton also goes out of his way to stress the great difficulty of obtaining books for export:[68]

It may be proper however to notice in this place that, exclusive of the difficulty there exists of obtaining such Chinese books, for their utility or curiosity may be deserving of a place in the Hon Company's library, much embarrassment is experienced in afterwards conveying them to our Ships, as the exportation of Chinese Books is positively forbidden by the Laws and regulations inforced at this Post.

Commercial intelligence was desperately wanted, but such intelligence was often sought in what could be learned about Chinese history, languages, technology and culture. In the series of exchanges just described, a subject of particular interest to the India House orientalists became a focus of inquiry: the relationship between "Babylonian" and Chinese writing. In 1797, the directors sponsored an expedition from their residency at Bussorah (Basra, Iraq) to the ruins of a city on the Euphrates River that French scholars had identified as Babylon (Hillah, south of Baghdad). The aim of the expedition was to collect and send on to India House inscriptions rumored to have been discovered in the ruins.[69] In 1800, nine large bricks containing inscriptions arrived at India House, the

[65] BL MSS EUR D562/16, "Extracts of Canton Consultations."
[66] BL MSS EUR D562/16, "Extracts of Canton Consultations."
[67] BL MSS EUR D562/16, "Extracts of Canton Consultations." (The Court's letter to Canton dated May 23, 1805.)
[68] BL MSS Eur D562/16, "Extracts of Letters Canton."
[69] Bengal public dispatch, October 18, 1797, quoted in Hager, Joseph. *A Dissertation on the Newly Discovered Babylonian Inscriptions*. Wilks and Taylor, 1801, p. xvii.

first "Babylonian" inscriptions in England, and became some of the first items displayed in the Company's library-museum. The directors also distributed some of the smaller stones, including one to the British Museum and one to Banks.[70] French antiquaries had speculated that "Babylonian" was a form of Chinese, and it was on this question that the directors requested input from French missionaries. A copy of the inscriptions had been sent out to a missionary informant a few years earlier, and a short reply was passed on, only noting that he promised to study the question. However, as Staunton noted in reply, the Chinese authorities had recently severely tightened the rules of communication and movement for Christian missionaries on the mainland, and he did not have hope of hearing any more from his source.

The closure of British access to the missionary network also made it difficult for Staunton to provide any new intelligence on the directors' final questions: "the date and origin of printed maps of China purchased by the Company [and] whether engraving and printing from metal plates is practiced in China." In answer, Staunton apologized that "The reasons stated in the last paragraph will account for the delay and difficulty we shall unavoidably experience in endeavoring to satisfy your Hon Court's enquiries on this subject." He thus could only speculate that "we are inclined to believe that the art of engraving and printing from metal plates is well known and occasionally practiced in China, in confirmation of which opinion we may add that several Chinese Books have been shown us, for the Printing of which it is affirmed metal types had been employed."

Eventually, a set of nearly 500 highly detailed botanical illustrations were deposited at India House, each with names in Chinese characters and English transliterations, some with Linnaean classifications.[71] A large collection of detailed paintings of over fifty Chinese ships drawn from the Canton harbor and its environs were also sent at around the same time. In addition, a set of large and very ornately carved silk lanterns were sent to the library and museum, where they hung in the main rooms for decades.[72] Other miscellany collected in the first decades included the "last will and testament" of an emperor;[73] "a commercial vocabulary in Chinese;[74] an ingot of silver;[75] the cap of a Mandarin; and the "shoe of a Chinese lady." As for books, a catalog of Chinese books in the Company library from sometime after 1816 lists eighty-six works, many in multiple

[70] "June 25 [EIC Gives a Babylonian Brick to Banks]." *The Gentleman's Magazine*, July 1, 1801.
[71] See BL Add.Or.1967–2007.
[72] BL Mss Eur D562/16 Extracts Canton Consultations, February 23, 1813.
[73] BL Mss Eur F303/3, June 1, 1821. [74] BL Mss Eur F303/5, 1830.
[75] BL Mss Eur F303/6, April 12, 1833.

volumes, and usually many copies of each title. The collection included many of the books that would be identified as important by the administrator and explorer (and future secretary of the Admiralty) John Barrow, who had accompanied Staunton on a diplomatic mission to Peking, in his *Travels to China* (1804).[76] In particular, the works of Confucius were collected in many different editions, including the "nine king [ching] or sacred classics" (or the "four books and five classics") forming a standard canon of Confucius, the I-Ching, a 134-volume encyclopedia from 1710 (abridged from a 6,000-volume work), and other works on subjects ranging from language, literature and poetry to astronomy, medicine, geography, law, mathematics and history.[77]

The Company establishments in Canton and Macao continued to secrete out books and seek expertise in Chinese languages and culture. The directors also supported a dictionary project, with a printing press set up in Macao, where the Company for a time had more leeway to gather and transmit information, and where early editions of Morrison's *Chinese Dictionary* were printed.[78] In the early 1830s, John Reeves sent a collection of natural history specimens from Macao as well.[79] In 1815 when another embassy to China was sent, led by William Pitt Amherst, the Company arranged for a naturalist (Clarke Abel, on the recommendation of Joseph Banks) to "get plants for the museum." The delegation was refused a meeting with the emperor, and though collections apparently were made, they were lost when a returning ship sank.[80] In roughly the same period, the Company also had allowed (though given little material support to) an Englishman, Thomas Manning, to travel to the interior and all the way to Lhasa.[81] In these ways over the first thirty years, the Company's library and museum slowly built up what would be, by then, Britain's largest collection of books, manuscripts, articles and specimens from China. But, as it would turn out, these

[76] On Barrow and China, see Choi, Ja Yun. "A 'Most Interesting Subject for the Investigation of the Philosopher': Conjectural History in John Barrow's Travels in China." *Journal for Eighteenth-Century Studies* 42, no. 3 (2019): 303–320.

[77] BL Mss Eur D562/16, Extracts of Canton Consultations.

[78] All in BL Mss Eur D562/16. See Chen, Songchuan. "An Information War Waged by Merchants and Missionaries at Canton: The Society for the Diffusion of Useful Knowledge in China, 1834–1839." *Modern Asian Studies; Cambridge* 46, no. 6 (November 2012): 1705–1735.

[79] Horsfield. *Catalogue of Mammalia*, 1851.

[80] Banks, Joseph. "Banks to Amherst on the Appointment of Able [sic] to Get Plants for the Museum." 8: 146–147, 1815. In *The Indian and Pacific Correspondence of Sir Joseph Banks*, vol. 8. Pickering and Chatto, 2014.

[81] See Banks, Joseph. "Banks to Manning on Latter's Application to EIC China Expedition." 7: 147, 1806. In *The Indian and Pacific Correspondence of Sir Joseph Banks*, vol. 7. Pickering & Chatto, 2013; Markham, Clements R. *Narratives of the Mission of George Bogle to Tibet, and of the Journey of Thomas Manning to Lhasa*. Trübner, 1876.

collections would be dwarfed by those that would be made once Britain went to war with China in the 1840s and 1860s.

The directors were also keen to acquire information and materials from the neighboring kingdoms and states beyond the Company's control. The Company's trade eastward of India across the Bay of Bengal and on to China was of critical importance. Of particular interest were China's cosmopolitan trading partners in Southeast Asia: Ava (Upper Burma [Myanmar]), Siam (Thailand) and Cochinchina (southern Vietnam). In these regions, the Company was in a very different situation than on the Indian subcontinent. Here, the Company was not yet a territorial power, and the information order was quite different. A mission to Siam and Hué during 1820–1821 was entirely unsuccessful in gaining trading concessions or relations with these states, but did result in new material being sent back to the museum: George Finlayson, a naturalist assigned to the mission, sent back mammals, birds, fish, reptiles and fossils collected.[82] Much more material was captured and brought back to Britain during the first Anglo-Burmese war of 1824–1826. Much of this ended up, through the prize agent process, in private hands. But many precious manuscripts, natural history collections and works of art were also separated out to be sent to the Company's library and museum.[83] And the directors hired at least one covert collector moving beyond Company territory: in 1835, for example, the directors hired as a "news-agent" Charles Masson, who had deserted the Company's army in the 1820s and had since traveled extensively in Afghanistan collecting information, artifacts and especially a vast collection of ancient coins, now in the British Museum. Masson was granted clemency and a small pension in exchange for the collections he had gathered in Afghanistan.[84]

[82] Finlayson, George and Sir Thomas Stamford Raffles. *The Mission to Siam, and Hué the Capital of Cochin China, in the Years 1821–2*. John Murray, 1826; see also Horsfield. *Catalogue of Mammalia*, 1851.

[83] Mercer, Malcolm. "Collecting Oriental and Asiatic Arms and Armour: The Activities of British and East India Company Officers, c.1800–1850." *Arms & Armour* 15, no. 1 (2018): 1–21.

[84] Wilson catalogued the Masson materials. See Wilson, Horace Hayman, and Charles Masson. *Ariana Antiqua: A Descriptive Account of the Antiquities and Coins of Afghanistan*. East India Company, 1841. Some 7,000 coins collected by Masson were only recently discovered in the British Library's India Office collections. Many others were sold by auction (many to the British Museum) in the 1860s. See Jansari, Sushma. "Roman Coins from the Masson and Mackenzie Collections in the British Museum." *South Asian Studies* 29, no. 2 (2013): 177–93. Also see Behrendt, Kurt. "Charles Masson and the Buddhist Sites of Afghanistan: Explorations, Excavations, Collections 1832–1835." *South Asian Studies* 36, no. 1 (2 January 2020): 107–9.

Chronology of Major Zoological Collections Acquired by the Museum at India House, London, Together with Indication of Related Wars or Other Events[85]

The South and Interior of the Indian Subcontinent

1799: Last Anglo-Mysore War: East India Company (EIC) gains large parts of southern India

1802: Eudelin de Jonville: A series of Insects from Ceylon, chiefly Lepidpotera, presented to the Indian Government on the transfer of Ceylon to the British Crown; A series of Drawings of Birds from Ceylon; Zoological specimens from Ceylon, chiefly Insects and Shells; with Drawings and descriptions

1803–5: Second Anglo-Maratha War: EIC gains small regions of central India and Gujarat

1804: Claude Russell, esq. Indian Serpents

1808: John Fleming, Esq. – Drawings of Birds

1808: The King of Tanjore's collection of drawings of mammals and birds

1808: Francis (Buchanan) Hamilton, M.D. Drawings of Mammalia, Birds and Tortoises from "Continental India." [Collections from the survey of Mysore]

1812: The King of Tanjore. Drawings of Mammalia and Birds from Southern India. Presented by John Torin, Esq.

1817–8: Last Anglo-Mahratta War: EIC gains large parts of central India

1817: Francis (Buchanan) Hamilton, M.D. Drawings of Mammalia and Birds. [Collections from the survey of Mysore]

1819: Francis (Buchanan) Hamilton, M.D. [Collections from the survey of Mysore]; Drawings of Mammalia, Birds, and Reptiles

1831: Colonel W. H. Sykes. The Collection of Natural History made during the Statistical Survey of the Dukhun [Deccan], consisting of specimens and descriptions of Mammalia, Birds, Fishes, Reptiles and Insects

1850: Colonel W.H. Sykes. A Collection of Reptiles, Insects, Mollusca, and miscellaneous Zoological specimens from the Dukhun [Deccan], preserved in spirit

[85] The chronology that follows lists the zoological acquisitions of the East India Company's museum, extracted from Thomas Horsfield's *Catalogue of the Mammalia in the Museum … of the East India Company* (W. H. Allen, 1851), and rearranged by region and date of arrival at the museum in London, with contemporary political or military context in italic. Note the list is not complete: very small donations, donations of unknown origin and mixed collections forwarded from the Bengal or Madras governments were excluded. Additionally, insect acquisitions are taken from: East India Company (English) Museum, London, London East India Company (English) Museum, Thomas Horsfield, and Frederic Moore, *A Catalogue of the Lepidopterous Insects in the Museum of the Hon. East-India Company*. W. H. Allen and Co., 1857. This chronology was originally published in Ratcliff. *The East India Company*, 2016.

Eastern Frontiers: Southeast Asia and East Asia

1811: EIC Invasion and Occupation of Dutch Java

1812:	Richard Parry, Esq. Drawings of Mammalia and Birds from Sumatra
1812:	Thomas Horsfield, M.D. Collections of Mammalia, Birds, Insects, Fishes, Reptiles] from Java "during the possession of that island by Britain" consisting of "a large series of Insects of all Orders; A large Collection of preserved Birds from Java, with Drawings"
1813:	Thomas Stamford Raffles, Lieutenant-Governor of Java-Specimens of Mammalia, preserved Birds and Insects from Java: Horsfield's Collection
1817:	Thomas S. Raffles. Mammalia and preserved Birds from Java: Horsfield's Collection
1819:	Thomas Horsfield, M.D. Collections of Mammalia, Birds, Insects, Fishes, Reptiles from Java consisting of "a large series of Insects of all Orders; A large Collection of preserved Birds from Java, with Drawings"
1820:	Sir Thomas Stamford Raffles, Lieutenant-Governor of Fort Marlborough-A collection of Mammalia, Birds and Reptiles from Sumatra
1821:	Sir Thomas Stamford Raffles, Drawings of Mammalia and Birds from Sumatra
1823:	George Finlayson, Esq. Surgeon and Naturalist to the Mission of John Crawfurd Esq., to Siam & Hué, the capital of Cochinchina. Collection of Mammalia preserved Birds, Fishes, Reptiles and Osteological Specimens made during the Mission.

1824–1826: First Anglo-Burmese War: EIC Gains Large Territories from the Burmese Empire Including Parts of Assam and Manipur (Now in Eastern India) and Arakan (Rakhine) and Tenasserim (Tanintharyi) (Now in Myanmar)

1833:	John Reeves, Esq. A Collection of preserved Birds from China, with specimens of Edible Birds'-nests. (Macao)
1840:	John William Helfer, M.D. A Collection of Mammalia and Birds from the Coast of Tenasserim. (Burma)

1840–1842: First Opium War: EIC Extends Trade into Southeast Asia and China

1842:	Bengal Government – in three shipments: The Entomological Collections made in Chusan, Canton &c. by Theodore Cantor, M.D., acting as Naturalist during the Chinese Expedition; Mollusca and other subjects of Natural History collected at Chusan, Canton &c.; Crustacea from Singapore and the China Sea
1849:	Lieut. James W. J. Taylor: Collection of Shells from Singapore and the Indian Archipelago

1854–5: Second Anglo-Burmese War: EIC gains large territories from the Burmese Empire Including Pegu

1854:	Theodore Cantor, M.D. – A large Collection of Birds, from Penang and the Indian Archipelago

Northern Frontiers

1814–1816: Anglo-Nepalese War (Gurkha War): EIC gains portions of Nepal and Sikkim

1827: Capt J. D. Herbert. Birds collected during his Geological Survey of the Himalayan Mountains

1832: Nathaniel Wallich, M.D. Skins of Mammals and Birds from Nepal

1837: John McClelland, Esq. Member of the Deputation to Assam for the purpose of investigating the Culture of the Tea Plant. Specimens of Mammalia, Birds and other subjects of Natural History, with drawings and descriptions

1840: Major R Boileau Pemberton. Specimens of Mammalia, Birds, and Insects collected during his Mission to Bootan [Bhutan] in 1837–8

1841: J.T. Pearson Esq.: A collection of Insects from Darjeeling

1841: C.W. Smith, Esq.: A collection of Insects from Chittagong

1842: Ganges Canal Project begins: EIC builds new infrastructure

1843: Hugh Falconer, M.D. [Ganges Canal Project], collection of birds from Northern India. [Large collection of fossils from the Siwalik Hills of Northern India]

1845–1846: First Anglo-Sikh War; EIC Gains Large New Northern Territory

1845: B. H. Hodgson, Esq. A large Collection of Birds from Nepal

1845: B. H. Hodgson, Esq. assistant to the British Resident in Nepal. A large Collection of Mammalia and Birds from Nepal and Tibet

1848: B. H. Hodgson, Esq. Mammalia and Fossils from Sikkim and Darjeeling; Several Birds from Sikkim and Darjeeling

1849: Colonel F. Buckley. A large Collection of Insects in all orders from the Himalayas

1850: Captain Richard Strachey. A large Collection of Mammalia and Birds from Ladakh and Kumaon [Large collection from Tibetan Boundary Commission.]

1854: Captain R.C. Tytler, Bengal Army. Several Specimens of Birds from Dacca

Western Frontiers

1839–1842: First Anglo-Afghan War

1842: The Bengal Government – Mammalia and Birds collected by William Griffith, Esq. during the Expedition to Afghanistan

1843: The Bombay Government – The Collection of Birds made during the mission of Capt. W C Harris to the Court of Shoa, Abyssinia

1846: Colonel W. H. Sykes. Specimens of Black and other Corals from the Persian Gulf

1851: The Bombay Government – Specimens of the Zoology of Mesopotamia, received from Commander Jones of the Indian Navy, consisting of Birds, Reptiles in spirit, and a few Mammalia and Fishes

Geography unlisted/Mixed collections

1824:	Lieut-General Thomas Hardwicke. A Collection of Birds, Mammalia and miscellaneous Zoological Specimens
1829:	Madras Government: A Collection of Mammalia, Birds and Insects made by the Company's Naturalist at Fort St. George
1832:	John George Children, Esq. Specimens of Insects
1833:	Madras Government: The Zoological Collections made by the late A. T. Christie, M.D., consisting of specimens in all classes of Zoology
1837:	Mrs. Impey – Indian Reptiles in spirit
1841:	Asiatic Society of Bengal. A collection of Mammalia, preserved Birds and Insects
1841 & 1843 :	John McClelland, Esq. Specimens of Mammalia, Birds, and Insects
1842:	J. T. Pearson, Esq. - Specimens of Mammalia and Birds
1843:	William Griffith, Esq. Specimens of Mammalia, Birds, Fishes and Reptiles
1844:	Asiatic Society of Bengal. A large Collection of Mammalia and Birds, with smaller Collections of Fishes, Reptiles and Insects, received by several separate dispatches
1845:	Matthew Lovell, Esq., Bengal Medical Service. Several Birds
1845:	J Bax, Esq. through Colonel Barnwell. A Collection of Birds
1846:	Asiatic Society of Bengal. Large Collections of Birds, received by several separate dispatches
1847:	Asiatic Society of Bengal. A Collection of Mammalia, Birds and Crustacea
1849:	Ezra T. Downes, Esq. Deputy Assay Master, Bombay Mint. Large Collections of Coelopterous and Hymenopterous Insects
1851:	The Bombay Government: Specimens of Birds, received from Commander Jones of the Indian Navy
1851:	W. E. Wood, Esq, Hon E.I.C. Medical Service. A specimen of the Adjutant or Gigantic Crane

Rangoon Relics for a Racquet-Ball Court

In the mid 1850s, Company garrisons were settling into occupation of the royal city of Rangoon (Yangon, Myanmar) after the Company prevailed in the second Burmese war of 1852–1853 and annexed Pegu and Lower Burma into what would be called "British Burma." Parts of the city were being razed to make space for barracks. One commander reported back to Madras that, during leveling of one of the temple sites in the Eastern Heights district, a large quantity of treasure had been unearthed.[86] The contents included gold pagodas, a bejeweled gold helmet and belt, gold-leaf manuscripts and "one gold bowl, with cover, containing a lot of charred human bones."[87] Brigadier Commander C. Russell forwarded a translation of the scroll but nothing else, requesting instead that his garrison be allowed to sell the grave-robbed items in order to pay for local improvements, partly waterworks but also, most especially, he hoped that "the proceeds [could] be appropriated for the erection of a theatre for the amusement of the European soldiers ... [and] for a Racket court for the officers."[88]

As British building and development projects expanded in the colonies, construction, and especially the frequent practice of leveling sites where old structures or burial mounds had been, now often unearthed such treasures. Coins and other antiquities were continuously being discovered in the colonies as more and more roads and canals were built, towns expanded, forests cleared and fields turned over. Buried treasure, for example, was increasingly subjected to a growing set of regulations. When a find came to the attention of the authorities, tax collectors, antiquaries and, increasingly, archaeologists often clashed with landowners over the rights of ownership of the treasure. Complicating things was the fact that all across the subcontinent, it was not uncommon for families to keep family heirlooms and treasures buried for safekeeping. An 1822 law detailing the rights to finders of a treasure trove required government to pay the finder half the value of the treasure. But this law did not distinguish between "hidden treasure" deposited by residents or their ancestors and a "treasure trove" deposited by unrelated people.[89] In 1851, the case of fifteen gold coins uncovered in a field in Kandesh (in northern Maharashtra) by one Patel seems to have forced a revision of that law to distinguish cases where owners or relatives of owners of buried treasure are present. Patel had

86 BL MSS EUR F303/81 ff. 24 (unnumbered) No. 56C.
87 BL MSS EUR F303/81 ff. 24 (unnumbered) No. 56C.
88 BL MSS EUR F303/81 ff. 24 (unnumbered) No. 56C.
89 "Copy of a Circular Issued by the Commissioner in the Deccan on the 27th July 1822," in MSS EUR F303/40 Bombay Political letter dated December 17, 1851 (no. 154), ff. 159–160.

produced evidence that there was "traditional information on his ancestors having buried some treasure in this field" and had obtained permission from an assistant magistrate to search for it. After the search, Patel reported that three pieces of gold had been found, but the magistrate suspected that more treasure had been found than reported, and conducted his own search, finding thirteen more gold coins. Although the laws of treasure also stipulated that finders' fees would be forfeited if unreported treasure were found, Patel was paid in this case.[90] In another case from Benares in 1851, a man identified as "Bulijore Sing" discovered "a lot of old gold coins ... while digging in or near his house." Sing reported this to the local police, but only after more than five months. The delay almost caused him to lose out on any payout, but the deputy commissioner in Benares judged

> that it would be politic for Govt to waive its right to the Treasure .... By doing so the people would acquire confidence and in the event of any coins or valuables turning up they would be inclined to come forward ... curious relics of antiquity would thus be brought to light instead of their being melted down by the finders.[91]

In this case, the finder was allowed to keep half the coins, and the rest went first to the hands of Major M. Kittoe, "Archaeological Enquirer," who produced a study and a chronological ordering of the coins, noting "many of the coins in this valuable collection appear to be new, at least they are not described by [James] Prinsep in his records in the JAS [*Journal of the Asiatic Society of Bengal*]." The coins were then sent to Calcutta, to the Asiatic Society of Bengal for inspection and exhibition, and only then were they sent on to the museum at India House.[92]

For the officer in Rangoon wanting a theater and ball court, the cultural and scientific significance of such treasure was entirely commonplace; the finds represented nothing more than a potential increase in the garrison budget. Having conscientiously had the scroll translated – perhaps presuming the information was, to the orientalists and administrators, the most valuable part of the find – he hoped to be able to convert the objects into cash. But the request to sell the items was not approved, and they made their way – like so many cultural and scientific resources – first to Calcutta and then, by 1856, to the India House museum.

[90] MSS EUR F303/40 Bombay Political letter dated December 17, 1851 (no. 154), ff. 153–156.

[91] Letter from the deputy collector at Benares to the officiating collector on August 5, 1851. MSS EUR F303/41 ff69.

[92] Letter from the deputy collector at Benares to the officiating collector on August 5, 1851. MSS EUR F303/41 ff69.

The 1830s were a rare decade without territorial wars, so there were no new surveys or wartime collecting expeditions. Still, the wars of the 1810s and 1820s continued to be a major force shaping the India House collection. The first Anglo-Burmese war of 1824–1826 ended with Company gains from the Burmese Empire all along the northeastern coast of the Bay of Bengal, including parts of Assam and Manipur (now in eastern India), Arakan (Rakhine) and Tenasserim (Tanintharyi, now in Myanmar). During the lull up to the first Anglo-Afghan war of 1839–1842, much of the material arriving at India House came from the northern edge of the Company's territories: birds from the Himalayas in 1827 (sent by J. D. Herbert, Geological Survey of the Himalayan Mountains); mammals and birds from Nepal in 1832 (sent by Wallich); mammals, birds, insects and drawings from Assam in 1837 (sent by John McClelland, collected during a deputation to Assam to investigate the culture of tea); and mammals and birds from Tenasserim in 1840 (sent by John William Helfer). Also in 1840 mammals, birds and insects arrived from Bhutan (again sent by John William Helfer, collected during a mission to Bhutan of 1837), as well as insects from Darjeeling (sent by J. T. Pearson) and Chittagong (sent by C. W. Smith) in 1841.[93] Birds, fish, reptiles and fossils from Siam and Cochinchina collected in 1823, and from farther east arrived rare and valuable edible bird nests and birds collected from Macao in.

Company wars came roaring back at the end of the 1830s, with renewed imperialist policies pushing aggressive dominance of trade with China and increasingly focusing on the security of the northern borders and the Russian Empire. By this time, the directors (or likely the India House curators via the directors) had secured funds to embed naturalists and collectors within the moving armies. By 1840 it would not be uncommon for specific instructions from the museum to be sent out to officers on campaigns. For example, a memorandum attached to the Tibetan Boundary Commission stressed "the importance which Government attach to the labours of the scientific department of this Mission" and listed "a few points which have an immediate reference to the interest of the Museum of Natural History in this House," asking for several dozen specific mammals to be collected.[94] The disastrous attempt to expand into Afghanistan (the first Anglo-Afghan wars of 1839–1840) carried with it the surgeon and naturalist William Griffith. Griffith had been assigned to the Army of the Indus "in a scientific capacity" and was primarily situated among the engineering corps in advance parties.[95] Griffith

[93] Horsfield. *Catalogue of Mammalia*, 1851. Also see UK NHM Z Mss. Ind.
[94] BL IOR: L/F/2/113: Finance & Home Committee, November 1847, cited in Desmond. *India Museum*, p. 61.
[95] Fleetwood, Lachlan. "Science and War at the Limit of Empire: William Griffith with the Army of the Indus." *Notes and Records: The Royal Society Journal of the History of Science*

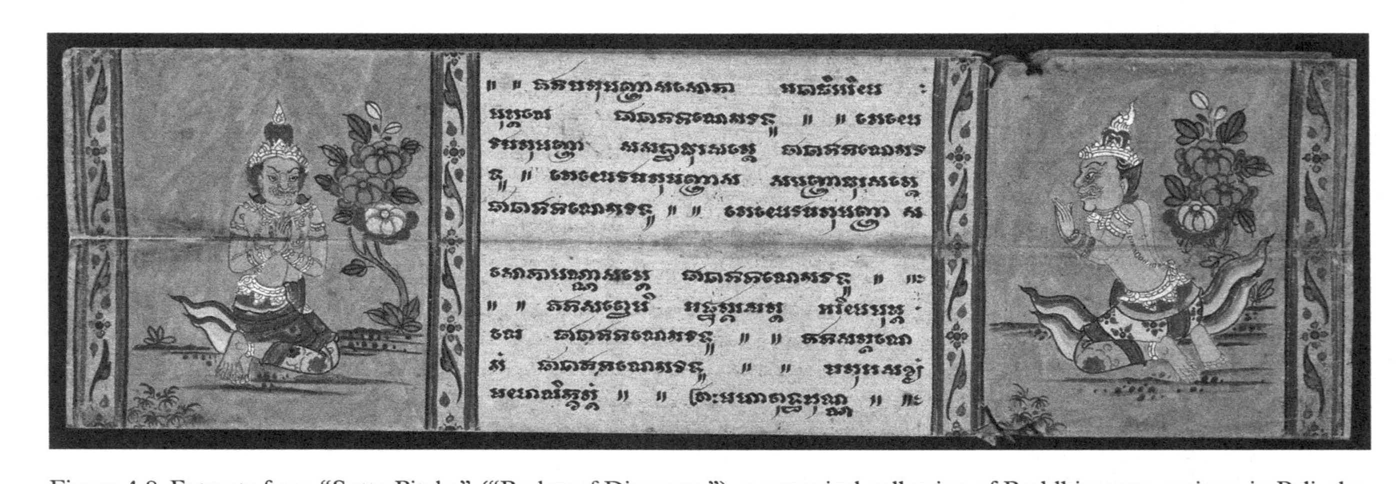

Figure 4.8 Extracts from "Sutta Pitaka" ("Basket of Discourse"), a canonical collection of Buddhist texts, written in Pali, the language of Theravada Buddhism practiced in much of Southeast Asia. Acquired in 1824 during the first Anglo-Burmese war. Now at the British Library (IO Pali 207, folio f.48). Courtesy of the British Library Board.

was thus granted rare access to areas of great interest to British administrators and naturalists alike. But it was also deemed too dangerous to go on collecting expeditions any distance from camp, making the actual work of collecting very difficult. Ultimately, Griffith would rely on a series of intermediaries, hiring out the work of collecting to locals and purchasing specimens from camp followers. Lachlan Fleetwood records at least twenty assistants hired by Griffith.[96] Many of Griffith's looted collections were lost, stolen or damaged (especially by camel transport), but his report and fourteen cases of specimens made it back to Calcutta and were sent on to India House in 1841. Griffith had hoped to return to London and to analyze his collections and publish the results. He had written to William Hooker at Kew Gardens that "my aim has been to amass materials for further study ... when a residence in Europe may enable me to avail myself of its splendid libraries and herbaria."[97] Yet Griffith never returned to Britain (he died in Malacca four years later). His Afghanistan collections, made at considerable cost to the Company, were, as Horsfield made clear, "bona fide the property of Government." Griffith had distributed other collections to individuals, but he had claimed these were "collections made away from the country in which I am employed by the Government .... Govt has been so extremely liberal to me that I should feel ashamed if people supposed I would dispose of any part of Govt collections on my own authority."[98]

Some of the zoological material went on display soon after arrival. It was also cataloged by Horsfield. However, the largest collection – Griffith's botanical collection – remained in the basement of India House, unopened for decades. The remaining papers in Calcutta, including maps and drawings, were eventually collected, edited and published by another surgeon-naturalist, John McClelland. Ironically, McClelland's justification for publishing the papers *before* sending them back to India House was exactly because of the acceleration of accumulation of Company science, an acceleration he believed the curators were not fit to handle: "the labours of the greatest Botanist that ever to set foot in India will be lost, perhaps for

75, no. 3 (April 2020): iii. Also see Lamond, J. M. "Afghanistan Collections of William Griffith." *Notes of the Royal Botanical Gardens Edinburgh* 30 (1970): 159–175.

[96] Fleetwood. "Science and War," p. 15. Rather than a lone collector exploring an empty territory, Fleetwood argues, wartime collecting was more like roving cosmopolitan "cultural borderlands," to use Fa-Ti Fan's phrase. See Fan, Fa-Ti. "Science in Cultural Borderlands: Methodological Reflections on the Study of Science, European Imperialism, and Cultural Encounter." *East Asian Science, Technology and Society* 1 (2007): 213–231.

[97] Griffith in a letter to William Hooker at Kew, August 6, 1840. Quoted in Fleetwood, p. 21.

[98] Griffith in a letter to William Hooker at Kew, August 6, 1840. Quoted in Fleetwood, note 96.

ever, swamped amidst the accumulated records of hundreds of men that are daily being added to their stores."[99]

The invasion of Afghanistan was an attempt to secure the far western borders of Company territory. At nearly the same time, the Company also pushed a trading dispute into military action at the far eastern edge of its range. During the first opium wars of 1840–1842, in which the Company sought to protect its trade, and especially the ability to supply opium to China, the Company also assigned a surgeon-naturalist to collect along with the campaign. Theodore Cantor, a nephew of Nathaniel Wallich, and a future collaborator with Horsfield on cataloging work at India House, went to India in 1835 and was hired in 1837 as surgeon attached to the Bombay Marine Survey (where he made a collection of fishes of the Ganges, sent on to India House). He was then sent to China with a Company regiment in 1840. Like Griffith, Cantor was instructed to collect for the government, meaning India House.[100] Also like Griffith, Cantor paid all expenses himself on the expectation that he would be reimbursed by the directors. Writing to Horsfield from Calcutta on April 20, 1841, he complains: "somehow or other I never received any assistance from Govt except a small quantity of spirits of wine [for specimen preservation], and even that rather late in the day. In a late letter from Govt, I am told they will pay the expenses I have actually incurred, leaving all other renumeration to the decision of the Court of Directors."[101]

In addition to having to put up his own money for supplies, Cantor would have taken a pay cut to go on "detached" duty as an assistant surgeon. He became ill at the end of the short collecting season ("nothing can be done at Chusan from the middle of October till the commencement of May") and sent his collections and drawings on to Horsfield, asking that he or "Mr. Hope" (Fredrick William Hope, clergyman, zoologist and first professor of zoology at the University of Oxford) would "do me that favor" and publish a summary and description of the collection in the *Transactions of the Entomological Society*.

[99] Griffith in a letter to William Hooker at Kew, August 6, 1840. Quoted in Fleetwood, p. 22. See Griffith, William and John McClelland. *Posthumous Papers Bequeathed to the Honorable the East India Company, and Printed by Order of the Government of Bengal . . .. Journals of Travels . . . William Griffith*. Bishop's College Press, 1847; Griffith, William and John McClelland. *Posthumous Papers Bequeathed to the Honourable the East India Company, and Printed by Order of the Government of Bengal: Palms of British East India*. Printed by C. A. Serrao, 1850.

[100] UK NHM Z MSS Horsfield. "Theodore Cantor to Horsfield," October 8, 1844. Cantor's insects together with his summary account arrived in 1841. See BL IOR/F/4/1949/84671; BL IOR/F/4/1991/88214.

[101] UK NHM Z MSS Horsfield.

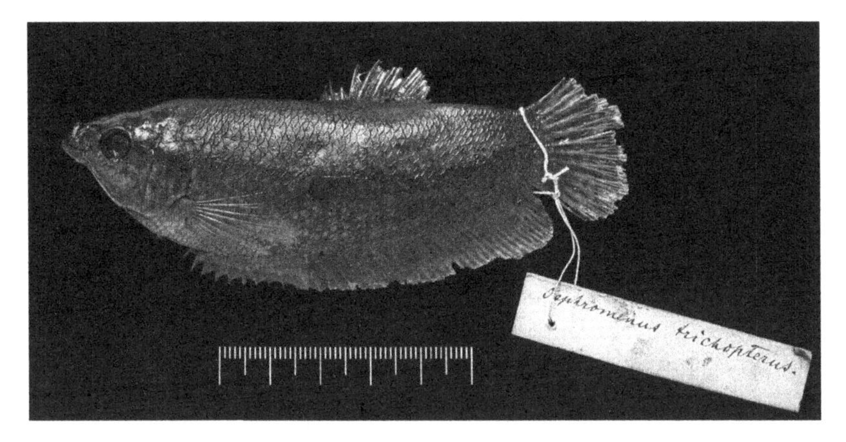

Figure 4.9 A holotype specimen of *Osphromenus trichopterus cantoris*, a freshwater fish, collected for the Company's museum by Theodore Cantor in Penang in 1840. Now in the Natural History Museum, London. By permission of the Trustees of the Natural History Museum, London.

"What I wish beyond all things is to have a general conspectus of the Entomology of Central China like that Hope has given of the Himalayah in Royle's work."[102] A complete series of the substantial collections and drawings was then forwarded to India House. The cataloging and publication preparation work was, as we will see was so often the case, taken up by a collective. William Griffith worked on the plants, Fredrick William Hope on the insects, Edward Blyth on the birds and William Benson on the mollusks. It was eventually published, at great expense and with color plates, by the Asiatic Society of Bengal.[103]

In 1842, Cantor was assigned as a civil surgeon to Prince of Wales Island (Penang), where he was also expected to continue his natural history collecting. Here he was handed the management of five hospitals in addition to a new sanitation department. Still, he sent onward to India House nineteen cases of specimens, containing 11,024 specimens collected during three years of service, including "specimens of edible birds'

[102] UK NHM Z MSS Horsfield.

[103] See, for example, Cantor, Theodore. *Catalogue of Reptiles Inhabiting the Malayan Peninsula and Islands*. Printed by J. Thomas, 1847. According to Turner, the high cost of the publication contributed to the demise of that periodical (*Asiatic Researches*). Turner, Ian M. "Natural History Publications Arising from Theodore Cantor's Visit to Chusan, China, in 1840." *Archives of Natural History* 43, no. 1 (April 1, 2016): 30–40. Also see Turner, Ian M. "Plant Species Described by William Griffith in 'Some Account of the Botanical Collection Brought from the Eastward by Dr. Cantor.'" *Edinburgh Journal of Botany* 72, no. 3 (November 2015): 413–421.

nests" of the "best kind," costing, he says, £300, and, in Case 12, a rare deposit of human remains: "Human Skulls ... murderers executed at Pinang December 21, 1843."[104] After a short stint as garrison assistant surgeon at Fort William, in 1848 he was sent back to war, this time to Ferozepur during the first Anglo-Sikh wars. He continued to send large collections back to India House until well into the 1850s.[105]

*

The territorial and trading expansion that followed the Napoleonic wars in Asia would result in another great influx of knowledge resources at India House. After an initial wave of wartime plunder such as that from the siege of Seringapatam, there would come more collections that now were the result of Company-led surveys and expeditions into newly acquired territory. The plunder-led phase of wartime collecting – where plunder would be gathered by the prize officers, sold at auction and only then were the proceeds distributed – was now followed by a much more organized and official form of collecting. The result was that even as collections were also growing all across Asia, such as those at the Calcutta Botanical Gardens, the cultural and scientific capital held at India House grew at an even faster pace during the Napoleonic wars and in the "little wars" of border aggression that followed.[106]

War-backed territorial expansion was critical to the growth of Company science in Britain. In addition to material accumulation in the context of war, areas beyond the Company's formal control were also targets of Company collecting, although, as we have seen, different means had to be employed in these regions. Consequently, as the next chapter will argue, the Company's missions and attempts at territorial expansion also becomes a driver of the changing material culture and practices of the historical and natural sciences in Britain in this period. Margot Finn has explored this connection with clarity, arguing that "colonial loot and military booty ... played an active role in inciting historical practice in nineteenth-century Britain."[107] Before

[104] UK NHM Z MSS Horsfield. Cantor to Horsfield, October 9, 1853. Cantor correspondence starts at No. 281.

[105] UK NHM Z MSS Horsfield. Cantor to Horsfield, October 9, 1853. Cantor correspondence starts at No. 281.

[106] Not all material collected in the above cases made its way back to London – far from it. Some of it can still be traced: Jonville's Ms. on the pearl fisheries remains at the Colombo Museum Library. Sivalingam, S. "Bibliography on Pearl Oysters." *Bulletin of the Fisheries Research Station* 13 (1962): 1–21.

[107] Finn, Margot C. "Material Turns in British History: I. Loot." *Transactions of the Royal Historical Society* 28 (December 2018): 5–32. Also see Gregorian, Raffi. "Unfit for Service: British Law and Looting in India in the Mid-Nineteenth Century." *South Asia: Journal of South Asian Studies* 13, no. 1 (June 1, 1990): 63–84; Lidchi, Henrietta

1800, semi-sanctioned practices of individual looting, collecting and personal enrichment were the norm. Such norms were beginning to change during and after the Napoleonic wars.[108] By 1855, the regulation of material designated "treasure" in formal terms had been replaced by a set of regulations and policies ostensibly meant to ensure that treasure became the property of British India. But even after such laws were in place, and after wartime looting had generally come to be seen as unethical in Europe, even more material flowed into Europe.[109] In no small part due to the growing pull of the library and museum at India House as a center of Company science, "national" treasure unearthed in the colonies, and often the most valuable and rare items, made their way back to India House to be eventually merged into *Britain's* state museums.

We have seen how the establishment of the Company's new library, museum and colleges was accompanied in the first decades by two related developments in the accumulation and management of knowledge resources at India House: internally, attempts at improving the organization and use of the materials, and, externally, projecting the Company as an institutionalized authority on knowledge of Asia. It is tempting to interpret these developments as yet another step in the Company's move toward becoming – or trying to appear to become – more state-like by adopting a state-like position of epistemological authority regarding Asia. This might be true were it not for the fact that, at the time, British offices of state were only just beginning to articulate and act upon those ideas themselves. It might be more accurate to interpret the Company, Parliament and Crown offices all undergoing in this period, at the level of information accumulation and management, a similar set of changes.[110]

However, within the next several decades, as political pressures put new stresses on the old form of the Company, part of these debates would turn attention to the India House repository and the Company's knowledge monopoly. In the aftermath of the loss of the India monopoly, within India House, an increasingly urgent concern for better

and Stuart Allan. *Dividing the Spoils: Perspectives on Military Collections and the British Empire*. Manchester University Press, 2022.

[108] See Lipkowitz. "Seized Natural-History Collections."

[109] Sandholtz, Wayne. *Prohibiting Plunder: How Norms Change*. Oxford University Press, 2007.

[110] Huw Bowen has argued that, in precisely this period, "the importance of the bureaucratic order and efficiency brought to East India House cannot be underestimated at a time when serious questions were being asked about the Company's fitness to govern India." See Bowen. *The Business of Empire*, p. 180. On corporations as models for state organization and function, see Ciepley, David. "Beyond Public and Private: Toward a Political Theory of the Corporation." *American Political Science Review* 107, no. 1 (2013): 139–158.

institutional self-knowledge now joined the ever-present worries over the limitations of the Company's understanding of the land and people under its domain. As Bowen has shown, there was also plenty of skepticism of the Company's system of conducting all business in writing and its ever-growing mass of documents at India House, and a persistent worry about the usefulness and quality of the information being accumulated. For example, some members of the Board of Control complained that most of the Company's correspondence tended to cover matters "that were extremely obvious and almost trifling." In 1823, Thomas Munro, former governor of Madras, called the archives and India House "a mass of useless trash."[111] Munro's larger point was to stress the limitations of knowledge gained through the written word when compared to understanding that comes with direct experience. With its specimens, samples, manuscripts and works of art, the library and museum contained a different kind of record of India, meaning, at the very least, that the directors did not have to rely "solely on the written word."[112] As we will see in the next few chapters, however, the debate over experience versus the archives only became more intense in subsequent decades. Furthermore, as the Company's relationship to the state further changed, and as new commercial and trading interests took hold, the discourse related to the utility of the library and museum – what it was for and whose interests it should serve – would also begin to change.

[111] Bowen. *Business of Empire*, pp. 180–181.

[112] As Sir John Malcolm worried was the cause of many of the Company's problems. See Bowen. *Business of Empire*, p. 179.

5 Systematic Possession

James Mill's Man in the Closet

Just after the 1813 charter renewal, the young writer James Mill took it upon himself to compose a new history of the Company's empire in Asia. At the time, Mill had been barely supporting himself and his large family through his prolific journalism work. Devoting precious hours to a *History of British India* was a gamble, especially for someone who had never worked for the Company or set foot anywhere in Asia. Writing from a liberal perspective that heavily critiqued the monopoly was even riskier. But the book turned out to be a great success for Mill; most importantly, it landed him, a year later, a coveted salaried position at India House as an assistant in the Committee of Correspondence. At that time, it was extremely unusual for someone without any experience in Asia to join the upper administration at India House. But the supposed expertise of the "British Indian" (i.e. a Briton who has spent time in Asia) is precisely what Mill criticizes in the opening pages of the work that made his India House career possible. Whereas – as in the case of Orme, Dalrymple, Wilkins, Marsden, Colebrooke and so many other orientalists and administrators – experience in India had long been seen as the essential basis for being considered an authority on India, Mill boldly asserted that it was now time for that quaint old idea to be retired. One line of the argument proposes that experience in India leads to bias and partiality, which leads to defective reasoning about India.[1] The other key to his argument was to do with the new collections in Britain:

> Whatever is worth seeing or hearing in India, can be expressed in writing. As soon as every thing of importance is expressed in writing, a man who is duly qualified may attain more knowledge of India, in one year, in his closet in

[1] For example, Mill argues that experience can exert an "immoderate influence, hang a bias on the mind, and render the conception of the whole erroneous." Mill, James. *The History of British India*. Baldwin, Cradock, and Joy, 1817, p. xv.

Figure 5.1 Leadenhall Street looking west toward India House, with the booksellers Parbury and Allen in the foreground.

England, than he could obtain during the course of the longest life, by the use of his eyes and his ears in India.[2]

Mill's preface infamously makes the case for the superior power of the home-country imperial archive, of not only the *possibility* of knowing at a distance but also the likely *superiority* of knowledge generated out of imperial centers. Mill's preface to the *History of British India* is a remarkable construction of a new metropolitan imperial authority based on the archival record. It is one prominent example of how the growth of collections and archives about Asia in Britain would begin to transform scientific practice. To some contemporaries, Mill's new imperial epistemology crystalized what had heretofore been a rather hazy vision of the utility of the kind of knowledge management institutions now growing at India House. And, although his epistemological critique of experience goes much farther than his contemporaries, it also came just at the start of a general trend in just that direction within government in Britain generally.[3] It is in this new perspective that some historians have seen the origins of a uniquely Victorian obsession with an imperial "total archive."[4]

This chapter and the next turn to the methods and practices of science at India House in the first three decades of the library-museum's growth. The first section focuses on the orientalists working at India House and Haileybury, and on how the material related to Asian languages, culture and history would be put to use for both specific administrative purposes and grand philosophical arguments. The second section turns to the naturalists at India House and the Company's colleges, and similarly explores the way Company science was engaged with both specific colonial projects and natural philosophical debates. For both orientalists and naturalists – that is, for both philosophical history and philosophical *natural* history – questions of classification and ordering were paramount. The unprecedented scope of information available would lead to an active search for new methods and practices. In nearly every discipline, the growing mass of information was seen as both a boon and a crisis. Orientalists, political economists and naturalists at work at India House and the colleges thus focused in similar ways on questions of systematics;

[2] Mill. *History*, p. xv.

[3] See, for example, Eastwood, David. "'Amplifying the Province of the Legislature': The Flow of Information and the English State in the Early Nineteenth Century." *Historical Research* 62, no. 149 (1989): 276–294.

[4] See Richards, Thomas. *The Imperial Archive: Knowledge and the Fantasy of Empire*. Verso, 1993. Also see, for example, Stoler, Ann Laura. *Along the Archival Grain: Epistemic Anxieties and Colonial Common Sense*. Princeton University Press, 2009.

that is, how to produce knowledge through the sorting, classification and comparison of information.

The increase in the quantity of knowledge resources in Britain was, in part, a consequence of the increasingly centralized organization of scientific labor across the empire. It was more and more common for naturalists and orientalists to argue that "theoretical" work was best pursued in Britain, while data collection should be the focus of the colonies. As Mill would put it: "The man best qualified for dealing with evidence is the man best qualified for writing the history of India. *It will not, I presume, admit of much dispute, that the habits which are subservient to the successful exploration of evidence are more likely to be acquired in Europe, than in India.*"[5] For example, for Mill, the work of making a "really useful history" (a scientific endeavor for him) now involved more than anything having the ability to process "evidence"; that is, the records and reports of "observers."[6] The scholar located at the center of imperial administrative accumulation should therefore act as a "judge" relative to colonial officers, who are like "witnesses":[7]

> He who, without having been a percipient witness in India, undertakes, in Europe, to digest the materials of Indian history, is placed, with regard to the numerous individuals who have been in India, and of whom one has seen and reported one thing, another has seen and reported another thing, in a situation very analogous to that of the Judge, in regard to the witnesses who give their evidence before him.

Mill's claim – that observers on the ground gathering particulars are less well adapted to "philosophize" – is not so idiosyncratic as one might think; at around the same time, similar debates were emerging among the British-based naturalists and their peers in the colonies, this time the question being who was qualified to identify new species (rather than mere varieties), with the metropolitan actors claiming only they had the necessary training and materials. Such distinctions were always contested by those based outside the metropolitan centers, and they were also sometimes drawn within metropole and province in the colonies as well. But the debate over the geography of scientific production was itself spurred by the relative growth of knowledge resources in the imperial home country, and that would, in turn, go on to support a Eurocentric distribution of scientific labor in the long term.

In the final section of this chapter, I turn to the place of Company science within the growing networks of civic science in Britain. Well beyond the confines of Leadenhall Street, Company science was in this period shaping the matter at hand available for knowledge production in Britain. Out of the materials extracted from India and gathered at India

[5] Mill. *History*, p. xiv (my emphasis). [6] Mill. *History*, p. xvii. [7] Mill. *History*, p. xvii.

House and elsewhere would be built an increasingly profitable web of what would now be called intellectual property resources, generally owned and traded by Europe-based actors. The folding of information about Asia into Britain's provincial systems of ordering and classification would contribute to the growth of European sciences at the time, while also generating social, intellectual and financial capital for the authors. This systematic possession of Asia in Europe was the stuff out of which not only careers and intellectual property but also whole disciplines and institutions could be made.

Philosophical Histories

Well before Mill's time, British orientalists from Jones and Wilkins onward were constructing relationship between Britain and Asia, between home and colony, that served and reflected political purposes in Britain. The publication of Wilkins's translation of the *Bhagavad Gita* in 1785 had been promoted by Hastings as a way of translating Indian culture for the understanding of Britons, a way of bridging physically disparate cultures supposedly united under the empire. Thirty years later, at Calcutta College, India House and Haileybury, the study of Sanskrit was now a cornerstone of both administrative training and philosophical study. At Haileybury, it remained the case that, most fundamentally, language skills were taught as part of the basic training for the Indian civil service. Sanskrit held out special promise, since the leading Sanskrit scholars of the day – Jones, Colebrooke, Wilkins and Hamilton – agreed that dozens of vernacular languages were derived from Sanskrit. In much the same way as Latin was studied as an entry into a range of Latin-based languages, so Wilkins and others now saw Sanskrit as key to the efficient mastery of multiple other languages of India. Sanskrit was also believed to be a powerful and flexible language, especially in relation to philosophical or scientific topics; without it, "the power of expressing abstract ideas, or terms in science, would be absolutely reduced to a state of barbarism."[8]

Like Latin and Greek, Sanskrit was thought by some to impart a particularly refined mentality upon the speaker. For the comparative philologist and Haileybury professor Alexander Hamilton, even more important was what the study of Sanskrit would reveal about the *mind* of the Sanskrit-speaker past and present. If, as Hamilton believed, languages were mirrors of the mind, then Sanskrit opened a path for the British to (finally!) comprehend their subjects; and it also raised, for Hamilton, the larger puzzle of how such a "perfect" language (suggesting

[8] Wilkins. *Grammar*, p. xi.

a highly advanced intellectual culture) could have developed historically under such "despotic" conditions of governance as were (widely assumed by the British to be) prevalent in ancient Asia. Studying Sanskrit, Wilkins argued, would attract, "uplift" and amuse Haileybury students, as well as fascinate the "lover of science, the antiquary, the historian, the moralist, the poet, and the man of taste."[9] The "extraordinary" language, was, in the famous estimation of William Jones, "of a wonderful structure; more perfect than the Greek, more copious than the Latin, and more excellently refined than either."[10]

Because of its antiquity, its widespread influence and its tantalizing similarities to Latin and Greek, Sanskrit philology was believed by some to hold the key to a new understanding of the history of civilization.[11] Hamilton and other Haileybury philologists argued that Sanskrit was the means to uncovering the genealogical connections between the world's civilizations. For example, in his review of Wilkins's *Grammar* in the *Edinburgh Review* in 1809, Hamilton uses the *Grammar* to make the case for a structural "analogy," as he called it, between Sanskrit and Latin, Persian, German and English, including in the review his own comparative wordlists to illustrate the argument.[12] Critically, this work required the construction of a history of the civilizations of the subcontinent, which in turn justified even more, and broader, antiquarian collecting. As Bernard Cohn has put it: "Each phase of European effort to unlock the secrets of the Indian past called for more and more collecting, more and more systems of classification, more and more building of repositories for the study of the past and the representation of the European history of India to Indians as well as themselves."[13]

Comparative philology was also important to the conjectural or philosophical histories of Adam Smith and later the Edinburgh professor James Ferguson (who influenced James Mill). Smith and others proposed models of different stages of society, moving stepwise from barbaric to civilized, and attempted to define metrics for different stages according to cultural markers such as the complexity of legal codes, or modes of agricultural production, or systems of governance.[14] These stage-based,

[9] Wilkins. *Grammar*, p. xi.

[10] Jones's *Third Anniversary Discourse*, quoted in Wilkins. *Grammar*, p. ix.

[11] Rendall, Jane. "Scottish Orientalism: From Robertson to James Mill." *The Historical Journal* 25, no. 1 (1982): 43–69.

[12] Hamilton, Alexander [attrib.]. "ART. VI. A Grammar of the Sanskrita Language." *The Edinburgh Review* 13, no. 26 (January 1809): 366–381, p. 381.

[13] Cohn, Bernard S. *Colonialism and Its Forms of Knowledge: The British in India*. Princeton University Press, 1996, p. 80.

[14] See Pitts, Jennifer. *A Turn to Empire: The Rise of Imperial Liberalism in Britain and France*. Princeton University Press, 2009, pp. 20 and 130–132; also see Rendall, Jane. "Scottish

or stadial, theories of history were, essentially, works of historical-political economy, and the political economists at Haileybury were tackling similar philosophical-historical questions. For Malthus, as with Ferguson and Hamilton, language was a window onto the mind, and understanding the natural philosophy of the mind was essential to any theory of political economy.[15] The first edition of his *Essay on the Principles of Population* began with two chapters on naturalist philosophy of mind. For Malthus, the critical question of how and whether societies can be "improved" came down to the question of how mind and body limited one another. He argued there was a natural limit on how much the mind can influence the body, and thus on how much improvement an individual and a society could make: "We can be quite sure," he wrote, "there is a limit to the improvement, though we do not exactly know where it is."[16] Referring to this edition, the natural philosopher and collector Alfred Russel Wallace called Malthus's *Essay* "the first work of philosophical biology" he had ever read.[17]

When Haileybury opened, although it may have been readily agreed that Sanskrit should be taught, the material for doing so was not available in England. There were no grammar books, no dictionaries, no workbooks. But among Wilkins's collections were manuscript copies of at least six of the manuscript grammars that the pandits of Benares had used, some dating back to at least the twelfth century. And so he set about compiling from these and other sources a *Grammar of the Sanskrita Language* for the use of the college (see Figure 5.2).[18] To print it, however (and to teach the students how to write Sanskrit), Devanagari typeface was needed. Wilkins designed, cut and cast the typeface himself Wilkins's *Grammar* was published in 1808 and immediately put to use in the Company colleges (giving Wilkins a nice side-stream of income).[19] Wilkins also worked with the printer J. & H. Cox to bring printing in various foreign typefaces to London. At several points in 1813, Cox would make use of the Company's collections of type, borrowing from the library "casts of Sanskrita Types" and "five Devanagari Copper plates and One Persian Plate" (March 17).[20] Wilkins also worked with Cox to produce textbooks for Haileybury and Addiscombe. Other professors

Orientalism: From Robertson to James Mill." *The Historical Journal* 25, no. 1 (1982): 43–69, pp. 52–55.

[15] Meiring, Henry-James. "Thomas Robert Malthus, Naturalist of the Mind." *Annals of Science* 77, no. 4 (October 2020): 495–523, p. 499.

[16] Meiring. "Thomas Robert Malthus," p. 503.

[17] Meiring. "Thomas Robert Malthus," p. 499. [18] Wilkins. *Grammar*, p. xii.

[19] BL MSS EUR F303/1.

[20] BL MSS EUR F303/1 (January 17, 1813; March 17, 1813).

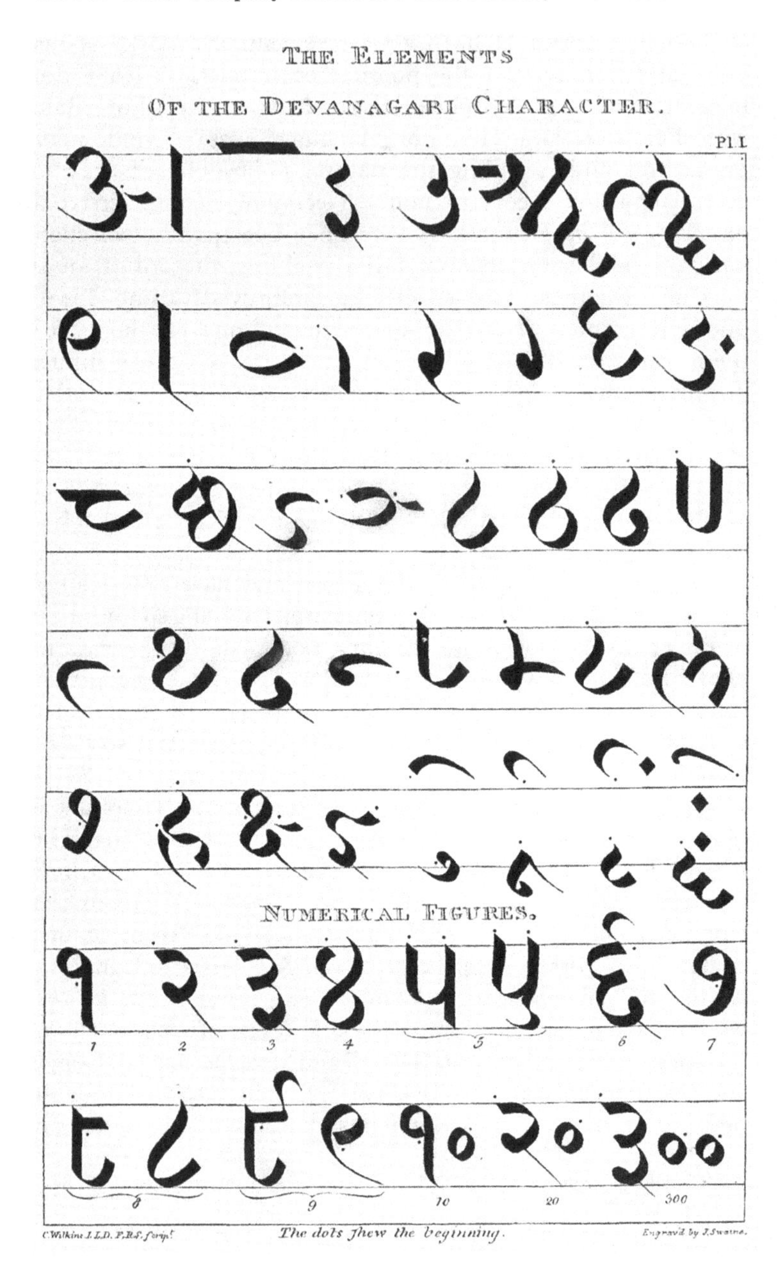

Figure 5.2 Wilkins's table of the elements of the Devanagari characters and numerals. From Wilkins, Charles. *A Grammar of the Sanskrîta Language*. London: Printed for the author by W. Bulmer, 1808. Courtesy of the Internet Archive.

were also busy publishing for the growing print market. In 1813, Cox printed for the Company 100 copies of the Addiscombe professor John Shakespear's *Hindustani Grammar*. Hamilton published the *Hitopadeśa* (1810), the first edition of a Sanskrit text to be published outside India, as well as a *Grammatical Analysis of the Sanskrita Hitopadeśa* (1810–11) and *Terms of Sanskrit Grammar* (1814).[21] A few years later Professor Charles Stewart published a descriptive catalog of the library of Tipu Sultan, as well as translations of some memoirs of Hyder Ali and Tipu Sultan.[22] Professor Dealtry published a mathematics textbook, *The Principles of Fluxions* (1810).[23] Professor Bridge published two volumes of *Mathematical Lectures* (1810 and 1811), as well as textbooks on mechanics, algebra and trigonometry.[24] Malthus was publishing articles on various political economic issues (the problems in Ireland, currency, principles of commerce, and currency and bullion) in the *Edinburgh Review*.[25] And as books flowed out from Haileybury professors, textbooks and other publications flowed in from the Company's library and its warehouses, where newly printed books for the library and colleges also arrived by the cartload from Bengal and Madras.

The world of Company politics was never far removed from the teaching and publishing of the Company's professors and curators. Malthus, through correspondence with former pupils such as Brian Houghton Hodgson, also sought to collect data from India and other regions with which to test his theory that civilizational development was dependent upon achieving agricultural surplus.[26] The *Histories*, *Travels* and *Journeys* of the surveyors, collectors and administrators were usually written or commissioned with broad policy aims in mind. The administrator, explorer and future secretary of the Admiralty John Barrow produced what Ja Yun Choi has argued is a distinctly "philosophical" account of China in his *Travels in China* (1804), which declares in the subtitle that the book attempts "to appreciate the rank that this extraordinary empire may be considered to hold in the scale of civilized nations."[27] Raffles's

[21] Rocher, Rosane. "Hamilton, Alexander (1762–1824), Orientalist." *Oxford Dictionary of National Biography*. Oxford University Press.

[22] Stewart. *A Descriptive Catalogue*.

[23] Curthoys, Mark. "Dealtry, William (1775–1847), Church of England Clergyman." *Oxford Dictionary of National Biography*. Oxford University Press.

[24] Cooper, Thompson. "Bridge, Bewick (1767–1833), Mathematician." *Oxford Dictionary of National Biography*. Oxford University Press.

[25] Pullen, John M. "Malthus, (Thomas) Robert (1766–1834), Political Economist." *Oxford Dictionary of National Biography*. Oxford University Press.

[26] Fennell, Shailaja. "Malthus, Statistics and the State of Indian Agriculture" *The Historical Journal* 63, no. 1 (February 2020): 159–185.

[27] Choi. "A 'Most Interesting Subject."

History of Java (1817) was undoubtedly intended to repair Raffles's somewhat battered reputation but it was also an impassioned argument in favor of maintaining British colonial interests in the region, a view that ran against those of many in both Parliament and India House. To make his argument, Raffles mobilized the information gathered over the past five years to paint a picture of a rich, fertile, productive, peaceful, pliant and strategically significant region.[28] Raffles's *History of Java* reads like a policy brief with its strong economic optimism. Even with Raffles's glowing optimism about the future value of having in Singapore a foothold in the region, the Company only very reluctantly (and after six years of legal wrangling with the Dutch) defended and ratified Raffles's acquisition of Singapore.

Buchanan's *Journey from Madras* was also written with a particular policy aim to hand. Buchanan's survey was the governor-general's chance to convince London that the war had been worth it, that these were possessions that would serve the Company's interests and aims.[29] But the argument for the value of this acquisition to the Company is cast in a very different light than that of Raffles's depiction of Java as a frontier of endless resources. David Arnold notes that Buchanan's instructions contain many references to the opportunities for "improvement"; and that Buchanan is asked to report in particular on the opportunities for improvement.[30] Thus, for example, with regard to the inhabitants, Buchanan's instructions were to note in particular the condition and treatment – the "protection, security, and comfort" of "the lower orders of the people." Under "manufactures and manufacturers" he is asked to pay attention to "how far the introduction of any of the manufactures of Mysore into any other of the Company's possessions might be productive of advantage, and respectively whether Mysore might derive advantage from the importation of the growth, produce or manufacture of Bengal."[31] Or, under "Farms," he is asked to comment on "how far the cultivation of the country may be improved."[32] He is even asked to inspect the cattle to consider possibilities for "the improvement of the breed."[33]

Buchanan responded with a multi-layered argument for the economic benefits – the better management of the "household" economy – that would be guaranteed by British rule. The overall depiction of the region provided by Buchanan is, however, and again in contrast to Raffles's Java, of a barren, unproductive landscape: traveling through Mysore in the dry

[28] Ratcliff, Jessica. "Hand-in-Hand with the Survey."
[29] Menon, Minakshi. "Transferrable Surveys."
[30] Arnold. "Agriculture and 'Improvement.'" [31] Buchanan. *Journey from Madras*, p. xi.
[32] Buchanan. *Journey from Madras*, p. ix. [33] Buchanan. *Journey from Madras*, p. lx.

season (but not distinguishing the two seasons), he found the landscape "sterile," lacking "verdure," having a "desert appearance."[34] The farmers he described as "indolent" and "slovenly."[35] In terms of "improvement," he presented the land itself as having great potential, and suggested it was "perfectly fitted for the English manner of cultivation" and should be "enclosed . . . . And planted with hedge-rows."[36] The great similarity between Raffles's *History* and Buchanan's *Journey* is that both lay primary blame for the present circumstances upon the prior political system and blame the countries' poverty, wants, waste and neglect on the policies of the vanquished states (and Tipu's reign is much more harshly depicted than that of the Dutch). Thus, through the surveys of the natural productions and farmlands (neglected), domestic culture (oppressed by monarchical laws), agricultural technology (backwards), commerce (repressively controlled), religions (silenced), monuments (destroyed) and so on, a case is made for a desperate need for a new domestic policy to rebuild the economy of the region.[37]

Thus, when published in 1817, Mill's *History of British India* was far from unique in being designed as a vehicle for particular political economic purposes. What does distinguish Mill's *History*, however, is both methodological and philosophical. Methodologically, the distinction lies, again, in his claim to authority by way of the imperial archive (in conjunction with a liberal education) rather than by way of experience (often also in conjunction with a liberal education). Philosophically, Mill's *History* stands out for its investment in a particular liberal utilitarian vision of the British Empire in Asia.[38] For Mill, history was a philosophical exercise, one that begins with theory and proceeds to marshal evidence in support of that theory. This results in (or, rather, begins with) a deeply negative depiction of the culture and society of "British India." Mill's remarkable claims about the backwardness of India, for example, draw heavily from his affinity with both the philosophy of mind and the historical traditions of the Scottish Enlightenment, in what Jennifer Pitts has called an "uneasy alliance of conjectural history and utilitarianism."[39] Even more than the epistemological arguments in the preface, Mill's *History* is now well known for the image of India that the book constructs, and for the widespread influence that image had for many years.

[34] Arnold. "Agriculture and 'Improvement.'"
[35] Arnold. "Agriculture and 'Improvement,'" p. 512.
[36] Arnold. "Agriculture and 'Improvement,'" p. 512.
[37] Vicziany. "Imperialism, Botany and Statistics," p. 633.
[38] See, for example, Majeed, Javed. *Ungoverned Imaginings: James Mill's The History of British India and Orientalism*. Clarendon Press, 1992; Schultz, Bart and Georgios Varouxakis. *Utilitarianism and Empire*. Lexington Books, 2005.
[39] Pitts. *A Turn to Empire*, pp. 123–125.

According to Thomas Trautmann, it is "the single most important source of British Indophobia and hostility to Orientalism."[40]

Throughout the book, Mill uses his own utilitarian rules for inserting "India" into a worldwide civilizational hierarchy. He makes many comparisons between not only "Indians" and "Chinese" but also "Indians" and "Africans," and "Indians" and "Mexicans." All of these societies and cultures, or more accurately all of the accounts, interpretation and reflections on these societies and cultures made by European travelers, were compared and ranked by Mill according to the strictures of liberal utilitarianism, wherein liberty of a particular form was calculated to generate the greatest happiness for the greatest number. The end result is to package "India" into a case study designed to prove the utility of liberalism *for Britain* and the necessity of imperial despotism *for India.*

The mission of Mill's "critical history" is to convince the reader that the societies of Asia are "uncivilized" or on the lowest rung of the ladder of civilization. But to do this he has to try to sweep away any argument for literary, scientific, moral, religious and artistic accomplishments – not an easy task when, for example, Sanskrit was then being studied with such reverence and enthusiasm. In chapter 10 of the notoriously bigoted "Of the Hindoos," Mill addresses the reasons scholars such as Jones, Wilkins and Colebrooke have so much praise for Asian cultures, arguing in essence that the scholars were simply overawed by the exotic.[41] In these and other ways, Mill disparages the orientalist histories, survey reports and administrator histories. But the rest of his research for the book also inevitably draws on the very scholarship that the preface criticizes. It is heavily dependent upon the work of Company officers or connected figures, including John Bruce (cited three times); Francis Buchanan (twenty-five times); Edmund Burke (twenty-one); John Barrow (eight); Clive; Colebrooke; Elphinstone; Teignmouth; Nathaniel Halhed (ten); William Jones (thirty-three); and James Rennell and Charles Wilkins.[42] For Mill, however, to use "flawed" sources such as he has, in order to build up a factual – a "really useful" – history is not contradictory. This just means that the sources must be evaluated and compared by a capable third party, someone qualified to act as a "judge" before these textual "witnesses."[43] Interestingly, Mill's methodology for his science of history is strikingly similar to methods becoming popular in astronomy at the

[40] Trautmann. *Aryans and British India*, p. 117.

[41] And at the same time he compares the Company orientalists' work to what he sees as the more clear and critical eye that Gibbon, from the safe distance of history, casts over early Jewish, Greek and Persian civilizations.

[42] From the index to Mill's *History of British India*, volume X (1858).

[43] Mill. *History*, p. xvii.

time for correcting for perceptual differences across observers – what would become known as the "personal equation" – averaged out through the method of least squares. Mill talks of "comparing the whole collection of statements with the *general probabilities* of the case . . . collected from the testimony of a great number of individuals."

The rhetorical use of this "scientific" history is very different from that of the earlier generation of orientalists. Recall that in Hastings's preface to Wilkins's *Bhagavad Gita*, he argued that "every accumulation of knowledge ... attracts and conciliates distant affections; it lessens the weight of the chain by which the natives are held in subjection; and it imprints on the hearts of our own countrymen the sense of obligation and benevolence."[44] This is precisely the "cosmopolitanism of sentiments" that Uday Singh Mehta ascribes to Edmund Burke as representative of the proto-"conservative" political ideology. In contrast, Mill, in rejecting experience and sentiments (and thus, says Mehta, any ability to truly engage and understand the strange and unfamiliar) in favor of the primacy of his philosophy of history, is engaged in a "cosmopolitanism of reason."[45]

The directors' decision to hire James Mill in 1819 was controversial. Not only was Mill's *History* critical of Company rule but Mill was also a vocal anti-imperialist generally, arguing that the colonies were only beneficial to "the few" – that is, a small group of investors – and were financially detrimental to Britain as a whole. But Mill was also convinced that utilitarian principles of governance were the only way to alleviate poverty in British India, and on this point he had key supporters.[46] Mill would become an immensely influential figure within the Company (and he pushed for expanding the roster of utilitarian political economists in the Company; he tried, for example, to get Ricardo to join the Court of Directors).[47] His long tenure influenced the future direction of British policy in Asia and its organization in Britain. But Mill's attempt to overturn the authority of experience, and of those with direct experience of India, was immediately debated and never uncontroversial, a reminder of the ever-present ideological hybridity within the ranks of the Company. Most famously, T. B. Macaulay attacked Mill's claims in "Mill on Government" (1829). Here Macaulay directly addresses the question of how political knowledge is acquired and what a "science" of government would look like. Arguing

[44] Mill. *History*, p. 13. [45] Mehta. *Liberalism and Empire*, pp. 22–23.

[46] His case was supported by friends David Ricardo and Joseph Hume (a schoolfriend who had joined the East India Company and returned wealthy). Canning was then president of the Board of Control and is said to have supported Mill against some Tory complaints.

[47] Ricardo, David. *Works and Correspondence; Edited by Piero Sraffa, with the Collaboration of M. H. Dobb*. Cambridge University Press, 1951. James Mill tried to convince Ricardo to join the Court of Directors (see v. 8, pp. 250–254, 262–263, 282, 292).

against Mill's abstract approach to the production of (politically useful) knowledge, Macaulay defends an inductive, empirical approach and argues for the utility of direct experience in acquiring politically useful knowledge.[48] It was also immediately critiqued as unjustly negative, and in later editions of the *History*, which the Company librarian was required to update, editorial additions would attempt to soften some of Mill's claims.[49]

Philosophical Natural Histories

> 1821, January 5 – Delivered to Dr Horsfield for delineation – Eight birds from Sumatra as specified on file[50]

While the orientalists and political economists worked to insert Asia into their historical orders of the world's civilizations, the naturalists were doing something similar with respect to orders of nature. The main focus of Company naturalists at India House in the 1820s would be to bring the Company's collection to bear on the contested question of the fundamental organization of nature, and whether or how it may be discovered. The period between 1820 and 1860 was the heyday of "philosophical natural history" in Britain.[51] Just around this time, natural history was taking on a much more geographical, or biogeographical, focus. The question of how distinct kinds emerge was increasingly answered with reasons of climate, altitude, topography and other geographical issues. Works like Alexander von Humboldt's biogeographical treatise on the Americas (and his proposed work in the Himalayas) not only enumerated species but also mapped their distribution and speculated as to the reason for their particular geographical range.

These discourses were in constant dialogue with the civilizational theories of the conjectural historians. Groups of plants and animals were described as populations, using statistics similar to those of census taking, sometimes organized into categories like "province," "county" or "nation." Both Darwin and Wallace were deeply influenced by Malthus. Interactions between groups and species were often described as wars or acts of colonization, of "obtaining the possession of the earth by

[48] Macaulay. "Mill on Government." *The Edinburgh Review* (March 1829). For a useful discussion of the debate, see Ball, Terence and Antis Loizides, "James Mill." *The Stanford Encyclopedia of Philosophy* (winter 2021 edition), Edward N. Zalta (ed.), https://plato.stanford.edu/archives/win2021/entries/james-mill.

[49] Mill, James and Horace Hayman Wilson. *The History of British India*. J. Madden, 1845.

[50] BL MSS EUR F303/3.

[51] See Rehbock, Philip F. *The Philosophical Naturalists: Themes in Early Nineteenth-Century British Biology*. University of Wisconsin Press, 1983.

conquest" and a "struggle against the encroachments of other plants and animals," as geologist and naturalist Charles Lyell put it.[52] Even more directly, the history, distribution and diversity of humankind itself was always part of the subject matter of natural history as well, and even the geographical origins of theories, philosophies and sciences were at the forefront of many of the naturalists' minds. For example, James Smith, longtime president of the Linnaean Society, in a survey of scientific methods concluded that "in those northern ungenial climates, where the intellect of man indeed has flourished in its highest perfection, but where the productions of nature are comparatively sparingly bestowed, her laws have been most investigated and best understood."[53] All of this, as Janet Browne has shown, was symbiotic with the Company's war-based imperial expansion.[54]

The cornerstone of classification in natural history had, since the late eighteenth century, been dominated by Carl Linnaeus's system of binomial nomenclature. Linnaeus was widely influential in Europe and his classification system was avidly taken up in colonial collecting. Linnaeus's "artificial" system was so called because it was not represented as capturing the essential distinctions between types in nature but instead was a useful human-made imposition of structure. It was based on distinguishing groups according to a few simple external characteristics, which themselves were chosen for ease of identification. The system of Linnaeus had, in the late eighteenth and early nineteenth centuries, been especially popular in Britain; the purchase of Linnaeus's own collection, and the founding of the Linnaean Society in London in 1788, extended that influence.

After the end of the Napoleonic wars, however, the Linnaean system increasingly came under attack in Britain. There was a general revival of philosophical debate over the aims and methods of species classification. Part of this revival of interest in systematics was related to identity and distinction among scientific laborers: at a time when the pursuit of natural history was being taken up by more and more classes of people, including army and navy subalterns and working-class men and women, a small group sought distinction as "scientific" or "philosophical" naturalists.[55]

[52] Browne. "Biogeography and Empire."
[53] Smith, James Edward. "A Review of the Modern State of Botany, with a Particular Reference to the Natural Systems of Linnaeus and Jussieu." In *Memoir and Correspondence of the Late Sir James Edward Smith*, edited by James Edward Smith and Lady Smith. Longman, Rees, Orme, Brown, Green and Longman, 1832, p. 445.
[54] Browne. "Biogeography and Empire."
[55] Secord, Anne. "Corresponding Interests: Artisans and Gentlemen in Nineteenth-Century Natural History." *The British Journal for the History of Science* 27, no. 4 (December 1994): 383–408. Also see Morrell, Jack and Arnold Thackray. *Gentlemen of*

Part of this involved drawing distinctions between naturalist in the colonies and those working at metropolitan or imperial centers. Along the same lines that James Mill had argued in the preface to his *History of British India*, heads of imperial collections, such as Joseph Hooker at Kew Gardens, argued that the naturalist's "philosophical" work, such as deciding when a new species had been discovered, should be left to the scientists in Britain and other centers of science.[56]

For those who were becoming unsatisfied with Linnaeus's *Systema Naturae*, the limitations of that system were now becoming clear because of the great new volume of material from all parts of the world being subjected to classification. The Linnaean system was, they claimed, both too "artificial" and too restrictive. Thus began an active search for a new theoretical foundation for describing the distribution and diversity of life on earth. In the early nineteenth century, the French theorist Antoine Laurent de Jussieu's proposed "natural" system revived debate about the validity and value of "artificial" versus "natural" systems of classification. The British botanist William Roscoe, for example, in 1815 defended the use of artificial systems on the grounds that the natural system, if indeed there was one, was currently unknown beyond "mere fragments," and quite likely unknowable:[57]

> [Nature's] vegetable productions are so numerous, their characteristics often so difficult to ascertain, they are related to each other by so many ties, that it is vain to expect that we shall ever be able clearly to define them, and accurately to seize upon the true distinctions; so as to combine the whole in the precise order in which they were primarily disposed by her hand.

Roscoe compared the rejection of Linnaeus's widely used artificial system in favor of Jussieu's new proposal to "those who, having a convenient and well roofed house, overturn it, in order to build one in the place of it of which they are unable to finish the roof."[58]

Science: The Early Years of the British Association for the Advancement of Science. Oxford University Press, 1981.

[56] Jim Endersby's study of Kew and Victorian science covers the "lumpers and splitters" debate in detail: Endersby. *Imperial Nature*. A summary of the arguments is found in Endersby, Jim. "Lumpers and Splitters: Darwin, Hooker, and the Search for Order." *Science* 326, no. 5959 (December 11, 2009): 1496–1499. Also see Mcouat, Gordon. "Cataloguing Power: Delineating 'Competent Naturalists' and the Meaning of Species in the British Museum." *The British Journal for the History of Science* 34, no. 1 (2001): 1–28.

[57] Roscoe in 1815, quoted in Novick, Aaron. "On the Origins of the Quinarian System of Classification." *Journal of the History of Biology* 49, no. 1 (2016): 95–133, p. 103.

[58] Novick, Aaron. "On the Origins of the Quinarian System of Classification." *Journal of the History of Biology* 49, no. 1 (2016): 95–133, p. 104.

At India House, philosophical natural history arrived in the wake of a large collection of specimens from Java. A year after Humboldt's last visit, and at just about the same time that Mill joined the Committee of Correspondence at India House, Thomas Horsfield arrived to take up the post of assistant curator and naturalist (see Chapter 4). Like Wilkins, much of Horsfield's time was spent managing the library, including supervising visitors and fetching books for visiting scholars and readers. As the naturalist at India House, Horsfield was also responsible for a vast amount of correspondence, of which only some materials survive. Some are from Company naturalists in the colonies asking for assistance, such as the letters of Edward Blyth, the influential zoologist and curator of the Asiatic Society of Bengal's museum, repeatedly complaining to Horsfield of a lack of communication from naturalists in Britain.[59] Others are to do with arranging visits to the India House collections or borrowing materials; for example, on November 23, 1820, William Buckland, the theologian and Oxford geology professor, sends a letter of thanks for a recent visit and a reference for an unnamed friend who would like to visit the entomology collections; on February 14, 1829, the geologist Charles Lyell, then in Paris, asks Horsfield to help an assistant of Baron Cuvier by allowing him access to any of the drawings of fishes at India House; on July 7, 1829, George Ord, the American naturalist and secretary of the American Philosophical Society, sends the first of a series of letters describing a trip to Paris, noting for his curator-friend such interesting facts as that the "[Parisian] men of science ... are not very communicative ... [but] the great collections are open to every one; and the libraries, which are extensive, may be enjoyed without either trouble or expense."[60] On August 1, 1835, a letter from David Don, botanist, professor at King's College and secretary of the Linnaean Society, asks Horsfield to go with him to see the collection at the Duke of Northumberland's Syon House. On October 3, 1837, John Joseph Bennett, assistant keeper at the Banksian Herbarium at the British Museum, asks Horsfield to exchange notes in preparation for a meeting with the head keeper, Robert Brown. On March 22, 1841, J. T. Peale of Government House Calcutta forwards a "circular" he had printed and distributed, which advertises the governor-general's interest in obtaining zoological specimens for the menagerie at Barrackpore, the Zoological Society and the East India Company's museum, clarifying that payment will be made on receipt of the live

[59] UKNHM Z MSS Horsfield. "Letters from Blyth, Calcutta." Correspondence with Francis Buchanan Hamilton, John Crawfurd and Theodore Cantor also survives in the UK Natural History Museum collection.
[60] UKNHM Z MSS Horsfield papers.

animals (while "only mere subsistence money" will be reimbursed "if they die on the road").[61]

In between all of this, from his office adjacent to the reading rooms, Horsfield was also at work on some of the most basic questions in natural philosophy: Is there an order or pattern to the organization of life, and if so, what explains it? Horsfield initially set to work displaying, describing and classifying the birds of the Company's museum. His primary source of reference was the Dutch aristocrat, bird collector and naturalist Coenraad Temminck.[62] In 1820, Temminck had founded the new National Museum of Natural History (the Rijksmuseum van Natuurlijke Histoire) in Leiden. (If Raffles hadn't convinced Horsfield to leave Java with his vast collection, much of it collected before the Company's brief takeover, before the Dutch returned, it is likely Horsfield's collections would have ended up in that institution instead of London.) Horsfield had first read his paper on the birds of Java at the Linnaean Society in 1820. A few years later, the Society itself splintered under the pressure of the nomenclature debates. Those who tended to be more radically opposed to the Linnaean system formed the Zoological Club in 1822. Horsfield was an early member and here he developed a friendship with William Sharp Macleay.

Macleay was a civil servant and entomologist, the son of Alexander Macleay, a collector of beetles and secretary of the Linnaean Society. The younger Macleay, working from his father's beetle collection, developed what came to be known as Quinarianism, which emerged as one of the most influential new "natural" systems of classification in Britain (and only Britain) in this period. In direct rebuttal to Roscoe, William Sharp Macleay believed that the search for a *natural* system should be the *primary aim* of natural history. He accused naturalists of "indolence," dismissing those who were content with filling out the Linnaean system as "mere *practical* botanists," like "the village herbalist," or worse: "The truth is that, like the religion of Mahomet, the Linnaean system has given rise in some parts of Europe to an unfortunate species of self-content, a barbarous state of semi-civilization, which is so far worse than absolute ignorance, that the existence of it seems to preclude every attempt at further improvement."[63] And so he went in search of the

[61] UKNHM Z MSS Horsfield papers.

[62] Horsfield, Thomas. "Systematic Arrangement and Description of Birds from the Island of Java." *Transactions of the Linnean Society of London* 13 (1822): 133–1; Gasso Miracle, M. Eulalia. "On Whose Authority? Temminck's Debates on Zoological Classification and Nomenclature: 1820–1850." *Journal of the History of Biology* 44, no. 3 (January 1, 2011): 445–481.

[63] Macleay, 1819/21, quoted in Novick, "Origins of Quinarianism," p. 104.

"natural" system. He believed he found evidence of it in the scarabs of his father's cabinet. One of Macleay's main critiques of the Linnaean system was that its linear, branching structure was unable to capture the gradual and continuous change from one group to the next that is observed in nature. Macleay's circular "Quinarian" system was, he claimed, able to capture this continuous spectrum of change. The basic assumption was that all natural groups form circular chains of members with what Macleay termed "affinities." Individuals had an affinity if they shared multiple points of subtle similarity (based on comparative anatomy), and the more points of similarity, the stronger the affinity. This is in contrast to "analogy," which also played an organizing role in the system, and which denoted a small number of major anatomical similarities. Furthermore (and controversially), all groups had five members. Thus, in a family of groups A–E, a chain of affinity would be described in which A had affinity with B, B with C, C with D, D with E, and E with A, closing the circle. Groups are nested within other groups in the same circular structure of affinities. The process of classification required detailed analysis of each specimen in order to rank the level of affinities and construct the circular chain. As a whole, as should be clear, the system was a radical departure from the system of Linnaeus.

All of this was first proposed in Macleay's *Horae Entomologicae* of 1819 and 1821. Volume 1 is a classification of *Scarabeus* according to the new system. The convoluted system of five-part nested circular chains worked beautifully in this limited case. Volume 2 is an audacious attempt at expanding the system to cover the whole animal kingdom. Importantly, as Aaron Novick has recently made clear, this second move was a conscious effort on Macleay's part to provoke challenges and tests to his proposed system.[64] For Macleay, this was how the discovery of the natural system must proceed. Remarkably, Macleay's difficult challenge was taken up by a number of contemporaries. Thomas Horsfield was one of them. Henry Colebrooke also adopted the Quinarian system in the 1820s. But, as with so many other areas, there was no single systematics dogma within the Company; other Company naturalists, such as Edward Blyth and Hugh Strickland, were harsh critics of the search for a "natural" system.[65]

[64] Novick. "Origins of Quinarianism."

[65] Colebrooke, Henry Thomas. "On Dichotomous and Quinary Arrangements in Natural History." *Zoological Journal* 4 (1828): 43–46. On Colebrooke, see Lowther, David. "Preliminary Analysis of the Hodgson Collection at the Zoological Society of London." *Archives of Natural History* 43, no. 1 (April 1, 2016): 90–94; Lowther, David A. "The Art of Classification: Brian Houghton Hodgson and the 'Zoology of Nipal'." *Archives of Natural History* 46, no. 1 (April 1, 2019): 1–23. On Blyth and Strickland, see Novick. "Origins of Quinarianism," p. 98.

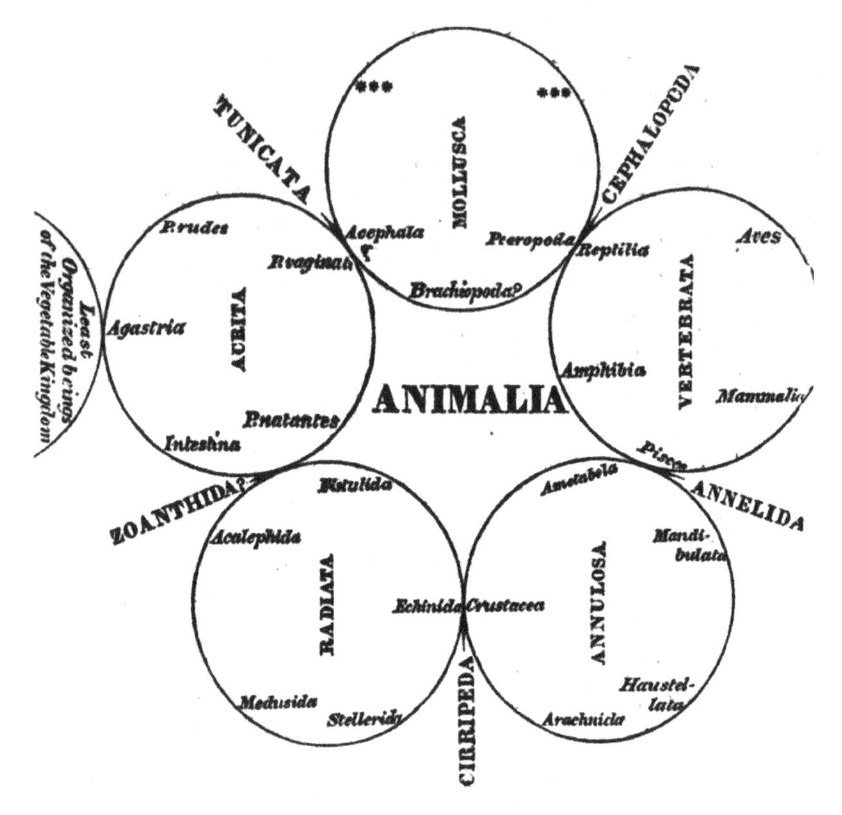

Figure 5.3 William Sharp Macleay's sample classification of the animal kingdom, showing "how the classes into which the animal kingdom may be resolved are thus found to return into themselves." Macleay, William Sharp. *Horae Entomologicae: Or, Essays on the Annulose Animals.* S. Bagster, 1819, pp. 317–318. From the Biodiversity Heritage Library.

In order to deal with the mass of specimens from Java, Horsfield had to divide up the work. He offered the opportunity to organize the insect collection to Macleay, and the plant classification to Robert Brown, future curator of the Botanical Department of the British Museum.[66] Both Macleay and Brown accepted, and Horsfield himself took up the project of classifying and describing the birds and mammals. So it was that Macleay next moved on to attempt to apply his system to the Company's insects from Java, what he called "the most valuable mass of

[66] See Horsfield. *Zoological Researches*, preface.

entomological information . . . a near-complete sample of forms" that had ever been collected "in the tropics."[67]

In justifying some of his bold rearranging, Macleay discusses at length the bloated state of Linnaeus's classifications, such as the genus *Carabus* to which was now assigned 1,600 species. "We every day hear of the difficulty of natural history having increased," says Macleay, "and doubtless it is increasing every hour; but this is owing to the number of new species which are pouring in upon us."[68] And yet, as he continues, the great advantage of this increase to science is that it had allowed the discovery of *natural affinities*, "which is now within reach of every person who does not allow himself to be frightened by the multitude of names which necessarily crowd the pages of the best modern works on natural history."[69]

Horsfield's *Zoological Researches in Java* (see Figures 5.4, 5.5 and 5.6) was also completed in 1825. Both works were published by the "Company's booksellers" Kingsbury, Parbury and Allen located just down the road from India House on Leadenhall Street, but the *Zoological Researches* was a very different type of publication than Macleay's *Annulosa*. It contained technically advanced illustrations (twenty-four of which were after drawings by William Daniell) and descriptions of seventy-six mammals and birds. As

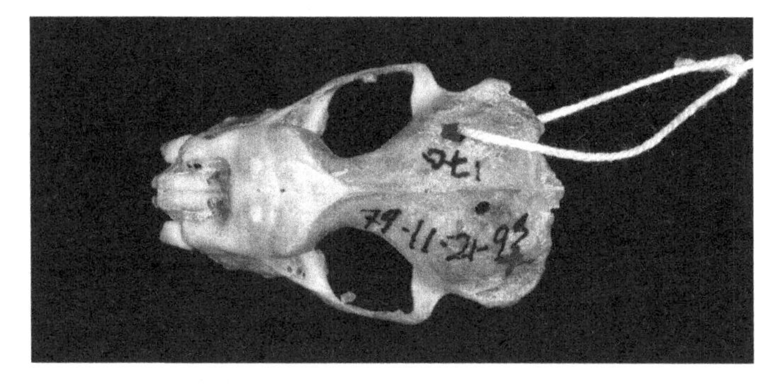

Figure 5.4 *Rhinolophus Lavartus* (horseshoe bat) type specimen collected by Thomas Horsfield in Java. Skull and label from the Natural History Museum, London (NHMUK ZD 1879.11.21.93). By permission of the Trustees of the Natural History Museum, London.

[67] Macleay, William Sharp and Thomas Horsfield. *Annulosa Javanica Or an Attempt to Illustrate the Natural Affinities and Analogies of the Insects Coll. in Java by Thomas Horsfield*. Kingsbury, 1825, p. vi.

[68] Macleay. *Annulosa*, p. 3.

[69] Macleay. *Annulosa*, p. 3. He concludes the introduction by noting: "So it is that, whether nature be regarded at the root or at the extreme branches of her tree, we always find her pursuing the same plan, and constantly displaying as much unity as beauty," p. 59.

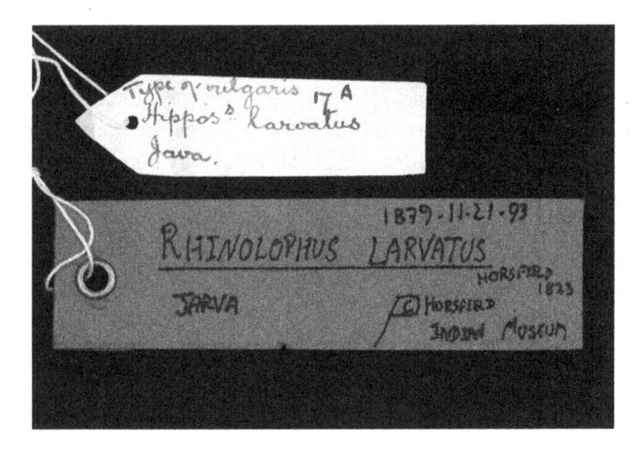

Figure 5.5 Label for *Rhinolophus Lavartus* (horseshoe bat) type specimen collected by Thomas Horsfield in Java. Skull and label from the Natural History Museum, London (NHMUK ZD 1879.11.21.93). By permission of the Trustees of the Natural History Museum, London.

the keeper of zoology at the British Museum John Edward Gray would later recommend for all zoological cataloging, the work was issued in a series, and each description was given its own unnumbered pages, so that the collection could potentially be arranged according to different systems.[70] The plates were lithographed, only the second English work of natural history to use this expensive new technique.[71] In the introduction, Horsfield introduced a few new genera and engaged critically at various points with the Linnaean models of his day and the endless debate over whether naturalists (particularly those in the field who collect) were becoming too quick to name variations as new species. Horsfield's style of objections to the Linnaean system can be seen in a lengthy discussion of the Javanese otter (*Lutra Leptonyx*):[72]

The Common Otter, the Javanese Otter, and the American Otter (including both the Canadian and the Brazilian Otter) are so nearly alike in external appearance, that the specific character drawn by Linnaeus for the Mustela Lutra, applies to

[70] McOuat. "Cataloguing Power."

[71] This is according to Christie's 2007 auction catalog, which cites Nissen IVB 453; Anker 212; *Fine Bird Books*, p. 82.

[72] Horsfield, Thomas. *Zoological Researches in Java, and the Neighbouring Islands*. Printed by Kingsbury, Parbury, & Allen, 1824. Reprinted: Horsfield, Thomas and John Sturgus Bastin. *Zoological Researches in Java, and the Neighbouring Islands*. Oxford University Press, 1990. The third part of the works of the Java collections, describing 100 specimens of plants, was not published until 1838: Horsfield, Thomas, John Joseph Bennett, and Robert Brown. *Plantae Javanicеæ Rariores: Descriptæe Iconibusque Illustrateæ, Quas in Insula Java, Annis 1802–1818, Legit et Investigavit*. G. H. Allen & Company, 1838.

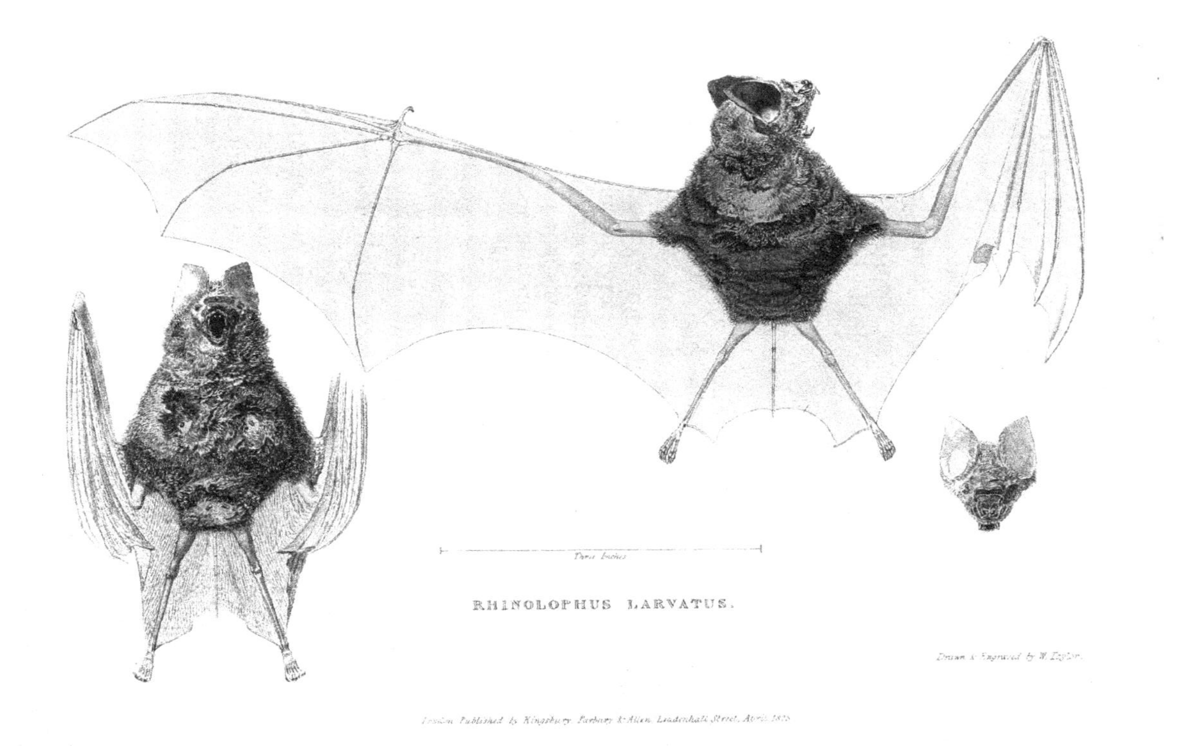

Figure 5.6 Illustration of a horseshoe bat (*Rhinolophus Lavartus*) from Horsfield, Thomas. *Zoological Researches in Java, and the Neighbouring Islands*. Printed by Kingsbury, Parbury, & Allen, 1824. From the Biodiversity Heritage Library.

them all. But as research is extended, and as new subjects are added to our Collections, a greater amplitude is required, both in the specific character and in the descriptions, in order to afford means to the naturalist to discriminate those species, which from an agreement in several external characters, are liable to be confounded.

In his 1828 *Descriptive Catalogue of the Lepidopterous Insects Contained in the Museum of the Honourable East-India Company* Horsfield takes on the task of further extending and testing the Quinarian system.[73] The introduction makes clear that Horsfield intends this work to be an extension of Macleay's *Annulosa Javanicae* and a further elaboration of the Quinarian system. In a long passage, he tries to explain the logic of the system, and the way in which each of the five "tribes" (as Macleay calls them) of a given group slowly edge into another, with the last of the group edging back into the first (see Figure 5.3):[74]

I have now traced the whole order of Lepidoptera in a rapid manner. I have attempted to show that it consists of five tribes, and that in the metamorphosis of each tribe, certain prominent or typical forms are manifested, indicating the subdivisions next in rank .... The gradual passage of one *tribe* into another, or the connexion of these higher groups by a natural affinity, has been only superficially stated; but it will be sufficiently apparent, I trust, that in the disposition of these tribes, I have attempted to follow the most gradual succession of nature; and I shall leave the proofs of this to the progress of the work itself. There is, however, one point regarding the connexion of the two principal tribes which presents itself for immediate notice. If the position above advanced be conformable to truth, we are now enabled to show with cogency, *that the whole order of Lepidoptera constitutes a series returning into itself* ... the circle is completed.

The centrality of classification and taxonomy to the scientific practices of European imperialism, and particularly the popularity of Linnaeus's system, has sometimes been interpreted by scholars as an attempt to impose an eminently practical but also hierarchical and gendered order upon the colonial world; the Linnaean system was, as one scholar puts it, a "grid that could aid in the rational ordering of the natural and social world."[75] On this reading, the draw of so many colonial naturalists to the Linnaean system was also a performance of Enlightenment reason (akin to the "cosmopolitanism of reason" Mehta

[73] Horsfield, Thomas. *Descriptive Catalogue of the Lepidopterous Insects Contained in the Museum of the Honourable East-India Company: Illustrated by Coloured Figures of New Species and of the Metamorphosis of Indian Lepidoptera, with Introductory Observations on a General Arrangement of This Order of Insects.* Vol. 1, parts I–II. Parbury, Allen, 1828.

[74] Horsfield. *Descriptive Catalogue of the Lepidopterous Insects*, p. 33.

[75] Baber, Zaheer. "The Plants of Empire: Botanic Gardens, Colonial Power and Botanical Knowledge." *Journal of Contemporary Asia* 46, no. 4 (October 2016): 659–679.

attributes to Mill) that also served to justify European superiority and imperial expansion. From this perspective, it is surprising, then, to find India House in the 1820s and 1830s to be a center for the pursuit of a radically different system, one that is rooted in "affinities" and circularity ("a series returning into itself").

But perhaps the most fundamentally "imperial" aspect of Linnaean classification is not the taxonomic structure but the application of the system itself, as an exercise of renaming, and thus taking a certain kind of intellectual ownership of the matter at hand. In this way, the Linnaeans and the Quinarians are equally rooted in, and helping to extend, the colonial political economy of science. Just as British scholars constructed both Anglicist and Orientalist versions of "India" for their stadial theories of history, so too did both Quinarians and Linnaeans voraciously order and name and take systematic possession of the natural history of Asia. Cumulatively, the result was that a robust business of knowledge production about Asia grew within, and for, Europe.[76]

Monopolies and Networks

By 1830, the material impact of the collections was also felt well beyond India House. Company science was also growing in tandem with the new professions, societies and intellectual networks of Britain's second scientific revolution. Between 1820 and 1840, the social organization of what was once called natural philosophy began to take on modern disciplinary distinctions. These changes happened in and around the growing collections of libraries and museums, as new subject-specific societies often formed around a perceived need to accumulate and manage information specific to their domain. In London alone, some of the new societies included: the Royal Horticultural Society (founded 1804, chartered 1861), the Geological Society (*f.* 1807, c. 1825), the Royal Astronomical Society (*f.* 1820, c. 1831), the Society for the Diffusion of Useful Knowledge (*f.* 1826), the Zoological Society (*f.* 1829), the Geographical Society (*f.* 1830, c. 1859), the British Association for the Advancement of Science (*f.* 1831), the Entomological Society (*f.* 1833), the Ethnological Society (*f.* 1843) and the Hakluyt Society (*f.* 1846). Privately organized and funded through individual membership subscription, Britain's thriving scientific societies figure prominently in the standard picture of a uniquely British form of private, civic scientific enterprise. Here, as has often been argued, is where Britain's unique culture of grand

[76] See Cohn. *Colonialism and Its Forms of Knowledge*, p. 77.

amateurs thrived, and space and resources were created for the dedicated practice of science in a time before university science faculties or other professional positions existed. Meanwhile, as Charles Babbage and other "declinists" worried about the status of British science often asserted, state support for science, even in the case of the Royal Society or the British Association, was minimal.[77]

However, in ways large and small scientific clubs and societies benefited greatly from the free labor of state and Company resources: Company servants, members of the military, politicians and government bureaucrats were all well represented in the new scientific society movement. Taking the Zoological Society, for example, the first president (and on many accounts the "founder") was the former lieutenant-governor of Java, Sir Stamford Raffles. As a group, the founding members listed in the Society's charter are remarkably diverse but the aristocracy and state servants are well represented. The list includes a landed aristocrat (Nicholas Vigors), a liberal Member of Parliament and "fashionable aesthete" (Charles Baring Wall – son of the private merchant Charles Wall), a tax bureaucrat (Joseph Sabine), a secretary of state (Henry Lansdowne) and a Whig politician who was both First Lord of the Admiralty and governor-general of India for the Company (George Eden). The role of wealthy amateurs and the political elite within scientific and literary societies would become the subject of much debate and much criticism.[78] Nevertheless, the existence of state connections through these elite participants was certainly there, bringing scientific societies at least partly within the orbit of political culture. In the case of the Zoological Society, these connections to the Crown would be especially fruitful, since it was through such channels that Queen Victoria was persuaded to grant the Society a portion of Regent's Park for their growing menagerie.[79]

As with members of the Royal Army and Navy, Company scientists and scholars were active in a wide array of clubs and societies. As will be seen in the next chapter, the society most closely related to the Company was the Royal Asiatic Society (*f.* 1824) (later the Royal Asiatic Society of Great Britain and Ireland), modeled on the Asiatic Society of Bengal, in which Henry Thomas Colebrooke, George Staunton, Charles Wilkins

[77] Hall, Marie Boas. *All Scientists Now: The Royal Society in the Nineteenth Century*. Cambridge University Press, 1984.

[78] See Ratcliff. "The East India Company." Also see, for example, the debate over the "political appointments" in the Royal Society raised by Charles Babbage: Babbage, Charles. *Reflections on the Decline of Science in England, and on Some of Its Causes*. B. Fellowes, 1830.

[79] Desmond, Adrian. "The Making of Institutional Zoology in London 1822–1836: Part I." *History of Science* 23, no. 2 (June 1, 1985): 153–185. Also see Grigson. *Menagerie*.

and Horace Hayman Wilson (Wilkins's successor) were all very involved.[80] To give just a few other examples, James Mill, T. R. Malthus and a group of friends founded the influential Political Economy Club in 1821. Mill was also central to the founding of the Society for the Diffusion of Useful Knowledge, and Mill and William Sykes were founding members of the Statistical Society in 1824.[81] Thomas Horsfield was also a founding member of the Zoological Society (along with Stamford Raffles) and the Entomological Society (where he was founding vice president).[82] John Forbes Royle (future India House curator; see Chapter 6) was a member of the Geological Society, the Linnaean Society, the British Association for the Advancement of Science and the Royal Society (where he was sometimes vice president). In addition to being a founding member of the Royal Asiatic Society, Royle also helped found the Philosophical Club of the Royal Society, the Royal Horticultural Society and the Royal Asiatic Society. But perhaps no Company servant was more involved in scientific society culture than Thomas Henry Colebrooke, who was, as Rosane and Ludo Rocher have shown, a "primary conduit" for connecting Company people and collections to the sciences in Britain.[83] Since his return to Britain, as the Rochers argue, Colebrooke's main concern "had been to integrate data on India into the purview of a vast range of scientific societies."[84] In addition to being the primary driver behind the founding of the Royal Asiatic Society, Colebrooke was also an inaugural member of the Royal Astronomical Society, where he contributed work on Hindu astronomy, and for whom he convinced the directors to house the records and reports of the Madras Observatory. He was also a founding member of the Zoological Society, helped get the literary magazine *The Athenaeum* off the ground and was an active member of the Geological Society, where he channeled specimens and reports from the Company's museum to the society meetings, as well as a member of the Linnaean Society, the Medico-Botanical Society of London and the Royal Institution.[85]

[80] On Staunton and the Royal Asiatic Society see Ong, Seng P. "Jurisdictional Politics in Canton and the First English Translation of the Qing Penal Code (1810)." *Journal of the Royal Asiatic Society* 20, no. 2 (2010): 141–165.

[81] On political economy versus statistics in this period, see Hilts, Victor L. "Aliis Exterendum, or, the Origins of the Statistical Society of London." *Isis* 69, no. 1 (1978): 21–43.

[82] See Hope's description of the early Entomological Society, writing to Darwin while he was on the *Beagle*: Hope to Darwin, January 15, 1834, Darwin Correspondence Project, "Letter no. 235," www.darwinproject.ac.uk/letter/DCP-LETT-235.xml.

[83] Rocher. *The Making of Western Indology*, p. 163.

[84] Rocher. *The Making of Western Indology*, p. 166.

[85] Rocher. *The Making of Western Indology*, p. 149.

For present purposes, it is important to stress just how central collections were to these new societies, the majority of which formed around an existing or proposed library or museum. Much of the early work of the Zoological Society, for example, was devoted to getting their collections off the ground. One of the very first orders of the first council meetings in 1825 was to secure arrangements with the keepers of the menageries at the Tower of London and Exeter in exchange for the temporary keeping of "such animals as may be presented to the society, until their own establishment is completed." The next several items record the status of their current small collection, including several new presents offered to the Society (two "rapacious birds" from Joshua Brooks and a deer taken by a naval captain from an island off the coast of Calcutta). The remainder of the business involved setting up the four main committees to manage, in addition to the finances, the Society's menagerie, the museum and the library.[86] Then, when the Zoological Society received its Royal Charter in 1829, the investments already made in accumulating a collection were key to the standing and status of the Society. The charter notes both that the Society has been formed for "the introduction of new and curious subjects of the Animal Kingdom" and that the members have already "subscribed and expended considerable sums of money for that purpose."[87] Similarly, the Geological Society's charter, adopted in 1827, stresses that the members have already "expended considerable sums of money in the purchase and collection of Books, Maps, Specimens and other objects and in the publication of various works."[88]

The Company's museum supplied materials to many of the new society collections. For years, it had been the practice to share so-called duplicate specimens with other collectors, as when, for example, in 1828 the directors permitted members of the Medico-Botanical Society of London to visit "the Herbarium of the Company . . . and [take] duplicates of such medical plants as are therein contained."[89] By 1830, the Company was sending duplicate specimens to the British Museum as well as other institutions such as at Oxford, Cambridge, the Zoological

[86] Zoological Society Library (ZSL): GB 0814 FA. Council Meetings. May 5, 1825.

[87] *The Charter, By-Laws and Regulations of the Zoological Society of London, inc. March 27 1829*. Waterlow and Sons, 1829.

[88] *The Charter, By-Laws and Regulations of the Zoological Society of London, inc. March 27 1829*. Waterlow and Sons, 1829.

[89] BL MSS EUR F303/5, June 4, 1828. On duplicates, from the UK NHM Z MSS IND Part I. Many of the Company collections had large quantities of each animal – for example, the Finlayson collection often has twenty, thirty or forty specimens of each animal. Also see "Catalogue of the Duplicate specimens of the Quadripeds and Birds collected in Sumatra by Sir Stamford Raffles," sample pages from the "Hardwicke." Also, the catalog of specimens from Wallich and R. Wright of Madras lists multiple specimens of most animals and plants.

Society and the University Museum in Geneva.[90] It was also donating manuscripts and works of literature to literary societies.[91] In general, and like other museums at the time, the Company's museum often operated less as a final resting place for inflows from the empire and more as a sorting house or sieve.[92] Well-preserved and rare items were first offered to (or requested by – as sometimes happened when the museum put duplicates on display) the "other national collections," as Horsfield referred to the British Museum and Kew. An agreement of formal regular exchanges of publications with the British Museum began in 1852. The many different grades of other unwanted specimens would be wound outward across Britain to Europe and abroad. While Kew, the British Museum and then, usually, Cambridge and Oxford were the first in line within a hierarchy of outflow recipients, a surprisingly broad number and type of institutions were also in line to receive the museum's donations.

An oft-repeated justification for moving collections from the colonies to London was climatic. Horsfield and other European curators worried that the specimens and manuscripts would rot in the heat and humidity of the tropics. They also argued that institutions in the colonies were not capable of proper preservation and care. Ironically, however, the Company's own voracious collecting would lead to the neglect and mistreatment of stored specimens. With the cellars in Leadenhall Street packed full by the 1830s, crates of documents, artworks or specimens would sometimes be left in the dockyard warehouses in New Street, vulnerable to "city dust, rats & other vermin."[93] Kew Gardens eventually rescued a huge quantity of materials from the basement of India House.[94] The Linnaean Society, the Zoological Society and the Geological Society were also regularly fed from the Leadenhall Street stores. The Horticultural Society helped to distribute seeds that came through the museum's doors. Smaller municipal societies also benefited, such as at Manchester, Liverpool, the Isle of Wight, Cornwall,

[90] Mss Eur F303/5, the British Museum on October 21 sent 52 mammals and 217 birds; on November 10, 106 insects from Java. On December 23, the directors sent duplicates to a number of other institutions: the museum in Geneva (for Dr. Candolle), the University of Oxford, the Zoological Society and the University of Cambridge.

[91] See, for example, a donation of "Babylonian" inscriptions to the Royal Literary Society. "Scientific and Literary," *The Athenaeum: Journal of Literature, Science ...*, no. 421 (November 21, 1835): 875.

[92] A private French collection is described as a "center of distribution" in Hoquet, Thierry. "Botanical Authority: Benjamin Delessert's Collections between Travelers and Candolle's Natural Method (1803–1847)." *Isis* 105, no. 3 (2014): 508–539.

[93] Royal Botanical Gardens, Kew. Herbarium presentations to 1900, vol. 1, February 5, 1858, ff. 249–251.

[94] See Cornish, Caroline and Felix Driver. "'Specimens Distributed.'"

Dublin, Boston, Philadelphia and even farther afield in Missouri. Most of the Company material organized by Joseph Hooker (see Chapter 7) was sent on to other repositories around the world. As we will see in the next chapter, half of the Siwalik Hills fossils of Hugh Falconer were sent directly to the British Museum, and into the early 1850s the Company produced plaster-casts of its share of fossils and shipped them on to Oxford and Cambridge, the St. Petersburg Academy of Science, the Military Academy at Addiscombe, and institutions in Australia, Sweden, the United States, Germany, India and beyond. Especially after 1833, war trophies from Company wars such as the Afghan and Opium wars would be donated to the new United Services Museum, an even larger military-linked museum, which had new display rooms in Whitehall near the War Office. Altogether, Ray Desmond reckons, sixty-four universities, museums, societies and individuals benefited from the Company's disgorging.[95] The East India Company thus became a prominent participant in an economy of barter, exchange, purchase and donation of material among hundreds of repositories across the world (Table 5.1).

Table 5.1 *Distribution list from April 8, 1850 of the Company-sponsored publication of Max Müller's* Rig-Veda-Sanhita: The sacred hymns of the Brahmans *(W. H. Allen & Company, 1849), illustrating the networks within which Company scholarship was embedded*

British Museum	Library of the Société	Ditto of Breslau
Library of the Royal Society	Asiatique Paris	Ditto of Marburg
Ditto Royal Asiatic Society		Ditto Tubingen
Ditto Royal Institution	Royal Academy Berlin	Ditto Göttingen
London Library	Ditto Munich	Ditto Leyden
Bodleian Library	Ditto Imperial Academy	Ditto [Lonnian?]
Library of Exeter College	St Petersburg	Ditto Uppsala
Cambridge Public Library		Ditto Chevalier Bunson
Advocates Library Edinburgh	Imperial Academy Vienna	Prof Bunnof Paris
Library of Trinity	Royal Library Copenhagen	Professor Larson Bonn
Coll Dublin	Oriental Society Germany	[Bentley] Göttingen
Public Library Paris	Library of University	Professor Roth Tubingen
Library of the French	of Berlin	Professor Stengeler Breslau
Academy	Ditto of Bonn	

[95] Desmond. *India Museum*, p. 53.

*

This chapter has explored some of the ways in which, after the establishment of the Company's library, museum and colleges in Britain, an imperial, Eurocentric geography of knowledge production would begin to take shape. Part of this had to do with physical ownership of and access to knowledge resources. The Company's remarkable ability to control access to Asia, and to dominate the accumulation of information about Asia in Britain, had, by the 1830s, given Company science a prominent role in shaping the material culture of science in Britain. The Company's influence was now exercised not only through restriction and protection but also through selectively opening access and sharing resources. The Company's formal monopoly was gone, but Company science now operated within a different social configuration of access and exclusion: the narrow social networks of club-society cultures of science.[96] This selective opening up also coincided, as the next chapter will make clear, with even more radical changes to the Company's remaining monopoly rights and its sovereignty with respect to the Crown. In consequence, even within Britain, there was a growing debate and disagreement over the nature and scope of access to the Company's library and museum, including accusations that the Company was maintaining an illegal knowledge monopoly.

But the establishment of British dominance within the colonial political economy of science also had to do with how the material was put to use, and in particular, at this moment, the systematic, intellectual possession of Asia through the placing of data about Asia within local theoretical and taxonomic systems. It would only be later in the nineteenth century, when modes and practices of European science began to establish a global presence, that the long-term consequences of the growing cultures of science in Britain would become clear. In the early nineteenth century, however, the philosophical and taxonomic work of Company science in Britain was – although certainly acquisitive and possessive – by and large a provincial, inward-looking world. Taxonomic debates were not aimed, in Horsfield's time, at capturing or overtaking the ordering and naming systems of Asia and imposing a new British order of things. Rather, the battle was much more provincial, between the Linnaean Society and the Zoological Club, or between those influenced by German *naturphilosophie* and those wedded to distribution

[96] On the similarities of monopolies and networks, see Schroeder, Ralph and Richard Swedberg. "Weberian Perspectives on Science, Technology and the Economy." *The British Journal of Sociology* 53, no. 3 (2002): 383–401, p. 392.

studies akin to that of Humboldt.[97] That provincial battle was nevertheless still over intellectual property claims (as they would be called today). And, with such a prize, the issue of priority of naming – and the general rule that the first "scientific" name given would remain the name – was equally critical.[98]

It is hard to find a more direct statement of the power of naming systems as generators of valuable property than William Kirby's address at the opening meeting of the Zoological Club, which Horsfield no doubt attended. Here, in a speech about the reasons for the new club, he also lays out in explicit terms how the philosophical issue of classification resolves into concrete material benefits:

> "*Nomina si pereunt, perit et cognitio rerum*," says Linne. Names are the foundation of knowledge; and unless they have "a name" as well as a "local habitation" with us, the zoological treasures that we so highly prize might almost as well have been left to perish in their native deserts or forests, as have grown moldy in our drawers or repositories. But when once an animal subject is named and described, it becomes a possession for ever, and the value of every individual specimen of it, even in a mercantile view, is enhanced.[99]

The Company's collections were now being used in policy debates, in the construction of philosophical theories and as the resources out of which the reputations and careers of Company scientists were made. And beyond India House, Company science was also coextensive with the new professions, societies and intellectual networks of Britain's so-called second scientific revolution. Out of these fragile networks would grow, over the next century, some of the key structures for the European dominance of the business of modern science.

[97] On the several versions of "philosophical naturalism" at work during this time, see Rehbock. *The Philosophical Naturalists*, introduction.

[98] In 1817 Temminck, for example, accused a contemporary of "plunder" and "the most insolent pillage." Gasso. "On Whose Authority?" p. 450.

[99] Kirby, William. "Introductory Address Explanatory of the Views of the Zoological Club Delivered at Its Foundation, November 29, 1823," *Zoological Journal* 2 (1826): 1–8, 5. Quoted in McOuat, "Cataloguing Power," p. 2.

6 Becoming National

A Radical Pearl Merchant's Demand

On April 22, 1834, the exact day that the new charter of 1833 took effect, one unnamed editor of a magazine wrote to the directors of the East India Company and demanded that they throw open access to the library and museum at India House:

> To H. St. George Tucker Esq., Chairman of the East India Company;–
> Sir, as that abominable monopoly expires this day, I request that the subordinate Government Board, called the Court of Directors of the East India Company, will direct that you and the other servants of the Crown, service in Leadenhall, facilitate my access to the public papers and books, the property of the nation, which have been so long buried in Leadenhall. If I do not receive immediate admission I shall apply to the superior authorities of the nation. Remember! Licenses and passports are out of date. Don't harden your hearts about what power is left.
> With sincere sorrow that House and all is not already at the hammer, your most obedient servant,
>
> (name withheld [Peter Gordon])[1]

The charter of 1833 dealt another blow to the Company's corporate sovereignty and monopoly privileges. Under the conditions set out in 1833, the remaining monopoly on the China trade was lifted, the Company was to cease *all* trading activities and all Company property was formally transferred to the Crown, to be held in trust by the Company.[2] This last directive meant that the Company's entire imperial infrastructure, from the presidencies to the Company ports to India House itself and everything within it, was now, on paper at least, a part of the British government. In practice, however, the Company retained

1 *Alexander's East India and Colonial Magazine* (1836), 1/5, p. 125. Only a few complete series of the periodical survive, including copies at the British Library and Glasgow.

2 "All real and personal property of the Company to be held in trust for the Crown for the service of India." "The Company to close their Commercial Business, and to sell their property not retained for Government." BL IOR/A/2/19, "Papers Relating to the Negation with His Majesty . . . 1833," pp. 506–507.

Figure 6.1 The East India Docks in 1806. The Company's control over shipping, and the associated dominance of the London docks, was a key target of the free-trade reformers. From Wikimedia Commons. Also see Green, Henry and Robert Wigram. *Chronicles of Blackwall Yard*. Whitehead, Morris and Lowe, 1881.

control over much of that property, and in many ways operations at India House remained unchanged after 1833. But, for the author of the above letter, the new ownership arrangement made all the difference: the Company's library and museum was now, so the author claimed, a "public" resource.

The author is very likely Peter Gordon a writer, adventurer and one-time pearl fishery manager in Madura, who had been in conflict with the Company for years over alleged mistreatment.[3] Gordon had been back in London since around 1830, and among other things was working on a set of volumes, coauthored with John Crawfurd, on the history of China.[4] Presumably it was this research that took him into the India House library in the first place. *Alexander's East India and Colonial Magazine*, which ran from 1831 to 1843, was relatively prominent, one of very few periodicals devoted to Britain's empire in Asia.[5] According to one periodical survey from 1838, its politics were "ultra Radical" and its circulation numbers small in comparison to the other principal journal on Indian topics, the *Asiatic Journal and Monthly Register* (which was "conservative, but not violently so").[6] In the years after the end of the Company's monopoly in 1833, in the pages of *Alexander's* Gordon would use the museum and library as a pressure point to expose all that was wrong, in the editors' views, with the remaining subcontracted and corporate-shareholder structured form of British rule in India. Gordon and other critics would use the issue of access to the Company's collections as a litmus test for the relationship between the British state and the Company. As the editors argued, if the Company had now truly relinquished its monopoly, and set aside any sovereignty claims with respect to the Crown, then the Company's library and museum should now be run exactly as a public resource; that is, like "the other National Museum," the British Museum and its library.[7]

[3] See Desmond. *India Museum*, p. 28.

[4] Murray, Hugh, Peter Gordon, and John Crawfurd. *An Historical and Descriptive Account of China: Its Ancient and Modern History, Language, Literature, Religion, Government, Industry, Manners, and Social State ….* The Edinburgh Cabinet Library, 18–20. Oliver & Boyd; Simpkin, Marshall, 1843.

[5] Its launch was celebrated by other periodicals such as the liberal *Weekly True Sun* and the *Literary Gazette*. See Anon. "The East India Magazine, or Monthly Register for British India, China, &c." *The Literary Gazette: A Weekly Journal of Literature, Science, and the Fine Arts* 724 (December 4, 1830): 784. On the history of *Alexander's* within the India periodicals in Britain, see Anon. "Critical Notices." *The British Friend of India Magazine, and Indian Review* (June 1845): 170.

[6] Grant, James. *The Great Metropolis*. T. Foster, 1837, p. 318.

[7] For example: "Hardwicke does not appear to have had any confidence in the management of the Oriental Repository at the India House, for he has bequeathed his treasures to the other National Repository – the British Museum." *Alexander's* (1836), 1/5, p. 318.

Figure 6.2 The Sale Room at India House, where until 1813 all goods from Asia (and until 1833 all goods from China) would be auctioned by the Company. By Joseph Stadler, 1808. Copyright British Library Board (asset P699).

This chapter turns to the changing relationship between the Company and its publics – in Britain and British India – and focuses on debates and practices related to ownership of, and access to, the Company's library and museum after 1833. As we saw in the last chapter, for one sector of the British public – naturalists, orientalists and other members of the scientific community – access to the Company's collections was growing ever wider during this period. But in terms of visitors to the brick-and-mortar spaces of Company science, access was much more restrictive than, for example, the British Museum. As we will see, while the legal ownership of the Company's knowledge resources could be transferred to the Crown with the passage of a new charter, just what it meant to be a "public" knowledge resource was up for debate. In this period, just as natural philosophy was resolving into separate disciplines with separate institutional structures (more on that in Chapter 7), the cultural space of knowledge production was separating into new and separate spheres: public versus private, national versus imperial, professional versus amateur. The Company's piecemeal absorption into the British state was not so much the erasure of a historical anomaly as it was part of the very process by which "states" and "publics" came to be more clearly defined against corporations and "private" interests.

Historians such as Phil Stern and David Ciepley have charted the changing history of the corporate–state relationship at the level of politics and economics.[8] In what follows, I consider how the public–private status of the Company was also debated and constructed in relation to science, education and access to knowledge resources. At a time when a coherent British imperial identity was only just beginning to crystalize, the extremely convoluted property relations for the library-museum (held in trust by the Company for the Crown, which in turn held it in trust for the people of British India) raised awkward questions about the very idea of an imperial public.

The Charter of 1833: The Library-Museum in the Era of Reform

Received ... 1000 cards of admission to library. Printed to order.

Library day book, May 19, 1835

The 1830s were a transformative period in British culture, and these changes would bring new pressures to bear on the practices of Company science at India House. The reformist Whig party had come into power in 1830. The Reform Act of 1832 significantly expanded voting rights. By the Slave Emancipation Act of 1833, slavery became illegal throughout the British Empire. Agitation to repeal import laws (Corn Laws) that favored agricultural landholders was growing, and the working-class Chartist movement was gaining momentum. Discourses and practices of knowledge production were also reshaped during the so-called Age of Reform. With liberal utilitarianism came a push for increasing access to, and state support of, science and education. Cheap periodicals such as those produced by the Society for the Diffusion of Useful Knowledge (SDUK) were aimed at facilitating the self-improvement and self-determination of the "middle ranks" at the center of Jeremy Bentham and James Mill's utilitarian social philosophy. The incorporation of the secular University of London in 1836 sought to break the hold of the Anglican Oxbridge colleges. Radical newspapers attacked "taxes on

[8] Stern. *The Company-State*; Ciepley. "Beyond Public and Private."

knowledge."[9] Scientific societies were also swept up in the vanguard. The British Association for the Advancement of Science (*f.* 1831) was established as an antidote to the perceived aristocratic elitism of the Royal Society. Subsequently the Royal Society, too, changed its rules of membership and governance in the late 1830s. Attention also turned to exhibitions, museums and galleries, and to the British Museum in particular, when in 1832 and 1835 a Parliamentary inquiry scrutinized the public utility of that institution.

It is in this context that, as the next charter renewal season approached, some shareholders began questioning whether the Company had done enough to make the library and museum accessible to the public. The India House library-museum had always been described by Wilkins and the Court of Directors as a "public" repository. Although free, access was restricted through a variety of different measures. Visitors were required to have a ticket of admission, and generally only directors had tickets to disburse. Wilkins also could admit visitors without tickets, but these were supposed to only be people "distinguished by rank or science."[10] Special arrangements would always be made for "persons of extraordinary high rank or status."[11] It was also rumored that anyone could bribe their way in with a small gift to one of the porters.[12] Descriptions of India House and its repository appear in London guidebooks in this period, and the wider reading public also came into contact with Company science through books that drew on the collections. Edward Moor's popular *Hindu Pantheon* (1810), for example, used illustrations of sculptures and drawings from the Company's museum (and Wilkins provided the Sanskrit labels).[13]

But India House was far less of a draw, and far less spectacular, than the many other cheap shows and rotating exhibitions for which London was becoming famous.[14] In the streets of London in this period, museum displays of foreign curiosities were of a piece with commercial displays of foreign commodities, and the experience of each shaped the experience of the other.[15] Across London, shows and exhibitions were integral to the

[9] See, for example, "Taxes on Knowledge," *The Leader*, London, 2(50), March 8, 1851.
[10] BL MSS EUR F303/35, Finance and Home Committee Minutes, July 16, 1817.
[11] BL MSS EUR F303/35, Finance and Home Committee Minutes, July 16, 1817.
[12] Desmond. *India Museum*, p. 1.
[13] Moor, Edward. *The Hindu Pantheon*. J. Johnson, 1810.
[14] An excellent overview is in Qureshi, Sadiah. *Peoples on Parade*, chapter 2. Also see Altick. *The Shows of London*.
[15] See, for example, Kriegel, Lara. *Grand Designs: Labor, Empire, and the Museum in Victorian Culture*. Duke University Press, 2008; Mathur, Saloni. *India by Design: Colonial History and Cultural Display*. University of California Press, 2007. On the textile collections of the Company in particular, see Driver, Felix and Sonia Ashmore. "The Mobile Museum: Collecting and Circulating Indian Textiles in Victorian Britain."

booming city's new shopping arcades and other innovations in consumer culture. For instance, at the Baker Street Bazaar, from the mid 1820s onwards, a gallery of shops sat adjacent to special exhibitions ranging from new carriage styles to livestock shows to, most famous of all, Madame Tussaud's waxworks.[16] Some popular shows about India or China could be found outside India House, such as the spectacular panorama of the battle of Seringapatam that was installed in 1800 at the Lyceum in the Strand, and could be experienced any day of the week for one shilling.[17]

Within India House in this period, consumer culture and new forms of exhibition were juxtaposed in a different way. The library-museum sat not alongside shops but the auction rooms where merchants would bid. The foot traffic that might lead to a side-trip into the displays was made up of Company employees and associates of all ranks, from sailors and warehouse workers to tea traders, shareholders and nabobs. The numbers of Company shareholders, pensioners, widows and current employees who might come to India House for business of some kind or another was by this time into the tens of thousands. Still, since no records of visitor numbers from this period survive (the earliest is the chairman telling a meeting of shareholders that about 4,000 visited in 1833), it is unclear how many people were actually wandering through the library and museum rooms when Wilkins began complaining in 1817 of the "immense crowds of all classes who have by various means obtained leave to visit the Library and Museum."[18] At that point, ticket-bearers were admitted any day of the week except Sunday. In the custom of the public museum before 1830, visitors would be accompanied by a guide, sometimes Wilkins or (later) Horsfield, other times an assistant. Although restricted to the guide's tour, visitors were allowed to handle objects, such as turning the crank that set Tipu's tiger growling, leaf through some of the manuscripts and even play some of the Malayan instruments on display from Raffles's collections. Wilkins and Horsfield also themselves fetched and returned books and manuscripts for library visitors, so the number of users is likely to have been relatively small. Peter Gordon, the radical anti-monopolist, gives one of the most detailed descriptions of the library visitor experience in 1835:

Victorian Studies 52, no. 3 (2010): 353–385; Patel, Divia. "'Made of English Thread': The Fabric of Empire." *Third Text* 33, no. 4–5 (September 3, 2019): 595–614.

[16] See University College London's Survey of London: https://blogs.ucl.ac.uk/survey-of-london/2019/06/14/the-baker-street-bazaar.

[17] See McAleer, John. "Exhibiting the 'Strangest of All Empires': The East India Company, East India House, and Britain's Asian Empire." In *The MacKenzie Moment and Imperial History: Essays in Honour of John M. MacKenzie*, edited by Stephanie Barczewski and Martin Farr. Springer International Publishing, 2019, pp. 25–45.

[18] BL MSS EUR F303/35, Finance and Home Committee Minutes, July 16, 1817. On the Chairman's estimate, see Desmond. *India Museum*, p. 28.

> Our first object naturally was to ascertain the contents of the library; the library is not sufficiently catalogued, therefore, an actual inspection of many books was the only means of ascertaining their contents. At the British Museum, the reading-rooms are furnished with such books as the readers are most likely to have occasion to refer to, and they are placed so that each reader can help himself to them; but, at the India House, all the books are *taabooed*; Dr. Horsfield alone can take a book from its shelf; hence, every reader is a constant source of trouble to the assistant; so much so, that, each reader cannot but feel a great degree of repugnance to go into the Doctor's room, and to disturb the studies of a man of science, for each book he requires, and to be quite embarrassed with the over strained politeness of the Doctor or the miserable economy of the Company which constrains Dr. Horsfield himself to perform the laborious and dirty work of a common porter, in taking down the books and bringing them to the reading desk. ... The Doctor simpers and says that nothing can be a trouble to a librarian.[19]

Back in 1817, Wilkins had requested that the directors *cut* the opening times, such that "the curiosity of the public may be liberally satisfied" without so much inconvenience for himself and the users of the library. The directors agreed to put new restrictions on visiting days, so that it was only open on Mondays, Thursdays and Saturdays between 10am and 3pm.[20] Although crowds were apparently unwanted by the librarian and visitors working in the library (who sometimes complained of the frequent growling and squealing of Tipu's tiger coming from the display galleries), both Wilkins and Horsfield also steadily expanded and changed the displays throughout the late 1810s and 1820s.[21] Such improvements were usually presented to the directors as being required by growing visitor numbers, such as when, in March 1818, Wilkins requested

> greater accommodation for the depositing of works which have greatly accumulated, and also for displaying certain works of natural history which have not been arranged for want of sufficient space ... which will contribute to make the Library and Museum more worthy of the attention of the numerous visitors, among whom are to be numbered Persons of rank and science of all nations.[22]

A little over a decade later, the issue of public access would re-emerge as part of the debates surrounding the charter renewal of 1833. In the Court of Proprietors (see Figure 6.3), shareholders raised questions about the running of the library and museum. The lack of museum guides or catalogs was brought up several times in the Court of

[19] *Alexander's East India and Colonial Magazine.* "A Visit to the India House Repository," 1836, 1/5 (January–May), pp. 129–130.
[20] BL MSS EUR F303/35 Finance and Home Committee Minutes, July 16, 1817.
[21] MacGregor. *Company Curiosities*, p. 176.
[22] BL MSS EUR F303/35, Finance and Home Committee Minutes, March 11, 1818.

Figure 6.3 Engraving of a regular meeting of the Court of Proprietors at India House. From *Illustrated London News*, May 4, 1844.

Proprietors in 1832.[23] Catalogs and guides were two things that the Parliamentary Commission discovered the British Museum was doing very well, and which visitors voraciously consumed. Cheaply available, these publications were regarded by the Commission as key to making the collections as widely accessible as possible. The *Penny Magazine* of the SDUK, a cheap periodical aimed at working- and middle-class "improvement," published guides to the British Museum, and the museum itself sold an affordable guide as well. The print representation of the collections was also seen by curators such as John Edward Gray, keeper of the British Museum zoological collections, as critical to the accessibility of the museum.[24]

[23] *The Asiatic Journal and Monthly Register for British and Foreign India, China, and Australia.* Volume 7 (April 1832), Parbury, Allen, and Company, p. 215.

[24] See Gray's interview in the 1835 Parliamentary Report on the British Museum, reprinted in Siegel, Jonah. *The Emergence of the Modern Museum: An Anthology of Nineteenth-Century*

The Company, meanwhile, still had no catalogs for any of its collections.[25] One proprietor, Captain Gowan, repeatedly compared the running of the Company's museum to that of the British Museum. Noting that the British Museum was now open to visitors six days a week, he proposed that India House should "meet them halfway" and open its museum to general admission three days per week. He and other proprietors also asked about the lack of labels on displays and the system of needing a card of introduction for admittance. Having heard that it was under consideration to transfer the whole collection to the British Museum (a rumor that seems to frequently crop up after 1833), Gowan seems to have been keen to make the case that the collection would be better cared for and more accessible in the hands of that institution.[26] Gowan also sometimes forcefully questioned the necessity and value of the Company running its own college, noting that even Oxford now had a Sanskrit Chair.[27] Reporting on these debates in the Court of Proprietors, *Alexander's East India Magazine* took the proprietors' complaints as evidence of the extremity of the Company's "despotic power of expulsion" and "corrupt patronage": "the Court of Directors persist in excluding the British Public from the Museum and Library at the India House; they also persist in refusing free access to their own constituents, and in resisting the expressed wish of the Superior Court for a descriptive catalogue."[28]

Somewhat surprisingly, the chairman and the Court of Directors (see Figure 6.4) did not reply to the shareholders by agreeing that the museum needed to improve access to the general British public. Instead, he defended the relatively closed nature of its collections as a necessity. The reason was that this was a specialized collection to be used for "serious study," and that if the museum were "thrown open," not only would things most likely go missing through "sleight of hand" (periodicals at the time often commented that the British public was uniquely untrustworthy when it came to acting properly in a museum) but it would also be much too disruptive to the "really useful" aspect of the collections, which was to serve the true orientalist, the scholar working on any and all

Sources. Oxford University Press, 2008, p. 123. Also see Gray, John E. "Some Remarks on Natural History Museums." *Analyst* 5 (1836): 273–280.

[25] *The Asiatic Journal and Monthly Register for British India and Its Dependencies*. Vol. 6 (November 1831), Black, Parbury, & Allen, p. 256.

[26] *The Asiatic Journal and Monthly Register for British India and Its Dependencies*. Vol. 6 (November 1831), Black, Parbury, & Allen, p. 256.

[27] *The Asiatic Journal and Monthly Register for British and Foreign India, China, and Australia*. Vol. 7 (March 1832), Parbury, Allen & Company, p. 212.

[28] *Alexander's East India and Colonial Magazine*, 1/5 (1836), p. 399.

Figure 6.4 Engraving of a meeting of the Court of Directors at India House. From *Illustrated London News*, May 4, 1844.

topics to do with Asia.[29] In short, and not surprisingly, the directors regarded the knowledge produced by "scholarly" orientalist work as coextensive with the useful knowledge required in governance: history, languages, literature, natural history and so on. After all, this very unique scholarly culture of British orientalism had, since its early days in the late eighteenth century, been totally intertwined with the Company and its shifting position within India. In this way, no matter its ownership status, the India House library and museum was *not* like the British Museum because it remained very closely tied to the actual work of state. The collections sat somewhere on a continuum between state archives (which themselves were only just then beginning to become organized into

[29] *The Asiatic Journal and Monthly Register for British and Foreign India, China, and Australia*. Parbury, Allen, and Company, April 1832, p. 216.

standalone institutions, let alone with conventions of public access) and public museums.

From the first plans for the India House library and museum back in 1798, it had always been discussed as a set of institutions intended to serve the interests of the people of *British India*, and it was also the British Indian taxpayer who funded the administration at India House. In these debates, however, the interests of the Indian taxpayer seem to have never been brought up directly. Serving the interests of the people of India had, in this case, always been understood as having *nothing to do* with providing the Indian public physical access (or even virtually in the form of catalogs or guides) to what was in effect their national museum and library. The assumption seems to have been that resources were to be put to use *for* the good of the Indian public, without being able to be used *by* them. This assumption is especially striking given that, during the Parliamentary inquiry into the British Museum in 1832, reformers were greatly worried about the inaccessibility of those resources to the typical taxpayer living outside London.

Liberal reforms bubbled their way into the Company administration through many of the changes made in the Charter Act of 1833. The starkest realignment of Company politics with the growth of liberalism came with the formal end of all Company trading, and the beginning of "free trade" (though still heavily mediated by laws and tariffs) between Britain and Asia. But radicals and reformers in Britain took up the cause of "improving" British India in many different ways during the charter renewal season. The proposed instruments of improvement were often not directly political or economic (though reformers worried greatly about the economic and political status of India) but fundamentally about education. Thus, promoting the growth of science and education in India became a key focus of the 1833 charter debates. For example, the remaining restrictions on "free" colonization – British travel to and settlement in India – were increasingly under attack by those who argued they also severely limited the free movement of ideas between Britain and India. The industrialist Dwarkanath Tagore and the writer and reformer Rammohun Roy were among a group of Calcutta businessmen and reformers who made a formal petition in 1829 for the end of such restrictions on free movement as well as free trade.[30] Both saw the opening up of India to more European business and enterprise as

[30] Zastoupil, Lynn. "Free Trade and a Reformed Parliament." In *Rammohun Roy and the Making of Victorian Britain*, Palgrave Macmillan US, 2010, pp. 111–128. For an extensive examination of how science embodied the contradictions of liberal imperialism in colonial and postcolonial India, see Prakash, Gyan. *Another Reason: Science and the Imagination in Modern India*. Princeton University Press, 1999.

a prerequisite for widespread educational transformation through informal interaction and communication. William Bentinck, governor-general of India during this period (1828–1835), echoed these views repeatedly. Bentinck's governorship is often taken as the beginning of a period in which liberal utilitarian views were particularly influential within British Indian politics, and he did push for similar kinds of education-driven reforms for India as Bentham and his allies were building in Britain. He avoided wars and sought financial reform in order to create a surplus for investment in education and improvement schemes. He promoted (in anticipation of Macaulay) English-language education over native-language education, started planning for a Ganges canal system and experimented with steam navigation to improve communication.[31]

Bentinck saw greater interpersonal mixing between Britons and Indians as key to raising the "mind" of India, which had been so he believed with a now-typical dogmatism, "buried for ages in universal darkness."[32] Although praising the "superior aptitude" and "thirst after knowledge" among Indians, he believed that it was only through anglicizing Indian culture that "improvement" in India could be effected. He argued that the "rigid preclusion of the free admission of Europeans to India" had "dammed up . . . the main channel of improvement to India."[33] Already by 1830 the size of the British communities in India had grown significantly, along with the scale of the Company's government. By 1830, the Company had nearly 40,000 individuals on its India payroll: 875 civil servants, 745 medical officers and over 36,000 "European" (mostly British) troops in its Indian army.[34] Bentinck also believed that the movement must go both ways, with more encouragement for Indian men "to go to Europe; [and] there to study in the best schools of all the sciences" and, above all, to see "what India *may become* by [seeing] what Europe, and especially England, is."[35]

During the charter debates of 1833, the directors therefore attempted to show that the Company was deeply invested in the growth of education, specifically in "European science" and literature, in British India. Noting that they had kept up the promise made in 1813 to give substantial

[31] Bearce, George D. "Lord William Bentinck: The Application of Liberalism to India." *The Journal of Modern History* 28, no. 3 (1956): 234–246, p. 235.

[32] This is in a letter soliciting support for the "Madras Steam Fund" to G. Norton, April 11, 1834, reprinted in *Alexander's East India and Colonial Magazine*, 1834, 11/12, pp. 498–505.

[33] Letter to G. Norton, April 11, 1834, reprinted in *Alexander's East India and Colonial Magazine*, 1834, 11/12, pp. 498–505.

[34] Bowen. *Business of Empire*, p. 262.

[35] And then he goes on to discuss the recent trip of Rammohun Roy who has "broken the ice . . . and I have no doubt other rich and well-educated natives are preparing to tread in his footsteps." Reprinted in *Alexander's East India and Colonial Magazine*, 1834, 11/12, pp. 498–505.

annual support to native education, the directors cited one report from 1826 that listed over sixty Company-supported schools, *madrassas*, colleges, book societies and presses. The majority of these institutions were, at that time, conducted in local languages and separate from Christian missionary schools. And, according to the directors, this investment was already opening up more pathways for natives to access skilled civil service careers. Through the disbursement of Company funding to smaller independent schools and colleges, "a considerable number of learned natives are retained, in their capacities as moulavees, moonshees, pundits and professors of the art of writing in native characters." Similarly, medical schools were "instructing native doctors in the science of medicine with a particular view of more effectively discharging their duties as vaccinators."[36] At the same time, with the charter of 1833 formally ending the ban of non-Christians and non-British employees in the Company's civil service, the Company's British-based colleges were now an obstacle to Asian subjects gaining some key civil service positions.

The subject of the Company's patronage of education and science in Britain came under scrutiny as well, and the record of the Company's education regime in the home country was shakier and harder to defend against the move for reform. Some close to the Company were highly critical of the financial and moral benefit of Haileybury for British India. The Member of Parliament and former chairman Richard Jenkins, for example, advocated for a revival of Wellesley's grand plan for Calcutta College, suggesting that this was in the best interests of the people of India. In the end, the Company just barely succeeded in keeping the Company's British colleges open. Any civil servants or military officers bound for India still had to obtain degrees from Haileybury or Hertford.

How to Break a Knowledge Monopoly

The Charter Act of 1833 transformed the nature of the Company. In the broadest terms, the Act significantly reshaped British colonialism into a now more familiar form of nation-based imperialism. In the process, both Britain and "India" resolved into more nation-like bodies. In the case of Britain, the arguably separate sovereignty of the Company was nearly entirely eroded: the Act explicitly designated India as a colony possessed by the Crown. And in the case of "India" – which at this time

[36] Great Britain. *Report from the Select Committee on the Affairs of the East India Company: With Minutes of Evidence in Six Parts, and an Appendix and Index to Each*. Parliament. 1831–1832. H. of C. Reports and Papers. 734, 735 I–VI. London, 1832, p. 399.

stretched northeast around the Bay of Bengal to include the Straits Settlements (i.e. parts of present-day Myanmar, Malaysia, Singapore) – it was for the first time united as a single administrative and legislative entity, with the creation of the governor-general of India position, and the raising of the Bengal government to the Supreme government of India, and the removal of the independent legislative powers of the Bombay and Madras presidencies. And as Tagore and Bentinck and other liberals had hoped, the Act also enabled, for the first time, the free movement of Europeans, and thus encouraged "colonization" in large parts of the new nation-like territorial entity as well as removing one of the Company's means of monopolizing knowledge of Asia.

Eventually, the Company's library and museum would also become more state-like and would merge with Britain's state collections. But the process itself was not at all smooth, would take many decades to complete and wouldn't really begin until after the abolition of the Company in 1858. One interesting immediate effect was the first disciplinary splintering of the museum collection, in this case "war trophies" being separated out from the rest of the collection and relocated. Standards and flags of vanquished powers such as those of Tipu Sultan were soon transferred over to a new naval museum at Haslar Hospital.[37] In another change, in 1838, public opening hours were eventually extended, with no tickets required on Saturdays between 11am and 3pm. At the same time, however, the directors continued to assert that the India House museum was *not* a public collection and that the business of empire came first. As the chairman wrote to a Member of Parliament enquiring about the museum opening times: "although the Museum in this house does not come under the denomination of a Public Institution, the Court feel happy in consenting to its being opened to the inspection of the public so far as may be practicable with reference to the business transacted under its roof."[38]

Another attempt at accommodating the public view within the necessities of the "business transacted under its roof" was a new policy explicitly barring visitors from copying or extracting "official documents" or other designated materials from the library, with the interesting exception of "documents of a literary or scientific character."[39] It seems to have

[37] BL IOR/H/787, no. 1–14: letters between Horsfield and others about transferring "colors," standards and so on to the Royal Naval Hospital, which was completed in 1836; "Letters about Transfer of EIC War Trophies to Royal Naval Hospital Chatham," 1836.

[38] BL Mss Eur F303/36–41, Court of Directors to Joseph Mane MP, May 15, 1838. And at a Finance and Home Meeting of October 31, 1838 the days open to the public were extended to Monday and Thursday.

[39] BL MSS EUR F303/36, Finance and Home Committee Minutes, April 4, 1838.

been Peter Gordon, the anti-monopolist, who prompted the Committee to clarify what visitors were and were not allowed to transcribe.[40]

The effects of the charter can also be seen in some changes to the collecting and patronage patterns for the library and museum. For example, after the rules for British travel to India became less restricted in the 1830s, the Company's control over exploration and travel writing would weaken, although expeditions into most inland areas still needed approval from the Company. For example, the well-known traveler and writer Emma Roberts, author of *Scenes and Characteristics of Hindustan* of 1836, made her career with the financial support of the periodical press (and only the tacit support of the directors). In 1839, she arranged a sponsored trip sponsored by the *Asiatic Journal and Monthly Register* who serialized her "Notes of an Overland Journey through France and Egypt to Bombay in 1839," and the directors supported her expedition with a substantial subscription.[41]

By far the biggest challenge the new charter created for the library and museum had to do with the new ownership status of the India House collections. Now that all Company property had been transferred to the Crown, the formal ownership status was clear, but the future physical location seems to have been in doubt. We have seen that some shareholders were now bringing up the possibility of merging the Company's collections with the British Museum. In the week that the new charter took effect, a London newspaper reported on the "important alterations" now taking place at India House since the new "arrangement" between the Company and government: some clerks and officers had been dismissed; others were now planning to retire; Haileybury College was (to the paper's surprise) *not* going to be broken up; "a great portion of the premises of the East-India House will become useless to the Company, in consequence of the loss of the exclusive trade with China and other commercial arrangements"; and "we have not heard whether the museum is still to remain in the East India-House, or the Library, which consists of a great number of scarce books and valuable manuscripts [will be moved]."[42]

Ultimately any decision on the location of the library and museum depended upon whether or not the British government agreed that the library and museum constituted part of the material necessary for the governance of India. This was an issue emerging from the *state-like* nature of the Company. However, there was also another set of issues emerging from the *corporate* nature of the Company: how the Company's privileged possession of two centuries' worth of commercial knowledge might

[40] BL MSS EUR F303/36, Finance and Home Committee Minutes, letter to Gordon, April 1838.
[41] BL MSS EUR F303, February 3, 1841.
[42] "Domestic News." *The Weekly True Sun*. April 20, 1834.

impact the opening up of the India trade to other British firms. Were critical trade secrets being kept locked up at India House? Was the Company doing enough to support the interests of British merchants and traders? Was the ability of British manufacturers to increase sales in India, or export raw materials for their factories, being hindered by the natural monopoly now maintained by the library and museum?

The potential commercial value of the knowledge resources held within the Company's library and museum would frame one attempt by another institution to take over management of the Company's collections. After 1833, amidst the discussion about the future of the India House library and museum, one possibility being considered was to transfer the whole of the collections to the Royal Asiatic Society.[43] Could the apparent or possible conflicts of interest over the Company keeping control of its library and museum be resolved by transferring it to a specialized third party such as the Royal Asiatic Society? With the prospect on the horizon, the members invited the writer William Cooke Taylor to remark on the "Present State and Future Prospects of Oriental Literature in 1835" to essentially make the case for the Royal Asiatic Society as the proper future home for the India House collections. Taylor opens his remarks with a nod to the utilitarian fashion of the moment:

> It may seem strange to connect Oriental commerce with Oriental literature, and many may deem the association unnatural; but no country in the world is more thoroughly utilitarian than England; in no other nation is it so difficult to introduce a new object of study ... without demonstrating its immediate pecuniary advantages.[44]

Taylor goes on to press the real economic utility of orientalist study to the nation. Essentially he takes the Company's still semi-sovereign defense of keeping the library and museum for reasons of (Indian) state and turns it back around to a properly (to him) *imperial* alignment: yes, the library and museum is essential to the work of the government of India, but as the Indian state is now formally subsumed under the Crown, its success and improvement, its relationship to the mother country, is now a matter of concern to all of the people of Britain, and – most importantly – a possible source of commercial gain for all.

Taylor also argues that the Royal Asiatic Society is deserving of the patronage and support of not merely scholars of the orient but also "all

[43] On the question of whether "the museum of the East India Company should be united with the Society's": *Alexander's* (1836), 1/5, p. 226.

[44] Taylor, William C. "On the Present State and Future Prospects of Oriental Literature, Viewed in Connection with the Royal Asiatic Society." *Journal of the Royal Asiatic Society of Great Britain and Ireland* 2 (1835): 1.

engaged in the commerce of the East, all who derive advantage from general traffic."

The RAS [Royal Asiatic Society] is designed to be the great storehouse of intelligence for all who desire information respecting the present state of trade and capabilities of all the countries between the Eastern Mediterranean and the Chinese seas. ... Were its advantages as clearly understood as they ought to be, it would have a branch in every port and a member in every counting house. In nothing more than in trade, and in no branches of trade than in those between England and eastern countries, has the truth of the aphorism been demonstrated, that "KNOWLEDGE IS POWER." ... But there are too many in the world ... who desire much to see the power increased and perpetuated, but neglect the knowledge which is its first element.

The Royal Asiatic Society was far from alone in gravitating toward work that was considered to be useful to British trade and industry, or in presenting collections as central to the development of commerce and trade. The spectacular expansion of industrial production in the first decades of the nineteenth century also coincided with a growing chorus of doubt and worry about the effects of such growth on the quality of design and the scope of industrial innovation. The potential for museums to support manufacturing and improve craft and industrial design was just then beginning to be promoted; similar claims were made, for example, in the House of Commons Select Committee Report on Arts and Manufacture of 1835.[45]

It is in this wider context that those who were really pushing the Company and the British state on this question of access to commercial knowledge at India House were often reformers and traders invested (either economically or ideologically) in the growth and expansion of trade with Asia. Peter Gordon was among the most vocal critics. It is unclear how invested Gordon really was in the future of the India trade (or of the Company's library-museum), but the Company's repository was especially useful to Gordon's crusade precisely because of the apparent contradiction of access to a publicly owned collection being managed by a corporation of shareholders ("publicly owned but kept very private").[46] For Gordon, it perfectly encapsulated how out-of-synch was the very existence of the Company with the era of free-trade reform; the library-museum is the one incongruously uncorrupted "object" hidden within the otherwise degenerate body of the Company, and as Gordon dramatically suggests, it will be the undoing of the whole thing:

The library at the India House has been hid from the notice of the Crown, the Parliament and the People in the immense mass of corruption which has hitherto

[45] Kriegel. *Grand Designs*; Mathur. *India by Design*.
[46] *Alexander's East India and Colonial Magazine*, Vol. 7, 1834, no. 49 (December), p. 571.

filled every apartment of the India House, but now that the Company of merchants has been compelled to give up its commerce, its museum stands forth as a conspicuous object; indeed it is so conspicuous that the Directors, who were merchant kings, seem likely to dwindle into puppet show men.[47]

Gordon's attention was also drawn to the Company's library and museum in the context of a records crisis at the level of the British government. In October 1834, the old Houses of Parliament burned to the ground. The fire was caused by an attempt to clear out some of the old archives. Bundles of wooden tally sticks used in accounting were added to the basement furnaces, causing them to overheat, and burning all the rest of the archives along with the entire building. In the wake of that fiasco, Parliament set up a Commission to investigate the state of public record-keeping in Britain, a move that would eventually lead to the establishment of Britain's first Public Records Office. Gordon, however, was focused on the differential treatment being given to colonial public records versus British public records by the Commission. Why, he asks, isn't government treating the Company's library and museum as the "national" collection that it is? Shouldn't the Public Records Commission survey include the India House archives and records, and, in fact, *all* of the libraries and archives in British colonies? Shouldn't the Commission reports be sent throughout the empire, and not given only, as the Commission had done, to libraries and archives in the UK? Gordon is essentially telling the government to act like the empire it now is and take more direct control of the records of its vast territories. (The Commission is called the "Public Records Commission of the United Kingdom" but Gordon refers to it as the "Public Records Commission of the Empire.")[48] Worst of all, he argues, is the situation at India House: "The libraries of Benares, Arcot, Tanjore, Seringapatam and Poonah have been plundered of their contents, which now lie rotting in the cellars of Leadenhall, corroded by damp and covered with dirt. The state of the archives, libraries, colleges, and schools of every portion of the British empire is constitutionally a proper subject for a grand jury to investigate."[49]

According to Gordon, Government, by its inaction, is allowing the Company to continue to hoard and neglect materials that the public now has a right to access. The "public" of the "Public Records Commission" is being (provocatively) understood as a trans-imperial

[47] *Alexander's East India and Colonial Magazine*, 1/5 (1836), p. 131.
[48] Anon. [Peter Gordon]. "Archives and Libraries." *Alexander's East India and Colonial Magazine*, November 1834, p. 477.
[49] Anon. [Peter Gordon]. "Archives and Libraries." *Alexander's East India and Colonial Magazine*, November 1834, p. 476.

entity, an imperial public, which has equal rights to knowledge of the records of its government:

In India, from time-to-time, severe threats are promulgated against officers, copyists &c who presume to reveal the secrets of the offices in which they are employed. We hope that every literary society in each colony will immediately apply to the British government for a copy of the publications of the Commission on the Public Records of the kingdom; the library at Arbroath [a royal residence in Scotland] has received a copy of the Board's publications, and surely the libraries at Colombo, Sydney, and Hobart Town have as good a claim as those at Arbroath and Lubeck, unless every British colony is to be branded as a Lubberland [i.e. lazy, unproductive].[50]

Government is here, by excluding colonial archives and libraries from the remit of "Public Records," failing to embrace the reality of the newly unified imperial sovereign and the new geography of the "public" under British law.

The Company, according to Gordon, is, in its turn, further exploiting this government inaction by actively mismanaging the collections. Gordon essentially claims that the Company is maintaining its knowledge monopoly through criminal neglect and mistreatment of the library resources, which amounts to a passive way of blocking public access. Gordon argues, for example, that the Company uses the library and museum as a personal treasury, selling and gifting items for profit. As proof of "the extreme impropriety of continuing to employ the India directors as the curators and conservators of a National Museum," Gordon relays a story about the directors misusing the "invaluable collection of Indian coins" and melting down ancient "Dariecs" (Daric coins) found buried near Benares.[51] Although no record of coin melting has been found, the library day books record numerous occasions where items were brought into the warehouse and then sent out as presents, or (less often) where precious items were taken out of display and made into gifts. Some of these gifts were very publicly made, such as the gold tiger's head and other precious items captured from Tipu Sultan that were presented to the royal family in 1832 just before the charter renewal debates began.

Even more, the pages of *Alexander's* detail the criticisms familiar from other visitor guides and the shareholder debates: lack of opening hours, few guides, no catalogs, unsystematic and unhelpful labels on objects, and the system of needing a card of admission provided by a director or curator. Gordon remains particularly focused on the lack of catalogs for the Company's library and museum. Gordon suggests that the lack of good documentation and access is willful and favors the ongoing corrupt rule of

[50] Anon. [Peter Gordon]. "Archives and Libraries." *Alexander's East India and Colonial Magazine*, November 1834, p. 482.
[51] *Alexander's East India and Colonial Magazine*, Vol. 7, 1834, no. 49 (December), p. 571.

the Company. Of the chairman, "his grand secret is to keep [Indians] in a state of profound ignorance . . . naturally enough this monster seizes upon all records, and locks up those which he does not destroy."[52] Noting that the Company has long experience in producing catalogs for its auction sales, Gordon argues that the Company's "Manuscripts, Books, Antiquities, Maps and Medals certainly deserve some degree of the care which is bestowed on compiling Catalogues of the Company's 'old musty' 'tarry flavoured' teas."[53] But as the Company refuses to make the collections accessible through proper documentation, Gordon sets out, as he tells his readers, to produce a catalog and guide to the library himself.

The first installments of his guide to "The National Library at India House" were published in *Alexander's* between 1834 and 1835.[54] Gordon begins by again reminding the readers that they should think of the Company's resources as a *public asset.*[55] The "guide" then goes on to detail the shoddy functioning of the library as a research institution, purposely underfunded, with the curators woefully overworked. But then, about two issues into the guide, in a dramatic turn, which Gordon relays in real time ("This article has been cut off, abruptly, by the receipt of the following letter") after the directors receive a copy of Gordon's *Guide*, they revoke his permission to access the library and museum. This "censorship" gives Gordon and the *Alexander's* editors much more fire, and they continue probing and attacking the Company library and museum from afar.

In Gordon's attacks on the Company's library and museum, there is always a lack of clarity in which people, he believed, rightfully own the India House collections. At one point, he argues that the right thing would be to "force" the "tyrants of Leadenhall" to restore the "records of India" to India. But at other points it is also argued that the Company (and thus the government of India) is so in debt to the British government that its records should stay in British hands.[56] Gordon's more common suggestion is that "The [India House] museum and library should immediately be removed to the British Museum where they would be accessible to every person."[57] The corrupt management of the Company is certainly seen as detrimental to the interests of India, but, to be clear, that is the case only because it stands in the way of liberal improvement that would

[52] Anon. [Peter Gordon]. "Archives and Libraries." *Alexander's East India and Colonial Magazine*, November 1834, p. 476.
[53] *Alexander's East India and Colonial Magazine*, Vol. 10, 1835, 7/12, p. 66.
[54] *Alexander's East India and Colonial Magazine*, Vol. 7, 1834, no. 49 (December), p. 571.
[55] *Alexander's East India and Colonial Magazine*, Vol. 10, 1835, 7/12, p. 66.
[56] Anon. [Peter Gordon]. "Archives and Libraries." *Alexander's East India and Colonial Magazine*, November 1834, p. 476.
[57] Anon. [Peter Gordon]. "Archives and Libraries." *Alexander's East India and Colonial Magazine*, November 1834, p. 471.

anglicize the whole system and, ultimately, increase British trade with India. The Company is therefore (as is a common trope in the Anglicist literature) painted as an old Asian despot itself, having adopted the old "corrupt rule of the ancient rulers of India":[58]

> Look at the rights of the cultivators of India – ask about the pergunnah rates – the tenure of land – the nature of slavery – the laws of caste – the rights of heads of families – the modes of trial – or any other important subject; and instead of finding it defined as by an English record, we find it uncertain as a Mahratta chieftain would desire, so that whenever occasion offered, he might intermeddle, and raise a dispute with his weaker neighbour.

In the end, Gordon's attacks seem to have had little effect on the question of whether the Company's collections should be moved out of India House and merged with the British Museum. The Royal Asiatic Society (which Gordon had claimed was almost as corrupt as the Company itself – it was "Philip sober" whereas the Company was "Philip drunk") was also unsuccessful in its bid to take control of the collections. However, in a move likely intended to appease such criticism, the directors did introduce one major change to the running of Company science at India House. In 1838, a new position was created: the reporter on the products of India. The first post-holder was John Forbes Royle. Royle was born in Cawnpore, India, was sent back to Britain for his education at Edinburgh High School and then Addiscombe, and returned to India in 1819. For nearly a decade he was the director of the Company's botanical garden at the former Dutch factory town of Saharanpur. He had arrived back in London in 1831 – yet another Company surgeon-naturalist with a rare collection in tow. Royle had wide-ranging interests, but he became known as a "pioneer of economic botany," as the distinct branch of botany focused on economically useful plants came to be called.[59]

The Office of the Reporter on the Products of India represented a new articulation of how the India House repository served not only British interests but also, so it was argued, the interests of the Indian taxpayer. Royle's office was meant to aid the "improvement" of India by, very specifically, increasing commercial trade between Britain and India. By 1850 this commerce-based economic route to India's improvement had, within the world of Company science in London anyway, begun to eclipse the earlier discursive focus on creating consilience on the grounds of artistic and literary exchange and translation. The market was to be the new meeting place and site of mutual benefit. And, as we shall see in

[58] Anon. [Peter Gordon]. "Archives and Libraries." *Alexander's East India and Colonial Magazine*, November 1834, p. 476.
[59] See "Dr. John Forbes Royle." *The Athenaeum; London* 1576 (January 9, 1858): 49.

Chapter 7, the Great Exhibition would allow Royle and his contemporaries to push this vision of the economic utility of museums to entirely new heights – with unexpected results.

Fossils in the Old Pay Office

Although possibly through no influence of Gordon and the radical anti-monopolists, the India House library and museum did, in the 1830s and 1840s, slowly open up access to the public, increase gallery spaces and finally begin publishing catalogs. Change came to the running of the museum and library in 1836, when Charles Wilkins, at eighty-six still in charge of the library, became ill and died soon after. Instead of succeeding Wilkins, Horsfield was given a separate position as naturalist and curator of the Company's museum, and an orientalist was sought to head the library. The Company succeeded in hiring one of the most prominent orientalists in Britain, Horace Hayman Wilson, who had recently been hired to the first chair in Sanskrit studies at Oxford. Wilson, the son of an accountant of the Company, went out to India as a surgeon in 1809. He had been trained at St. Thomas's Hospital, London and, as he explains it, turned to the study of languages on his outward voyage when a fellow passenger from India began teaching him Hindustani.[60] Once in India, he was hired by Dr. John Leyden, assay master at the Calcutta mint, as his assistant. Wilson took over from Leyden in 1816, and remained assay master until he returned to Britain in 1833. By 1811 Wilson had joined and (with Colebrooke's aid) become secretary of the Asiatic Society of Bengal. In 1813 he published his first translation, *The Cloud Messenger* (a translation of Kalidasa's *Meghadhuta*), and in 1819 he published a *Sanskrit–English Dictionary*. Wilson was also involved in the education debates of the 1820s, was on the board of the School Book Society and the Hindu College, and was secretary for public instruction in Calcutta.[61] While in Calcutta, Wilson had already delved into Company collections, publishing the first catalog of the vast (and controversial) Mackenzie Collection. The massive catalog was printed at great expense at the Asiatic Society's press in 1828.[62] He was known for his opposition to the requirements of Christian instruction, evangelization and attempts to regulate Indian religious practices (he even opposed Bentinck's very public

[60] Courtright, Paul B. "Wilson, Horace Hayman (1786–1860), Sanskritist." *Oxford Dictionary of National Biography*. September 23, 2004. Oxford University Press. Also see Wilson, Horace Hayman. *Essays Analytical, Critical, and Philological on Subjects Connected with Sanskrit Literature*. London, 1864.

[61] Courtright, Paul B. "Wilson, Horace Hayman (1786–1860), Sanskritist."*Oxford Dictionary of National Biography*. September 23, 2004. Oxford University Press.

[62] Wilson, Horace Hayman. *Mackenzie Collection: A Descriptive Catalogue of the Oriental Manuscripts and Other Articles Illustrative of the Literature, History, Statistics and Antiquities*

and popular measure to abolish *suttee*), all of which put him in favor with many of the Indian scholarly and political elite in Calcutta but made him less popular among some circles in London. All of this makes it even more surprising that Wilson would in 1832 be elected as the first Boden professor of Sanskrit at Oxford – a new position intended formally to support the spread of Christianity in India. When Wilkins died, the directors offered him the post of Company librarian, which he took on in addition to his professorship, cementing a new link between Oxford and the Company.

The hiring of Wilson and Forbes Royle and other expenditures on the museum and collections at India House are especially notable in this period after 1833, during which time internal "reform" was slashing the number of clerks, which sunk from around 200 in 1828 to 56 in 1844 (the smallest administration since 1765).[63] These were also years of financial strain, as Company profits declined and its expenditure, especially on border wars, continued to grow. But the positions devoted to information management continued to multiply. By at least 1847, India House also had a "statistical office" with eight clerks.[64] There are other signs that the charter of 1833 stimulated an expansion of investment in the sciences at India House. The astronomer and natural philosopher John Herschel was asked to be a general scientific advisor to the Company in 1838.[65] Although Herschel declined, from the late 1830s onward the directors were much more often seeking advice from the Royal Society, and the exchange of publications and reports (especially astronomical, magnetic and meteorological records) between the India House library and the library of the Royal Society was increasingly frequent.

The increase in curator positions after Wilkins's death would have been a welcome addition given the growing scope of the Company's collections. Throughout the 1830s and 1840s, the India House collections expanded much faster than in the previous decades. With so much material coming in, much of the curators' work involved managing the inflows. Wilson was dealing with a steadily increasing amount of library material, particularly printed matter sent from India. By 1830, four daily English newspapers and

of the South of India Collected by the Late Lieut.-Col. Colin Mackenzie, Surveyor General of India. Asiatic Press, 1828.

[63] Boot, H. M. "Real Incomes of the British Middle Class, 1760–1850: The Experience of Clerks at the East India Company." *The Economic History* Review 52, no. 4 (1999): 638–668, p. 639.

[64] Subramanian, S. "A Brief History of the Organisation of Official Statistics in India during the British Period." *Sankhyā: The Indian Journal of Statistics (1933–1960)* 22, no. 1/2 (1960): 85–118.

[65] Herschel, John. "Herschel to Roberton on Advising the CoD." June 8, 1838. In *A Calendar of the Correspondence of Sir John Herschel*, edited by Michael J. Crowe. Cambridge University Press, 1998, p. 189.

multiple weeklies in both Bengali and English were being printed in Calcutta alone, and the India House library was meant to receive and retain copies of all of metropolitan newspapers and journals published in India. Printing technology was now in the hands of not only the presidency and missionary presses but also book publishers and state-civic associations such as the Calcutta School Book Society and the Bibliotheca Indica series published by the Asiatic Society of Bengal. And even in Britain, the number of books coming out on topics related to Asia or to the Company was higher than ever (though nowhere near the number of Asia-printed works being gathered at India House).

Separately from the imports received and the purchases made by the librarian, the Court of Directors exercised one form of cultural patronage by way of subscription for publications. The directors were approached by hopeful authors who wrote to them and usually included a prospectus or a printed copy of their work. The most successful applications for support would result in the Company purchasing 100 copies, although this was quite rare. More often a large subscription would be forty copies of a work; less fortunate authors would be granted only six, or often just one subscription. As the century progressed, the number of authors who approached the Company slowly increased. Some of the successful authors were well-known, successful writers; others were Company servants or were connected to the Company in some way.

In 1844, the first year exact visitor numbers are available, 16,003 persons are recorded in the visitor books, and the numbers would grow slightly until the Great Exhibition of 1851, when the numbers would jump significantly (from 18,623 in 1850 to 37,490 in 1851).[66] Despite the criticisms of reformers and anti-monopolists, the Company's collections were still largely considered private property; the now formally national character of the Company's collections seems not to have made a very wide impression. The *Saturday Journal* reports on the extended opening times with a brief review:

> A SMALL museum at the East India House is now open freely to the public on Saturdays. The day is rather an awkward one for the majority of London sight-seers; but as the museum is, of course, private property, that is, the property of the East India Company, it is a privilege to be admitted to see it on any day that the directors may choose. It would add considerably to the privilege if the objects in the museum were labelled with a few descriptive particulars, which might inform the visitors, not merely of names, but of history, meaning, or use.[67]

[66] Desmond. *India Museum*, p. 41.

[67] "Museum of the East India Company." *The London Saturday Journal* 2, no. 29 (July 20, 1839): 37–38.

Example of the Court of Directors' patronage of new publications. In 1839–40, the Court agreed to the following subscriptions (as described in the Library Daybooks):

Forty Copies for the Use of the Court

The British Empire in the East by Count Bjirnst[Jena]
Achean and the Coast of Sumatra by John Anderson
A Personal Narrative of a visit to the Ghizni, Kabul and Afghanistan by E. J. Vigne
A Series of Prints from Drawings made in the Holy Land, Egypt, Arabia & Syria
Rough Notes of a Campaign in Sinde and Affghanistan in 1838/9 by Major James Outram
A Narrative of the Campaign of the Army of the Indus in Sind and Kabool in 1838/39 by R. H. Kennedy MD
The War in Affghanistan in 1838 Capt. Henry Havelock
The Oriental Portfolio for 1841 by Messrs Smith Elder & Co.

Twenty Copies

The Persian Moonshee by the late Francis Gladwin esq.

Six Copies

A series of views in Affghanistan taken by Sir Keith Jackson during the recent Campaign

Three Copies

An engraving from the Portrait of the Honble Mountstuart Elphinstone printed for the Oriental Club
An Engraving from the Portrait of His Grace the Duke of Wellington as Master of the Trinity House by Thomas Boys

One Copy for the Library

Travels to the City of the Caliphs by J. R. Wellsted esq.
Print of the Procession of Her Majesty through the Gallery of the House of Lords by J Starling
Engraving of His Majesty Mohammed Shah of Persia J. H. [Turgg]
Engraving of a Full Length Portrait of General Lord Viscount Combermere
"Analogy of the old and new Testaments" [a prospectus sent requesting the patronage] by Mr. J Whewell

The *Penny Magazine* also published a guide to the India Museum in 1841 after learning that the India House collections were now open to the public without an admission card from 10am to 3pm on Saturdays.[68]

[68] "The East India Company's Museum." *Penny Magazine of the Society for the Diffusion of Useful Knowledge, Mar.1832–Dec.1845* 10, no. 587 (May 29, 1841): 208.

In 1851, Henry Greene Clarke, who also published pamphlet guides to other national collections (including the British Museum, the National Gallery and the Naval Gallery at Greenwich Hospital), published a short guide to the museum at India House, which depicted the galleries as needing more English-language labels, difficult to access, poorly lit and sometimes small, cramped and "subterranean."[69] Guidebooks from the 1840s and 1850s usually stressed two things about the Company's library and museum. First, it was one of very few London museums that could be accessed free of charge, especially in East London.[70] Second, generally India House was still in 1850 the only place where one could see a large amount of material from India, Southeast Asia and China on permanent display.

One material effect of the charter of 1833 was that some of the most spectacular spaces inside India House – including the auction rooms and pay offices – were now redundant. By the mid 1840s, the museum was expanding into new rooms in India House. From the time Horsfield took over as curator, a steady stream of museum construction, expansion and improvements were presented to the Court of Directors, which almost always approved the expenditures. In 1837, Horsfield had five large glazed cases constructed for specimens of birds and other specimens from Assam and Madras.[71] The floor space of both the library and the museum continued to expand and, while Wilson kept building bookshelves, Horsfield kept building cases.[72] The old pay office, one of the largest rooms on the ground floor, was now refitted on the model of the British Museum but "of course to a much more limited extent" to house large archaeological finds ("various articles of ancient Hindoo Sculpture") and natural history specimens. Additional rooms above the original library and museum rooms were given over to an expanded natural history display, particularly focused on Horsfield's collection of birds and insects.[73]

In these ways, within India House, by 1850 the everyday work of imperial administration ran alongside the everyday work of museum management. Every September the library and museum would close to the public and extra hands would be brought in, cleaning the rooms and

[69] Clarke, Henry Greene. *The East India Museum; a Description of the Museum and Library of the Honourable East India Company, Leadenhall Street.* H. G. Clarke & Co., 1851. See Desmond. *India Museum*, pp. 40–41.

[70] "India Museum, East India House." *East India Magazine* (March 21, 1841): 219. The London General Railway, Steam-Boat, and Omnibus Guide. "Museums," January 1857.

[71] See, for example, IOR Finance and Home Minutes in BL Mss Eur F303/37: F&H May 28, 1837; F&H November 7, 1838.

[72] BL Mss Eur F303/36–51; F&H February 16, 1839; July 22, 1840.

[73] BL Mss Eur F303/36–51; F&H September 10, 1845; BL MSS EUR F303/49; F&H September 13, 1848 – all of the physical work of expansion over the past five years added up to £853.11.9 (1843–8), as tallied by the Clerk of the Works.

all the cases, doing repairs, and preparing and mounting specimens.[74] Such maintenance continued on a smaller scale all year long. Printed matter had to be sent to the binders. Specimens also had to be stuffed, pinned or otherwise prepared. Some of this – likely the insect preparations – Horsfield did on his own; in addition to paper, ink and candles, the library and museum also regularly purchased large quantities of pins. By the early 1850s, the Company's entomological collection was, according to some entomologists, beginning to rival that of the British Museum.[75] Throughout 1830–1 several hundred specimens were sent to taxidermists, for example the ornithologist and preparer John Gould. Gould (whose *Birds of Asia* of 1850 would be dedicated to the Company) would sometimes be paid for his taxidermy work with bird specimens selected from the Company's collections.[76]

Perhaps the most elaborate new natural history preparation and display at India House involved the Siwalik Hills fossils from northern India (see Figures 6.5 and 6.6). The fossil collection made by the naturalist Hugh Falconer (1808–1865) and engineer Proby Cautley arrived during the 1840s. Soon after graduating from the University of Edinburgh in 1829, Falconer went to London to assist Nathaniel Wallich on his cataloging of the herbaria from Calcutta. At the same time, he also assisted in the cataloging of John Crawfurd's fossil collections that had been extracted from the banks of the Irrawaddy during a diplomatic trading mission to Ava (Myanmar), which Crawfurd had donated to the recently established Geological Society. Falconer succeeded Royle at the Saharanpur Gardens (see Figure 7.2) in 1832, and would take up a professorship of *materia medica* at the Calcutta College, and superintendence of the Calcutta Botanic Gardens, when Wallich retired in 1848. Some fossils had been uncovered by the Bengal engineers during work clearing riverbeds for the construction of the Doab canal. Between 1834 and 1839, and relying on local sources – Falconer mentions in particular scrutinizing *Ferishta's History of Dekkan* (a Company-sponsored English translation of an influential seventeenth-century Persian work) and tracking down a rumored gift to a raja of a giant elephant tooth – Falconer and Cautley, an engineer, surveyed the Siwalik Hills at the base of the Himalayas for fossil remains.[77] Their discoveries would be by far the largest yet found by Europeans in Asia; huge deposits of

[74] F303/37 August 23, 1843.
[75] Douglas, John William. "Entomological Localities." *The Zoologist* 10 (1852): 3517.
[76] See BL Mss Eur F303/5, especially 1831–1832. The payment in specimens to Gould happened on September 10, 1832.
[77] Ferishta was Muhammad Qasim Hindu Shah, a sixteenth-century Persian historian. Falconer, Hugh. *Palæontological Memoirs and Notes of the Late Hugh Falconer: With a Biographical Sketch of the Author*. R. Hardwicke, 1868, p. xxviii.

fossil remains of many reptiles and mammals.[78] From Calcutta, the *Nautilus* shipped on July 1, 1841 with 187 crates of fossils.[79] Falconer arrived back in London in 1841 with eighty more crates, and Cautley sent along another twenty-two cases, which arrived in August 1844. The shipments were so heavy that special arrangements had to be made with HM Treasury in order to allow the arrival to be duty free.[80]

In agreement with the directors, the British Museum was given a large set of Falconer and Cautley's collection, and Falconer's work on his collection in the British Museum is well known.[81] But Falconer (with the aid of Horsfield and Wilson) was at the same time supervising the preparation and display of an extensive and only partially duplicate set of fossils at India House. The directors had seen to it that "a room will be appointed in this house for the reception of the collection of fossils made from the Himalayan Range and for its use in the suitable arrangement of the Collections during his furlough." (The same letter also informed Falconer that his botanical collection from Saharanpur was in transit "for the court and will be taken charge of by Dr Royle and Dr Horsfield.")[82] With the aid of Falconer, Wilson convinced the Court to hire the same assistants Falconer had employed in the British Museum, to continue the delicate and laborious process of "freeing similar specimens from the earthy matter by which their true forms are, in a great measure, concealed." The stonecutter Frederick Pullman and two assistants spent at least a year at India House working on extracting the fossils.[83] And once this was done, he was retained for another few years in order to make plaster-casts of the fossils for distribution to other museums and universities.[84] Pullman was recommended to the directors as highly experienced, having, for example, "cleared the mastodon's head" (the large mastodon skull was one of the most famous of the British Museum collections), and at the same time cheaper than either

[78] Chakrabarti, Pratik and Jaydeep Sen. "'The World Rests on the Back of a Tortoise': Science and Mythology in Indian History." *Modern Asian Studies* (January 1, 2016): 808–840.

[79] UKNHM DF 105/4–30. July 28, 1842 from Cautley at the Doab canal to Prof. Forshall at the British Museum ends with "P.S. the 187 chests appear to have been sent to England in the 'Nautilus' last year, therefore if you have not got them, they may be found in the India House." This series also mentions the giant tortoise and mastodon skull of the India House.

[80] BL MSS EUR F303/38 Revenue Committee August 28, 1844, arrived from Calcutta on the *Windsor*.

[81] Some record of the agreement between the British Museum and the Company is found in UKNHM "BM (Natural History) Collections 1838–1929," 1929 (1838): DF 105/4–30.

[82] BL MSS EUR F303/37; August 23, 1843, Court of Directors.

[83] BL MSS EUR F303/38; F&H July 23, 1845; F303/38 F&H August 6, 1845.

[84] BL MSS EUR F303/38; F&H July 8, 1846.

"artists" or "lapidaries." Still, when the directors questioned the need for India House to enter into fossil cast production, Horsfield stepped in to explain that such casts would be a kind of currency with which the Company could purchase missing specimens for its own collections:

> [Pullman's] labors have produced a series of specimens which the geologists of Europe will appreciate and be compelled to refer to, from their absence from the cabinets of the Continent or the British Museum. Nevertheless, the series is not complete, and gaps can only be filled up by Casts of Specimens from Continental collections or from the British Museum, but these casts can most readily be obtained by an interchange of Casts from the Company's rare specimens.[85]

By 1850 sets of casts were regularly being requested by different institutions and sent out to them; for example, the University Museum Stockholm and the naval museum at Haslar Hospital received "a complete series of Casts of the Himalayan Fossils, in five cases," while the Ludlow Museum of Natural History received "a Cast from the Head of the Rhinoceros ... from the Siwalik Hills."

All of this growth, together with the greater opening to the public, presented by Wilson as "the improved condition of the library," as he put it, suggested it was time for a catalog of the printed books.[86] Printed in a run of 1,000, this was the first catalog for the Company's library or museum. It would soon be followed by numerous other partial catalogs of the museum. The first major published museum catalog, Horsfield's catalog of mammals, was printed in 1851, just in time for the opening of the Great Exhibition. A catalog of the Company's collection of birds followed in 1854, and an insect catalog was published in 1857.[87] Meanwhile, in 1841, Wilson had produced a descriptive catalog of the coins and antiquities from Afghanistan in the Company's collections. Many of these had been plundered and purchased before the first Anglo-Afghan war by the agent Charles Masson.[88] Wilson had also continued to publish Sanskrit translations, as well as, in 1840, two books on Hindu history and religion.[89] Catalogs, books and expanding exhibits were just

[85] BL MSS EUR F303/38; F&H July 4, 1846.

[86] BL MSS EUR F303/38; F&H April 24, 1844. One thousand copies would be printed at a cost of £145.0.0 by "Messrs J & H Cox the Company's Printers." BL MSS EUR F303/38; F&H December 17, 1844.

[87] Horsfield. *A Catalogue of the Mammalia*; Horsfield, Thomas and Frederic Moore. *A Catalogue of the Birds in the Museum of the Honorable East India Company*. Vol. 1. W. H. Allen & Co., 1854; Horsfield, Thomas and Frederic Moore. *A Catalogue of the Lepidopterous Insects in the Museum of the Hon. East-India Company*. Vol. 1. W. H. Allen and Co., 1857.

[88] Wilson. *Ariana Antiqua*.

[89] Wilson's publications include: *Select Specimens of the Theatre of the Hindus* (2 vols., 1826–7), a translation of the *Samkhya karika* of Isavarakrnsa (1837), *The Vishnu Purana:*

a few of the ways in which the knowledge resources of the Company were increasingly on the move.

*

After the end of the Company's monopoly and the transfer of the library and museum to Crown ownership in 1833, the Company's scientific and educational institutions in London could not easily be resolved into a British "public" institution. The 1833 charter was just the beginning of a long and convoluted process that would remake the Company's knowledge monopoly into a public knowledge resource in Britain. In this way, in the first decades of the Age of Reform, the brick-and-mortar library-museum at India House was not only a mediator between visitors and the idea of "India" in Britain but also a mediator between reformers and the idea of "empire." Historians have argued that, in the eighteenth century, the Company's expanding empire was significant in forging a sense of "Britishness."[90] Now, as the monopoly was unwound, the question of how the Company's knowledge resources should be managed became one of the many issues through which the sense of "British imperialness" and in particular the idea of an imperial public would be established.

The Company's piecemeal absorption into the British state was not so much the erasure of a historical anomaly as part of the very process by which "states" and "publics" came to be more clearly defined against corporations and "private" interests.[91] While these issues stemmed from the old state-sovereign elements of the Company, other issues stemmed from the corporate-shareholder structure of the Company. As British manufacturers and merchants sought to increase the India trade, the Company remained a target of criticism, and in this context the directors were accused of hoarding knowledge essential to the development of free trade between Britain and India. The new position of the reporter on the products of India was likely intended to address some of this criticism by actively focusing on disseminating commercially useful information. But

a System of Hindu Mythology and Tradition (1840), *Lectures on the Religious and Philosophical Systems of the Hindus* (1840), *Sketch of the Religious Sects of the Hindus* (1846) and various volumes of collected essays and occasional translations. He had begun a translation of the Rig-Veda but did not complete it before his death. See Courtright, Paul B. "Wilson, Horace Hayman (1786–1860), Sanskritist." *Oxford Dictionary of National Biography*. September 23, 2004. Oxford University Press.

[90] See, for example, Mackenzie, John M. "Empire and Metropolitan Cultures." In *The Oxford History of the British Empire: Volume III – The Nineteenth Century*, edited by Andrew Porter and Wm Roger Louis. Oxford University Press, 1999, pp. 270–293.

[91] Stern. *The Company-State*; Ciepley, David. "Beyond Public and Private: Toward a Political Theory of the Corporation." *American Political Science Review* 107, no. 1 (2013): 139–158.

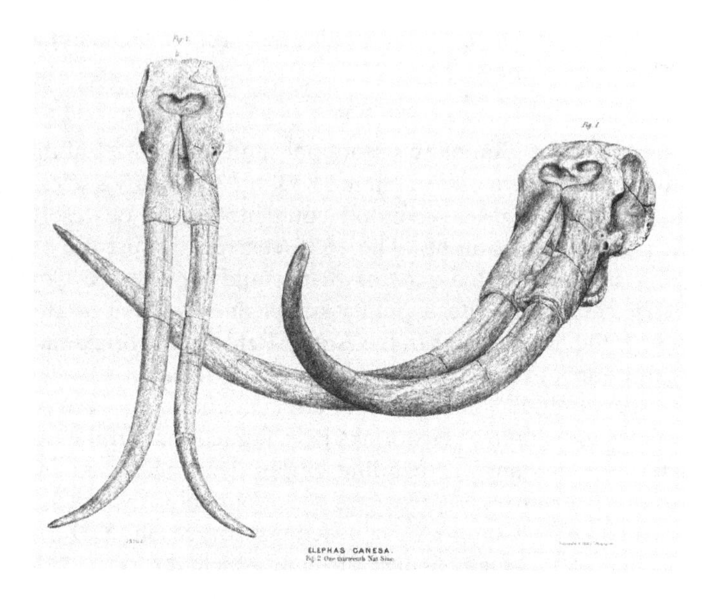

Figure 6.5 Illustration of a reconstruction of a Stegodon skull, which Falconer classified as "Elephanta Gansea," plate 23, in Falconer, Hugh and Proby T. Cautley. *Fauna Antiqua Sivalensis, Being the Fossil Zoology of the Sewalik Hills, in the North of India*. Smith, Elder and Co., 1846. Also see Falconer, Hugh and Charles Murchison. *Description of the Plates of the Fauna Antiqua Sivalensis*. R. Hardwicke, 1845.

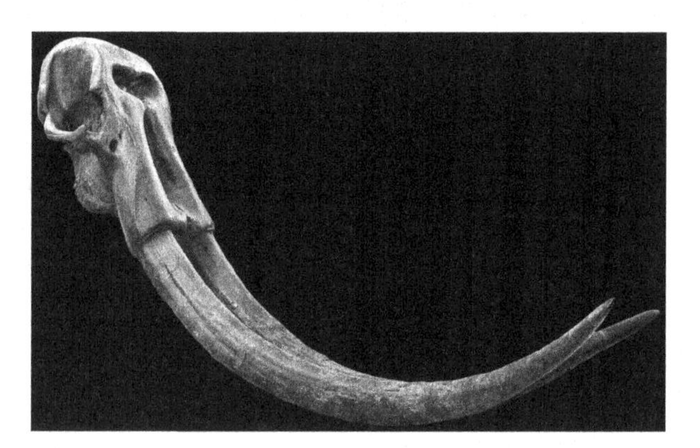

Figure 6.6 Reconstructed fossil skull of a Stegodon, an extinct genus of proboscidean, collected by Proby Cautley and Hugh Falconer in the Siwalik Hills in the late 1830s. By permission of the Trustees of the Natural History Museum (PVM 3008).

in practice, as we will see, Royle's position at India House was equally tied to a "commercial public" as well as a "scientific public," two categories that very often overlapped. To be sure, from the radical perspective of *Alexander's*, the shared club culture of the monopolists and the scientists was nothing to celebrate. Its pages ridiculed the tight connection between the Company ("Philip drunk") and the Royal Asiatic Society ("Philip sober"): the former were "men incorporated for the wicked and corrupt purpose of plundering India"; the latter were "the self-same men meeting in their better moments to make some return for the evils they inflict upon the people of India."[92] Clubs and societies were also the point where scientists in Britain increasingly were able to plug directly into the Company's knowledge resources and build upon the resources provided by the Company's former monopoly and still paid for by British Indian taxpayers. In the contemporary discourse on the utility of these rapidly expanding venues for self-improvement and rational entertainment, the incongruity of British India's "national museum" being located thousands of miles away at India House could not have been more acute.

[92] *Alexander's East India and Colonial Magazine* (1836) 1/5, p. 219.

7 The Commercializing Mission

Margaret Tytler's Model India

The Bengal Government engages Miss Tytler to prepare a set of models of Indian manufactures and agricultural implements for transmission to the East India Company's Museum in London.

March 17, 1824[1]

In 1819, Margaret Tytler, a Scotswoman with a self-described "love of science," accompanied her brother, Dr. John Tytler, on his first assignment with the Bengal Medical Service, where he was assigned to Patna.[2] Between 1821 and 1823, Tytler embarked on a remarkable project to construct highly detailed scale models of social and economic life in the region (see Figure 7.1). Visiting farms, forests, homes, workshops, mills and distilleries, Tytler employed local artists to draw and measure the tools and instruments of different manufacturers, then hired local makers of toys and statuettes to produce scale models (usually at half an inch to a foot) out of carved stone, wood and leather. Her process was empirical and observation-based. She describes, at one point, being sickened by the fumes in the distillery where she went to measure an apparatus. She also describes being "instructed" by a local distiller, who she calls Mushoo, in the gathering and preparing of herbs to make toddy. Tytler claimed the models are so exacting that, after constructing the model still, she claims, "by means of this model I distilled a tumbler full of spirits."[3] In all, she made over sixty precise models of a wide range of workers and their instruments, from bird catchers to millers of grains

[1] BL IOR/E/4/711 (Bengal public letters), p. 271.

[2] National Museum of Scotland Ms.069 (411): Margaret Tytler, "Catalogue of Models." Other model sets are now at the Victoria and Albert Museum and Kew Gardens. Thanks to Caroline Cornish for information on the Kew models. On modeling craft traditions in India and their appropriation in the later nineteenth century by European collectors and tourists: Smith, Charlotte H. F. and Michelle Stevenson. "Modeling Cultures: 19th Century Indian Clay Figures." *Museum Anthropology* 33, no. 1 (2010): 37–48.

[3] Tytler, "Catalogue," no. 57.

Figure 7.1 Ebony model showing a method of catching birds, produced in Bihar Patna c. 1815–1821, commissioned by Margaret Tytler. By permission of the National Library of Scotland (item reference: A. UC.832.77).

and oils, to weavers and makers of gold and silver thread to butter churners and poppy lancers. After one set was produced for the Bengal government, Tytler was asked to produce a second set specifically for the India House museum.

Tytler's models were early forerunners to a major new focus for the India House museum in the 1850s: the production of detailed industrial intelligence about India's trades and industries, intended for the use of British manufacturers and traders. As we have seen throughout the book, although Wilkins's 1798 proposal to the Court of Directors was nearly completely full of items of commercial value or related to understanding Asian trade, the actual shape that the library-museum took was much broader and, at least until the 1840s, bore little relation to Wilkins's proposal. By the late 1850s, India House would display many hundreds of models similar to those of Tytler's. As an article in *The Times* would put it in 1857, an hour spent in the new "model room" would "convey clearer ideas of Indian life and Indian customs than would be gained by the perusal of many dreary volumes."[4] This is just one way in which, in the final decades of the Company's existence, the discourse of knowledge production at India House was increasingly about growing and facilitating trade. This focus was amplified, as we saw in the previous chapter, in the new office of the Reporter on the Products of India, and especially in connection with the Company's central role in the Great Exhibition.

To be clear, however, focusing on commercial utility in no way meant the abandonment of engagement with questions of broad historical and philosophical interest discussed in the previous chapters. On the contrary, the rise of the economic museum movement in this period involved those very same questions, but now also applied in new ways to questions of how to develop the British–Indian trade. For example, Tytler's models were accompanied by a manuscript catalog that sets the whole project within a stage-based understanding of civilizational development similar to that of Adam Smith. On the surface, Tytler appears to have produced a relatively straightforward collection of models illustrating common commercial and agricultural practices in Patna. But Tytler's catalog also folds her observations into a broad theoretical framework of understanding of both economy and empire. She has chosen and ordered the particular models in order to illustrate the progress of civilizational needs according to the stadial theories of Smith, whom she quotes repeatedly. She also references a range of other European influences such as Erasmus Darwin's "Love of the Plants." The catalog at times works hard to create a "model" of rural Indian society designed to fit the Smithian economic-material stages of civilization. But it is also clear that Tytler's worldview has been shaped by her time in North India. For example, she discusses articles from the *Asiatic Researches* as well as the "Treatise on Agriculture" by one "Mater Jeet Singh Rajah Tikaree" (probably Mitrajit Singh Tikari Raj, a prominent zamindar).[5] Although in this way Tytler's analysis is a cosmopolitan production, it also sets this new world within what would be, to the intended audience, a very familiar intellectual register.

This chapter follows the economic turn in Company science at India House in the decades after 1833. The first section considers new institutional developments in the connection between the India House library-museum and collections-based science institutions in the colonies. Increasingly, the India House library and museum would be represented as at the top of a hierarchy of Company science establishments, reaching

[4] *The Times* (London), April 7, 1858, p. 10, quoted in McAleer, "Exhibiting the Strangest of All Empires," p. 40.

[5] See Chatterjee, Kumkum. *Merchants, Politics, and Society in Early Modern India: Bihar, 1733–1820*. Brill, 1996.

from London to the presidency governments and out into the rural divisions and settlements. The chapter then turns to the growing economic focus within the India House library-museum. The Company itself was no longer directly participating in trade, but it was responsible for the agricultural, industrial and other trade-related policies for British India. Part of the new responsibilities of the Reporter on the Products of India position were meant to aid the administrators in such areas of state. But the turn to a science of trade and industry was also, in part, the result of the directors more fully embracing the mission of making the library and museum useful for the (British) public. Altogether, with a new, more clearly defined role as a mediator of industrial, educational and scientific relations between the home country and the colonies, these developments combined to bring new energy and purpose to the library and museum at India House. In almost exactly the same moment, however, the decisive undoing of the Company was brewing, fermented not by the free-trade liberals in Britain but instead by the disaffection and defiance among British Indian subjects.

The Library-Museum and Company Science in British India after 1833

In 1840, amid an endless stream of incoming and outgoing books and documents, the library day books briefly record the beginning of the Victorian era with a notice of the deposit of a gilt-framed color print depicting the young queen opening Parliament for the first time in 1838.[6] By this time, the Company's collections at India House were overflowing, and Horsfield sought extra storage space in the cellars and attics. More and more, India House was redirecting donations and deaccessioning portions of its collections to other libraries and museums, which were mushrooming up all over what was, by now, the biggest city the world had ever seen. One London guide from 1851 lists eighty libraries.[7] Another from 1853 describes seventy-seven "literary and scientific institutions," virtually all of which held some kind of collection.[8]

Perhaps the constant commerce in and out of the Company's stores is one reason why the curators could worry about a lack of storage space in one memo and, in another, demand more material from India. Various

[6] BL MSS EUR F303/6, May 30, 1840.

[7] Weale, John, ed. *London Exhibited in 1851: Elucidating Its Natural and Physical Characteristics, Its Antiquity and Architecture; Its Arts, Manufactures, Trade and Organization, Its Social, Literary, and Scientific Institutions and Its Numerous Galleries of Fine Art.* John Weale, 1851.

[8] Atkin, George. *The British and Foreign Homœopathic Medical Directory and Record, 1853.* Aylott & Company, 1853, p. 18.

policies were now in place to encourage the regular movement of materials from the colonies to the imperial center. Significantly, it became routine for the Company to cover any port duties on items sent to India House for the museum or library. By at least the 1830s anything designated a "specimen of natural history" was guaranteed a duty-free import.[9] As the Royal Navy would do for statues, tombs and obelisks shipped to the British Museum from Egypt, the Company regularly footed the transport bill for massively heavy imports, such as the cache of ancient sculptures from Amaravati, or the semi-regular shipments of cases of books weighing over 500 pounds, destined for its library and museum.[10] Despite, therefore, the loss of the last vestiges of formal monopoly power in 1833, the Company's ongoing natural monopoly ensured that, Company science in Britain would emerge as an even greater center of accumulation and production, and its place within British scientific networks would continue to expand.

Printed matter from the colonies arrived less as gifts or the results of particular surveys or expeditions and more with the regularity of scheduled export commodities. By the 1840s, the library was trying to keep up with the outputs of British India's rapidly growing (and closely monitored) periodical press. Each quarter, presidency governments sent out collections of the latest periodicals. Both native-owned and government-aligned or missionary presses were churning out new books in dozens of languages in Calcutta, Bombay, Madras and elsewhere.[11] At Wilson's repeated insistence, sets of many of these works were sent back to London. The Calcutta Book Society, for example, supplied not only regional schools and colleges but also the Company's library and the students of Haileybury with vocabularies, primers and works of literature in Hindustani and many other languages. Many hundreds of copies of works would be exported to London, and much of this was distributed to Haileybury for use in classes or as student prizes. But many more were sold by the Company to bookshops such as Allen & Co. In fact, the Company's bookselling activities

[9] See, for example, duties and customs lists printed in many colonial newspapers, such as "Abstract of an Act for Granting Duties of Customs." *Singapore Chronicle and Commercial Advertiser*, May 1, 1834.

[10] See, for example, BL IOR Finance and Home Committee minutes, December 19, 1837, reporting the arrival of a box of "pine cones" as well as three cases of books from Calcutta weighing 566 pounds, the whole cost to be cleared by the Company and the items deposited in the library. On the shipping of the Amaravati materials, see Howes, Jennifer. "Colin Mackenzie and the Stupa at Amaravati." *South Asian Studies* 18, no. 1 (2002): 53–65.

[11] See Stark, Ulrike. *Empire of Books, An: The Naval Kishore Press and the Diffusion of the Printed Word in Colonial India*. Orient Blackswan, 2009; White, Daniel E. *From Little London to Little Bengal: Religion, Print, and Modernity in Early British India, 1793–1835*. Johns Hopkins University Press, 2013.

were becoming so regular in the later 1840s that it could be seen to violate the legal ban on Company direct engagement with trade. However, the practice seems to have been to only sell on imported books at cost, as the directors were keen to stress. As was made clear to one assistant librarian: "we are not to be Booksellers, nor have we ever been."[12]

The curators at India House were now much more active and aggressive in monitoring collections and finds via the Indian periodical press and pushing for the shipment of finds back to Britain. The Court of Directors (or, more likely, the museum staff via the chairmen) kept close tabs on the movements of these collections through various ports, noting, for example, when only portions of a collection were received in London while other parts had been siphoned off into collections in the colonies.[13] The India House curators' more aggressive attempts at policing the flow of knowledge resources were in part a matter of the increased self-regard as the center of the Company's sciences but also in response to the increasing pull of the growing centers of accumulation in the colonies. The number of museums, botanical gardens, libraries and scientific societies in British India had grown rapidly in the last few decades, and that growth was accelerating markedly in the early 1850s. In the wake of the 1833 charter, dozens of colleges and *madrassas* were established in British India and many of these had their own museums and botanical gardens.[14] One of the largest college museums was at the Medical College of Calcutta. In terms of large urban institutes, the Bombay Literary Society (*f.* 1804), Madras Literary Society (*f.* 1812) and Singapore Institution (*f.* 1826) all supported libraries and museums. When the Geographical Society of Bombay (*f.* 1832) formed, the first order of business was to begin establishing a library and collection of maps, manuscripts and instruments.[15] The fate of numerous colonial collections would be tied to the fluctuations of the Company Museum's appetite.[16]

[12] See later description of earlier practices in BL MSS EUR F303/81, August to September 1864.

[13] BL IOR: E/4/763 India and Bengal Despatches, September 16, 1840, ff. 155–172. Quoted in Desmond. *India Museum*, p. 55.

[14] The new education institutions after 1835 are described in: "EXTRACT from the REPORT of the GENERAL COMMITTEE of PUBLIC INSTRUCTION of the PRESIDENCY of FORT WILLIAM in BENGAL, for the Year 1836." Appendix I in Leveson-Gower, Granville George. *Select Committee of House of Lords on Operation of Act for Better Government of H.M. Indian Territories First Report, Minutes of Evidence, Appendix; Second Report; Third Report; Index*. House of Commons Papers, Sessional Papers, March 1852.

[15] For a critique of the government's lack of support for the collections, see *Alexander's East India and Colonial Magazine*, 1836, 1/5, p. 303.

[16] See Ratcliff. "The Company's Museum." Also see Miguel Puga, Rogério. "The First Museum in China: The British Museum of Macao (1829–1834) and Its Contribution to

Perhaps no colonial institution was more closely tied to the India House library and museum than the Asiatic Society of Bengal (ASB), the collections of which would eventually form the first national museum in British India in 1875. The dynamic between the Company's Museum in London and the ASB was, during this period at least, quite the reverse of the case of the London societies just described. Nathaniel Wallich had become the first curator of the ASB's collection in 1814, and (later exports to London aside) he had strongly argued for the importance of establishing libraries and museums *in* the colonies.[17] As Wallich had expected, the ASB's collection grew quickly:[18] "from *China*, from New South Wales, from the Cape, and from every quarter of the Honourable Company's possessions, specimens of natural history, of mineralogy and geology, have flowed in faster than they could be accommodated." But as collections grow, so do their costs, and by 1836 the ASB, now in debt, was considering putting the collection up for sale. Having grown too big to be funded by a voluntary organization, the collection was proposed as a foundation for the creation of the Government of India's first national museum. The problem, however, was that India *already had* a national museum; it just happened to be in London. When approached with the request to fund a national museum by establishing government support for the ASB collections, the governor-general of Bengal, the same George Eden who had co-founded the Zoological Society, passed the decision on to the Home Government, since the Court of Directors were already supporting a library and museum "at considerable expense." Although Eden believed that "such institutions in Europe, however perfect, do not supersede the necessity of providing similar in India," he sensed it was unlikely the Court of Directors would then fund a similar institution in India.[19]

Upon hearing that the question would be passed back to London, the ASB requested an interim 200 rupees per month in order to keep its collections in basic maintenance. When the question of a national museum in India finally came in front of the Court of Directors, they replied in 1839 that the Court would not object to that small sum "for the cost of preparing specimens and maintaining the collection in order." It was silent on the idea of a national museum, although it also sanctioned the Bengal government to spend some small amount of funds on purchases for the ASB museum, so long as "on all such occasions, you will

Nineteenth-Century British Natural Science." *Journal of the Royal Asiatic Society* 22 (2012): 575–586.

[17] Nathaniel Wallich to the Royal Asiatic Society, February 2, 1814, in Mitra. *Centenary Review of the Asiatic Society of Bengal*, p. 35.

[18] Mitra. *Centenary Review of the Asiatic Society of Bengal*, p. 37.

[19] Mitra. *Centenary Review of the Asiatic Society of Bengal*, p. 39.

forward to our Museum [that is, the one at India House] a selection from the articles which may have been so procured."[20]

Meanwhile, back in London, Horsfield kept surveillance on the ASB's *Transactions*, which published lists of new donations. He would at times aggressively seek the transfer of material collected by officers on duty. He also interpreted the ASB's new collection grant as a contract binding it even more clearly to a subordinate position under the London museum. As he reiterated in a letter to the ASB chasing up a missing section of materials from Bhutan and Nepal:[21]

> We now call your attention to several points respecting the relation in which the Asiatic Society is placed towards the Company's Museum in England in consideration of this grant .... For any naturalist or officer who may accompany any mission or deputation on behalf of Government, the most full and complete series resulting from his labors ... the most valuable and interesting results of scientific deputations and missions on behalf of Government ... are to be dispatched to England for the Company's Museum by the earliest opportunity.

But the transregional picture of Company collections is more complicated than Horsfield's demands upon the Calcutta museum might suggest. For one thing, the ASB's collections were growing overall, as were those of the branch societies in the Madras and Bombay presidencies. Madras, too, was by the 1830s receiving small but regular government subsidies, and it, too, was regularly sending materials back to London. By 1851, Madras had a Government Central Museum, whose 40,000 annual visitors dwarfed the visitor numbers of the India House museum.[22]

Within the Company's expanding territories, Calcutta, in particular, emitted its own gravitational pull, attracting many donations from across the Company's territories. This would, for example, be the case for parts of the so-called British Museum in Macao, a library and natural history museum founded by three East India Company supercargoes in 1829.[23] The collections of the library and museum grew largely from the donations of Company personnel, but private British traders and Portuguese locals also contributed. It grew steadily for several years until 1833 when the East India Company finally lost its China monopoly, when the supercargoes packed it up, along with the rest of the Company's offices, for transfer back to Calcutta. Some of the material was then donated to the ASB; for example, the trader Robert Inglis donated his collection of birds to the Society's museum. But much of what would become the

[20] Mitra. *Centenary Review of the Asiatic Society of Bengal*, p. 40.
[21] Dispatch, September 16, 1840, reprinted in Desmond. *India Museum*, p. 56.
[22] Founded by Edward Balfour around 1851 on the occasion of a large donation of geological specimens. Balfour reports 40,000 annual visitors in the 1853–1854 report.
[23] Puga. "The First Museum in China."

well-known parts of this collection – John Reeves's collection of fishes, for example – would soon make their way to India House in London. As the *Journal of the Asiatic Society of Bengal* noted sourly, "It had been proposed to transfer the whole [Macao] collection to Calcutta, and as far as concentration is beneficial, it is to be regretted that this munificent intention had been abandoned."[24] By the 1840s, the ASB collection was massive, and the Society itself a hive of activity. A typical meeting such as that of May 5, 1847 begins, as reported in the *Bombay Times*, with a list of gentlemen balloted for election from all over British India, from Pondicherry to Singapore and also as far away as the United States (F. E. Hall Esq. of Harvard University).[25] Then letters were read. The Court of Directors thanked the Society for recent "important contributions to the Museum of the India House in the Zoological Department," from the Marine Department a recent meteorological register for "Kyook Phyoo" (Kyaukpyu) in Burma; from the surveyor-general's office a recent meteorological record from Calcutta; a query from the secretary to the Military Board, requesting information about timber trees in Bengal. Then contributions for publications received through Captain Newbold were listed, including: a notice by Hekekyau Bey, late president of the Ecole Polytechnique of Cairo, on the temples and emerald mines of the eastern desert of Egypt; a notice from Mr. [Brian H.] Hodgson from Darjeeling submitting several essays on natural history; from Captain Hutton, of Mussoorie, a note on the *Ovis Ammonvides* of Hodgson; from Major Showers, of Moorshadabad, a copy and translation of the Persian inscription on a gun of Aliverdy Khan; from the Reverend Mr. Mason, a note on the Landshells of the Tenasserim Province, and a report from Dr W. B. O'Shaughnessy of the Mint, forwarding an assay of gold dust from the Boundary Commissioner of the Punjab. Then the curators and librarians submitted their monthly reports of additions to the collections and the meeting was adjourned save the final comment: "We should not omit to notice that the tables were as usual covered with numerous and beautiful specimens of objects of Natural History."

The Science of Trade and Industry

After 1833, as we have seen, some politicians, shareholders and would-be India traders accused the directors of maintaining, in effect, a natural

[24] Quoted in Puga. "The First Museum in China," p. 585.
[25] "Bengal [Report on ASB Meeting]." *The Bombay Times and Journal of Commerce (1838–1859)*, May 22, 1847.

monopoly on knowledge of Asia by way of the library-museum. However, among scientists, savants and a certain sector of the British public, access to Company science was now wider and more active than ever. On the night of March 24, 1835, for example, King's College held an outdoor lecture series in the Strand by the inventor and professor of experimental philosophy Charles Wheatstone, who used a variety of instruments to illustrate a lecture about sound and vibrations of air.[26] The centerpiece of the demonstration was one of the Javanese musical instruments brought back to India House by Stamford Raffles. Several complete Gamelan sets survived the ship fire that destroyed the rest of Raffles's exports from Java, and these had been on display at India House since the 1820s. Wheatstone had asked to borrow the instruments in order to study their acoustical qualities, particularly their "remarkable employment of reciprocal vibration."[27] The Raffles collection of Javanese instruments was something of a regular guest on the thriving popular lecture circuit – back in 1828, Michael Faraday had used a set at the Royal Institution during a Friday night lecture on "vibrations producing sound." In this instance, Faraday had wanted to use a "China Gong supposed to be in the E.I. Company's Museum." Wilkins, a member of the Royal Institution himself, had informed him that "we have never been in possession of such a specimen" but suggested Lady Raffles might have something to loan.[28] In another Royal Institution lecture in 1830, Faraday had displayed the Company's new "geodetic instrument" that was to be used in the Survey of India.[29] In 1841, glazed tiles in the Company's museum were loaned to the Society of Arts.[30] Models of Indian boats were loaned to the Society of Engineers. In 1853, a set of model ploughs were loaned to the Royal

[26] He writes about his study of this instrument in Wheatstone, Charles. "On the Resonances or Reciprocated Vibrations of Columns of Air." *Quarterly Journal of Science, Literature and Art* 3 (March 1828): 175–183.

[27] Davies, James Q. "Instruments of Empire." In *Sound Knowledge: Music and Science in London, 1789–1851*, edited by James Q. Davies and Ellen Lockhart. University of Chicago Press, 2017, pp. 145–174, p. 145.

[28] Wilkins to Faraday at the Royal Institution, 6 February 1828. "Faraday0346" in *εpsilon: The Michael Faraday Collection.* Also see *Quarterly Journal of Science* 25 (1828): 173. On Wilkins's membership, see Faraday to Earl Spencer, February 8, 1832. "Faraday0543" in *εpsilon: The Michael Faraday Collection.* Professor Faraday also worked for the East India Company on designing lightning conductors for powder magazine stores (BL IOR/L/MIL/5/413, Coll 313: 1838–41) and see IOR/E/4/797, Dispatches to India and Bengal, October to December 1847.

[29] "Royal Institution. [Faraday Inspects EIC Geodetic Instrument]." *The Athenaeum; London* 133 (May 15, 1830): 297–298.

[30] Aikin, Arthur. *Illustrations of Arts and Manufactures: Being a Selection from a Series of Papers Read before the Society for the Encouragement of Arts, Manufactures, and Commerce.* J. Van Voorst, 1841, p. 35.

Institution for a lecture on "ploughs of the world."[31] Company curators also sometimes gave lectures at the Royal Institution, such as in June 1838 when the newly hired India House curator John Forbes Royle was the speaker at Faraday's Friday evening lecture series, entertaining the audience with a discourse on "The vegetation of the Himalayan chain in connection with climate."[32] Royle also lectured on the *materia medica* of India at King's College, for which he sometimes borrowed specimens and samples from the Company's museum.

These are just a few of the contexts in which objects from the Company's collections were loaned out for public lectures or events at universities or scientific societies. From the 1820s onward, as discussed in Chapter 5, Company science was strongly represented in Britain's civic clubs, institutes and societies devoted to science and education. And, from the 1830s onwards, it is in these settings, among the literary and scientific public who filled public lectures and private clubs, that the Company's collections and their curators played an active role in shaping a more public discourse about Asia in Britain. In the 1830s, the brick-and-mortar location of Company science, within the galleries at India House, was, by all accounts, devoid of interpretation or instruction (i.e. no catalogs, useless labeling). It was instead beyond Leadenhall Street, in the thriving public and civic spaces for research and education, where Company science participated in seemingly endless exercises in comparing, contrasting and ranking the similarities, differences and historical connections between "India" and "Britain" or "Asia" and "Europe." The Javanese instruments, for example, were integrated into a comparative study of musical knowledge across the world, while at the same time Wheatstone, a prolific inventor, especially of electrical apparatus related to telegraphy, also used the instruments to demonstrate physical theories about the nature of sound.[33] Royle's lectures combined the history of medicine with the latest knowledge on the "laws which regulate the geographical distribution of plants," illustrated with his encyclopedic collection of contemporary Indian medicines.[34] His ultimate aim was to aid in the discovery of new (to Europe) medicines and treatments, but his method was deeply historical, and he also claims to have been able to "pick up one or two of the lost links in the history of the science": using new translations

[31] BL Mss Eur F303/36–51, February 15, 1853.
[32] Darwin, Charles, Frederick Burkhardt and Sydney Smith. *The Correspondence of Charles Darwin: Volume 2, 1837–1843*. Cambridge University Press, 1985, note 1, p. 89.
[33] Davies, James Q. "Instruments of Empire," p. 153.
[34] Royle, J. Forbes. *An Essay on the Antiquity of Hindoo Medicine: Including an Introductory Lecture to the Course of Materia Medica and Therapeutics, Delivered at King's College*. Allan & Co., 1837, pp. 26, 34.

of Sanskrit canons by his India House colleague Horace Hayman Wilson, Royle plotted connections between his bazaar collections from India and ancient Greco-Arabic medical treatises (and thus Britain's own scientific history).[35] In a similar vein, Henry Thomas Colebrooke's work on the history of Hindu mathematics and astronomy, which he presented at many scientific society meetings, was devoted to examining Hindu contributions to Arabic (and thus European) history of science.[36] The glazed tiles were one of numerous museum items lent to the Society for the Encouragement of Arts and Manufactures (the Society of Arts) for the purpose of illustrating the history of different trades and manufactures. Arthur Aikin, a chemist and lecturer at the Society of Arts, found the Company's museum especially rich in "models, in specimens, and in products illustrative of the arts and manufactures of India, and other oriental countries" and regularly requested Wilkins's help in locating objects from the museum to use in his presentations.[37] For example, the tiles, from the ruins of a fifteenth-century mosque in Gauda, as well as specimens of bricks, were used in a historical survey on pottery.[38] Aikin's influential lectures on the history of arts and crafts also used, for a chapter on bone-based manufactures, a Chinese lantern from the museum; in a section on the history of iron, manuscripts and specimens from the Company's library; and, in a section on the history and art of paper making, samples of bark-cloth paper from Java and wood and cotton paper from Kashmir.[39]

Other Company offices and departments, at India House and throughout the presidencies, had always been focused on gathering and analyzing information on the economy and productivity of the colonies. The Company's vast tax collection apparatus, surely the most information-intensive sector of Company administration by this time, depended upon fine-grained revenue surveys that attempted to measure village finances, landholdings, agricultural productivity, manufacturing and consumption. At the Company's library and museum, however, until the late 1830s, the projects and publications that came out of India House or

[35] Royle. *An Essay on the Antiquity of Hindoo Medicine*, p. 25.

[36] Colebrooke, Henry Thomas (trans.), Brahmagupta, and Bhaskara. *Algebra, with Arithmetic and Mensuration, from the Sanscrit of Brahmegupta and Bhascara*. John Murray, 1817. Also see Raina, Dhruv. "The European Construction of 'Hindu Astronomy' (1700–1900)." In *Handbook of Hinduism in Europe*, vol. 2., edited by Knut A. Jacobsen and Ferdinando Sardella. Brill, 2020, pp. 123–151.

[37] Aikin, Arthur. *Illustrations of Arts and Manufactures: Being a Selection from a Series of Papers Read Before the Society for the Encouragement of Arts, Manufactures, and Commerce*. J. Van Voorst, 1841, p. 3.

[38] Aikin. *Illustrations of Arts and Manufactures*, p. 35.

[39] Aikin. *Illustrations of Arts and Manufactures*, pp. 217, 289, 358.

Haileybury were no more focused on, for example, manufacturing or agricultural productivity, than those of other institutions such as the British Museum or Royal Society. The distinctly trade-related collection of Tytler's kind was still rare in the 1820s.[40] A few other donations and gifts similarly connect directly to the ongoing attempts by British merchants and manufacturers to gain ground in the Indian market. For example, in June 1813 "sixty-one specimens of sacrificial and domestic utensils [in wood] used by the Hindus; transmitted to England as patterns for the manufacturers" arrived.[41] In the 1820s, Henry Thomas Colebrooke had successfully instituted a study of weights and measures used throughout the Company's territories.[42] In general, though, Company collecting remained much more wide-ranging, opportunistic, war-driven and unstructured, and the pursuit of economically useful knowledge more diffuse, until well into the 1830s.

Several factors contributed to this change. One is, in both Britain and the colonies, the steady acceleration in the rate of accumulation and the multiplication of related institutions. By the late 1830s, disciplinarily specific collections were becoming common, and (as we saw after the charter of 1833) India House faced the question of whether it made sense to maintain a Company museum in its broad organization encompassing all things Indian. Already, by 1838, the curators tended to immediately pass on botanical donations to Kew Gardens or the Linnaean Society for storage and cataloging.[43] A second factor is the expanding influence of liberal utilitarian political views and imperial policies, in particular the discourse of a particularly *commercial* form of improvement.

[40] Models were becoming more popular as gifts or items offered for sale to the directors, although none appear to have been so systematically constructed and described as Tytler's. For example, in 1844, a large model of the "car of Jagaunath" (Jagannath, sometimes presented in English literature as "bloodthirsty Juggernaut") bought from one Mrs. Elizabeth Nash for £10. (See BL MSS EUR F303/38, F&H minutes, June 20, 1844.) In 1848, a Colonel stationed in Lahore commissioned a highly detailed scale model of the city as a present to the Court of Directors. The delicate construction was damaged en route to Bombay, rebuilt and then sent on to London, where it was also put on display in the Old Pay Office. (See BL MSS EUR F303/39, February 16, 1848, no. 16; BL MSS EUR F303/40; see bound volume 30 of 140.) In 1850, a Mrs. C. T. Sealy of Bengal sent to the Court four boxes of "clay figurines" as a present to the Museum. (See BL MSS EUR F303/40, March 21, 1850, no. 27.)

[41] BL MSS EUR F/303 2, June 26, 1813.

[42] The result was Kelly, Patrick. *The Universal Cambist, and Commercial Instructor: Being a Full and Accurate Treatise on the Exchanges, Monies, Weights and Measures of All Trading Nations and Their Colonies; with an Account of Their Banks, Public Funds, and Paper Currencies*. Author, 1821. See Rocher. *Colebrooke*, p. 156.

[43] In Hooker's massive *Flora Indica* of 1855, the preface addresses exactly the reorganization of the sciences being driven by a kind of information overload, and centralization but then separation of disciplines. Hooker also notes that the Company declined to support the project.

A third, closely related, factor is the erosion of the Company's privileges and therefore its autonomy with respect to other commercial interests. It had always been the case that British manufacturers were more interested in pushing for increasing Britain's exports to Asia. After 1833, manufacturers could and did exert a much more direct force upon trade and trading policy.[44] This was the period in which the factory-produced machine loom cotton textiles from Lancashire were just beginning to challenge the long-held dominance of Indian textiles in the global trade. Commerce with Asia was now thrown open to a raft of new players, and the meaning and the value of "commercial knowledge" was rapidly transforming. The British takeover of the Indian textile and manufactures market, and the shift in Indian exports from manufactures to raw cotton that would feed the machine looms in Britain, was still decades away. But there are signs that, as early as 1834, the directors were taking up the idea of developing cash crops for European export with a new level of commitment. Since the late 1790s, some members of the Board of Control (often with the support of Joseph Banks) had periodically tried to push the development of cash crop plantations in India, but there had so far (with the exception of opium for the China trade) been little incentive for the Company to try to make a major shift in Indian agriculture in this direction.

In short, after 1833, the interests of British industry were able to make a stronger impression upon colonial policy, and this was reflected in the new branches of specialization within Company science. Projects came and went for plantations or harvests of the anti-malarial bark of the chinchona, indigo, teak for shipbuilding and, most of all, cotton. From 1836 onward, experiments and trials aimed at expanding cotton production for export were conducted in all three presidencies. A decade on, however, little had changed: in 1847 the majority of cotton production in India was small-scale and sold domestically.[45] And finally, at the same time, the measures of "improvement" for British India were increasingly tied not only to education or even the spread of Christianity but also to economic and material change.[46]

[44] Webster, Anthony. "The Strategies and Limits of Gentlemanly Capitalism: The London East India Agency Houses, Provincial Commercial Interests, and the Evolution of British Economic Policy in South and South East Asia 1800–50." *The Economic History Review* 59, no. 4 (November 1, 2006): 743–764.

[45] "Return of Papers in Possession of East India Company, Showing Measures Taken since 1836 to Promote Cultivation of Cotton in India." *19th Century House of Commons Sessional Papers* 42, no. 439 (January 1, 1847). https://parlipapers.proquest.com/parlipapers/docview/t70.d75.1847-024275.

[46] Adas. *Machines as Measure of Man*; also see Morus, Iwan Rhys. "Manufacturing Nature: Science, Technology and Victorian Consumer Culture." *The British Journal for the History of Science* 29, no. 4 (December 1996): 403–434.

Nowhere was this new discourse more visible than in the growing "economic museum" movement. London's first self-described "economic" museum was the Museum of Economic Geology (*f.* 1835), which opened as a small gallery under the direction of the Ordnance Survey and by 1851 had moved to a purpose-built gallery in Jermyn Street. The Horticultural Society, specializing in commercially useful plants, was the driving force behind the establishment of Kew Gardens as a national botanical garden in 1840, opening up the former royal gardens and plantations to the public, and at the same time developing new varieties for export and commercial development. As we saw in Chapter 6, even the Royal Asiatic Society was, by 1835, expressing its own mission and importance in commercial terms. Some of the first economic museums were established by agricultural societies in British India, such as the Agricultural Society of Western India and the Agricultural and Horticultural Society of India, both of which announced new economic museums in 1847. One article in the *Bombay Times* captures the vision of these collections and their role in "improvement":

> Few things could be devised more likely to assist . . . in the extension of improvement. A collection could be formed of all the artificial and mineral productions of the districts – of our dyes, drugs, gums, pigments, ores, gems marbles, limestones, and building stones, of our arms, tools and implements of all sorts, such as would be fruitful alike of instruction and amusement. At home, collections such as this become a sort of show-room for the benefit of the selling and buying classes of the community. [Visitors] have placed before them things long desired had they known where or how they were to be procured. Our bazaars and godowns are full of such things, and people are every month sending home [to Britain] for articles which, had they known it, have long been quite within their reach in Bombay. As illustrative of the productions of the industry of the east, Bombay could quickly form a collection not to be surpassed in any part of the world.[47]

[47] "The Agricultural Society [Economic Museum Plan]." *The Bombay Times and Journal of Commerce (1838–1859)* (November 27, 1847). Also see, for evidence of the plan going forward: "Public Meeting in Honor of the Governor [Clerk, Patron of Agricultural Society]." *The Bombay Times and Journal of Commerce (1838–1859)* (March 22, 1848). This museum would eventually feed into the Victoria and Albert Bombay; see Bombay (Presidency), Government Central Museum and George C. M. Birdwood. *Catalogue of the Economic Products of the Presidency of Bombay: Being a Catalogue of the Government Central Museum, Division 1, Raw Produce (Vegetable)*. Printed at the Education Society's Press, 1862. On the Agricultural and Horticultural Society museum, see mention of it in "Why Does Not India Supply Us with Cotton: Singapore Cotton." *The Bombay Times and Journal of Commerce (1838–1859)* (April 14, 1847). Also see Tavolacci, Laura. "Vegetable Gardens versus Cash Crops: Science and Political Economy in the Agricultural and Horticultural Society of India, 1820–40." *History Workshop Journal* 88 (October 1, 2019): 24–46.

The key difference between these "economic" museums and other kinds of museums is that, as the *Bombay Times* article explains, the aim is specifically related to communicating *commercial* or *market* knowledge: household consumers will discover new things to buy for their home; manufacturers large and small will discover new materials or designs for their workshops or factories; traders will discover new wares to export.

Back in India, Royle (the future Reporter on the Products of India) had been director of the Company's botanical garden at the former Dutch factory town of Saharanpur (West Bengal) (see Figures 7.2 and 7.3). From that base, and while also acting as medical director of two local hospitals, Royle had amassed a large collection of information and specimens related to the relatively unexplored but politically significant Himalayan region at the northern edge of British influence, and at the geographical dividing line between China and India. Royle hired local collectors to make field trips and investigated the plants and medicines sold locally.[48] He was focused in particular on Indian *materia medica*, which he studied with an eye to correlating with ancient Greek medical traditions. The herbarium Royle had brought from Saharanpur contained about 10,000 specimens. Royle's herbarium was part of a boom in Company botanical collections in London. A few years earlier, another massive herbarium – in fact, the entire collection of the Calcutta Botanical Gardens up to about 1827 – had arrived back in London as well. Nathaniel Wallich, who had taken over as director of the Calcutta Botanical Gardens, returned to London with this vast collection while on a health leave and with the intention of publishing a catalog of the herbarium. Although Wallich had returned to London earlier than was strictly allowed, the directors supported Wallich, providing him an apartment in Soho to work in.[49] The secretary, writing to Bentinck on Wallich's return to India, noted that Wallich "has labored hard in the distribution of the collections from India which have been sent throughout Europe and which have gained for the East India Company in the scientific world, as the professors say, *immortal* fame."[50]

Royle's collections from Saharanpur were initially deposited in the Company warehouses, and were later donated to the Linnaean Society, where Royle was an active member (along with the Royal Society, the

[48] Royle, J. Forbes. *An Essay on the Antiquity of Hindoo Medicine: Including an Introductory Lecture to the Course of Materia Medica and Therapeutics, Delivered at King's College*. 1837, p. 25.

[49] When the directors offered a set of specimens to Cambridge professor J. D. Henslow, they were to be collected from Wallich at 61 Frith Street, Soho, London. Court of Directors, East India Company to J. D. Henslow. *Epsilon: HENSLOW-105*. December 31, 1829.

[50] Auber to Bentnick, October 13, 1832, letter 504, in Philips, C. H., ed. *The Correspondence of Lord William Cavendish Bentinck*, vol. 2. Oxford University Press, 1977, pp. 917–920.

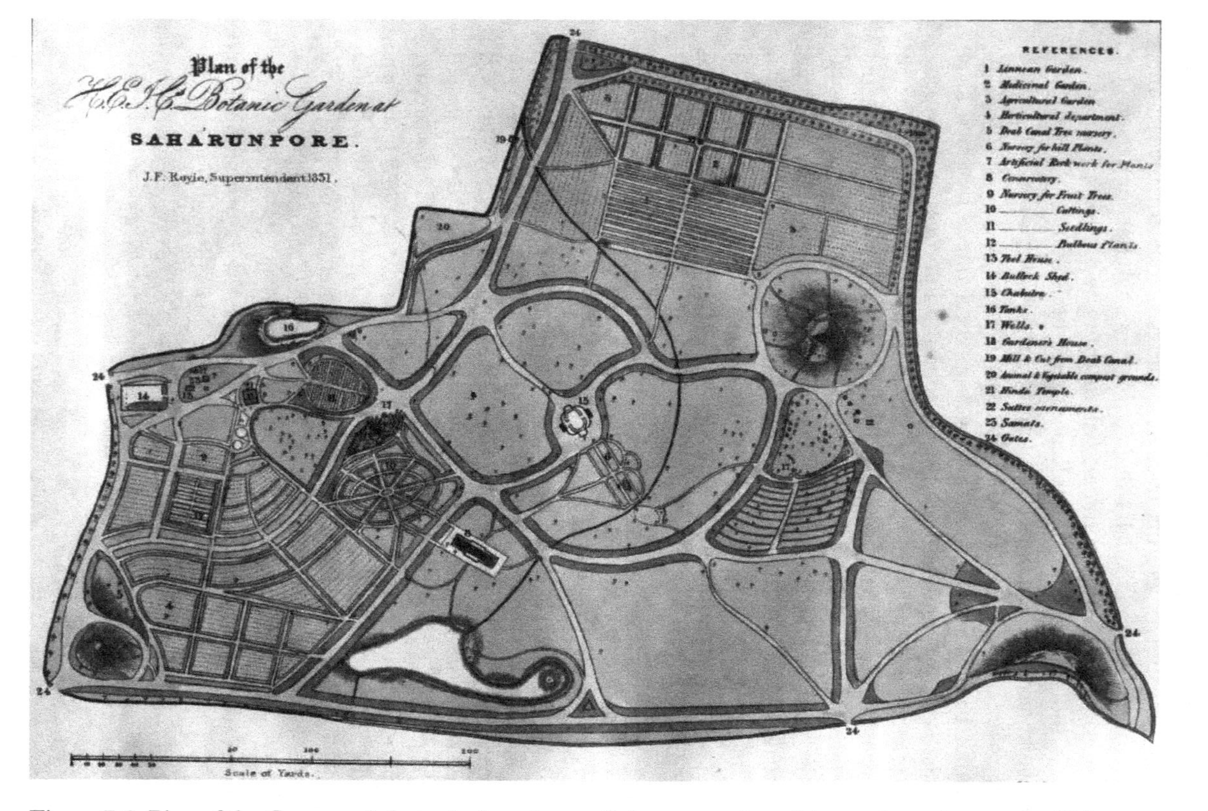

Figure 7.2 Plan of the Company's botanical gardens at Saharanpur, describing such sections as the "Linnaean garden," "Medicinal garden," "Agricultural garden" and "Doab canal trees nursery." In Royle, J. Forbes. *Illustrations of the Botany*

Figure 7.3 Illustration of the Cassia or Senna plant, from Royle, J. Forbes. *Illustrations of the Botany and Other Branches of the Natural History of the Himalayan Mountains: And of the Flora of Cashmere by J. Forbes Royle*. Vol. 1, Wm. H. Allen, 1839. https://doi.org/10.5962/bhl.title.449.

Geological Society and the Horticultural Society). Starting in 1832, working from his collection, Royle began organizing the publication of a biogeography of the Himalayas. He enlisted the help of many others, making it a collaborative project typical of these large works of natural history. Royle drew on the wide network he had cultivated while at Saharanpur, including: De La Beche (geology), Proby Cautly and Hugh Falconer (fossil drawings), Reverend Hope and Mr. Ogilby (zoology) and, for the botanical section, Robert Brown, George Bentham, Augustus de Candolle, W. J. Hooker and Don Lindley. The publication to come out of this, the *Illustrations of the Botany and Other Branches of the Natural History of the Himalayan Mountains*, was issued in eleven parts between 1833 and 1839. Many of the original illustrations were produced in India by Indian artists either before his return or while he was in

London, sent along by Wallich after his return to Calcutta.[51] As with many other works of Company science at the time, it was published by Allen & Company of Leadenhall Street.

With a focus on the geography of plants, the *Illustrations* was undoubtedly influenced by Alexander von Humboldt's work on South America. Royle's *Illustrations* was much more than a catalog of plants; it was also a treatise on the influence of climate and altitude on plant distribution, containing suggestions of the historical relationship and contemporary similarities between "European" and "Asian" species. It also incorporated Royle's interest in the history of Indian *materia medica*, giving not only a geography of plants but also a historical record of local plant knowledge. For example, it contains an index of plants mentioned in the work that are also mentioned in Persian and Arabic sources (together with the name in Latin and Arabic).

In 1835, Royle was involved in the Royal Asiatic Society's push to gain public funding (if not also the Company's museum), and he was the driver behind the Society's new "agricultural committee." In 1837, Royle took up the post of professor of *materia medica* at the new King's College in London. His first set of lectures were on the "Antiquity of Hindoo Medicine" in which he explored the history and properties of common medicines found in the bazaars of North India.[52] When, in 1838, Royle was also hired into India House as the first Reporter on the Products of India, his new office was, at first, devoted exclusively to botany (and was sometimes then referred to as the "botanical office" at India House).[53] According to Company surveyor T. B. Jervis, both Royle and Wilson were of a new generation that saw the Company's science as particularly "muddling" and "unsatisfactory" with "no practical utility."[54] From this time onward, the trajectory of Royle's work

[51] See Archer, Mildred. *Natural History Drawings in the India Office Library*. Published for the Commonwealth Relations Office by H. M. Stationery Office, 1962.

[52] See Forbes Royle, John. *Essay on the antiquity of Hindoo medicine, including an introductory lecture to the course of materia medica and therapeutics, delivered at King's College*. W. H. Allen, 1837.

[53] On Royle, also see Desmond. *The European Discovery of the Indian Flora*; Desmond. *India Museum*. Also see "Dr. John Forbes Royle." *The Athenaeum; London* 1576 (January 9, 1858): 49. Stearn, William T. "Royle's 'Illustrations of the Botany of the Himalayan Mountains'." *Journal of the Arnold Arboretum* 24, no. 4 (1943): 484–487.

[54] This emerges in a tussle between George Everest and Thomas Best Jervis over the position of surveyor-general of India. See Everest, George. *A Series of Letters Addressed to ... the Duke of Sussex, as President of the Royal Society, Remonstrating against the Conduct of That Learned Body*, 1839, p. 30. Everest is defending the utility of Company science and turns the tables on the Royal Society, accusing it of "a spirit of selfishness and monopoly ... which prompts that body perpetually to form a little clique, knot or coterie of a particular set at their head, within the compass of which all is gold, pure, refined gold, [and] without it all, dross, mere dross," p. 9.

turned sharply toward investigating economically significant natural resources and making the Company museum better serve the needs of British industry and the interests (as understood in Britain) of Indian trade and agriculture. Biogeography – understanding what kinds of plants would flourish under which climatic and soil conditions – was, for Royle, the critical starting point for any plantation or agricultural experiments. After his biogeography of the Himalayas, his publications turned sharply toward economically significant topics, including a survey of the "productive resources" of India, a manual of *materia medica*, several books on cotton cultivation in India, a catalog of Indian woods for craft and manufacturing, and a catalog of the "fibrous plants" of India, for use in making rope and cord.[55] Each addressed a pressing natural resource issue of the time. British manufacturing had become hugely dependent upon American cotton produced by slave labor, exposing the British economy to a potentially crippling weakness should the abolition movement succeed in America or should a war break out with the former colony. Wood for manufacturing, ships and furniture was in high demand, and supply concern would lead to the Company's first forest protection laws.[56] Cordage was vital to shipping, and Britain relied significantly on imports from Russia, an uneasy situation given the accelerating expansion of the Russian Empire in the 1840s leading up to the Crimean war.[57]

It is important to stress that, at the time, in practice, there were always deep institutional connections between "economic" botany, biogeography and the philosophical natural history explored in Chapter 5. For example, Royle's office was an important resource for Darwin in the years leading up to his *Origin of Species*. In the audience at Royle's Royal

[55] Royle, J. Forbes. *Essay on the Productive Resources of India*, 1840; Royle, J. Forbes. *On the Production of Isinglass along the Coasts of India with a Notice of Its Fisheries*, 1842.
Royle, J. Forbes. *A Manual of Materia Medica and Therapeutics: Including the Preparations of the Pharmacopoeias of London, Edinburgh and Dublin*. J. Churchill, 1847; Royle, J. Forbes. *On the Culture and Commerce of Cotton in India, and Elsewhere*. Smith, Elder & Co., 1851; Royle, J. Forbes. *Descriptive Catalogue of the Woods Commonly Employed in This Country for the Mechanical and Ornamental*. Holtzapffel, 1852; Royle, J. Forbes. *The Fibrous Plants of India, Fitted for Cordage, Clothing, and Paper. With an Account of the Cultivation and Preparation of Flax, Hemp, and Their Substitutes*. Smith, Elder & Co., 1855; Royle, J. Forbes. *Review of the Measures Which Have Been Adopted in India for the Improved Culture of Cotton*. Smith, Elder & Co., 1857.

[56] Rodrigues, Louiza. "Commercialisation of Forests, Timber Extraction and Deforestation of Malabar: Early Nineteenth Century." *Proceedings of the Indian History Congress* 73 (2012): 809–819.

[57] "The Culture and Commerce of Cotton in India, and Elsewhere." *The Edinburgh Review* 102, no. 207 (July 1, 1855): 40–59. The review notes that Royle's work on cotton came out at a time when anxiety over relations with the United States was worrying textile producers. This review also covers Forbes's *Fibrous Plants of India*.

Institution lecture in 1838 was the young Charles Darwin, who had just returned from his voyage on *HMS Beagle*, and who had written to Royle to introduce himself a few weeks before the lecture.[58] Some of Darwin's thoughts on the lecture are recorded in his notebooks, where he notes Royle's argument that "Botanical Provinces will turn out not nearly so confined as now thought" and takes down the evidence presented of similarities in certain genera across continents and vastly different geographies.[59] The year before, Royle, as secretary of the Geological Society, had worked with Darwin on one of Darwin's earliest publications for the Society's journal, on coral formations and points of elevation in the Pacific.[60] The two remained in correspondence through the 1840s. Darwin also corresponded with others associated with India House, including Horsfield, Edward Blyth (the former curator of the ASB's museum) and the surveyor and paleontologist Hugh Falconer.[61] But it was Royle to whom Darwin most often turned for answers to natural historical questions regarding India, evidence of how the Office of the Reporter on the Products of India sat at the intersection of economic botany and philosophical natural history.

Royle's office was an important resource for Darwin. He enquires about, among other things, a map of coastal elevations that appears in the Calcutta journal *Gleanings in Science*,[62] who he should send some unusual seeds to[63] and if he could borrow copies of a journal published in India, the *Transactions of the Agri-Horticultural Society of India*. Regarding the latter, Royle was unable to loan it at the time (he seems to have been using it during preparation of a report on cotton), so the courier Darwin

[58] Darwin writes to Royle, May 24, 1838, that Darwin would like to attend a lecture by Royle on the geography of plants of the "Himalayah." Darwin Correspondence Project, "Letter no. 415," www.darwinproject.ac.uk/letter/DCP-LETT-415.xml.

[59] See notes to "Letter no. 1109," www.darwinproject.ac.uk/letter/DCP-LETT-1109.xml.

[60] Darwin Correspondence Project, "Letter no. 383A," www.darwinproject.ac.uk/letter/DCP-LETT-383A.xml. See "'Elevation and Subsidence in the Pacific and Indian Oceans': On Certain Areas of Elevation and Subsidence in the Pacific and Indian Oceans, as Deduced from the Study of Coral Formations. By Charles Darwin." *Proceedings of the Geological Society of London* 2 (1838): 552–554.

[61] Edward Blyth offers "Miscellaneous Notes for Mr Darwin," especially on domestic animals (rabbits, pigeons, cats, etc.) of Asia, citing Royle's work. Edward Blyth to Darwin, August 22–23, 1855. Darwin Correspondence Project, "Letter no. 1983," www.darwinproject.ac.uk/letter/DCP-LETT-1983.xml. When Dawin is working on his manuscript of the *Origins*, "ask Royle/Falconer" is scribbled in some margins regarding questions about the survival of British botanical species in India. Hooker to Darwin, November 9, 1856. Darwin Correspondence Project, "Letter no. 1983," www.darwinproject.ac.uk/letter/DCP-LETT-1983.xml.

[62] John Grant Malcolmson to Charles Darwin, July 24, 1839. Darwin Correspondence Project, "Letter no. 528," www.darwinproject.ac.uk/letter/DCP-LETT-528.xml.

[63] Darwin to John Lindley, April 8, 1843. Darwin Correspondence Project, "Letter no. 668," www.darwinproject.ac.uk/letter/DCP-LETT-668.xml.

had sent returned empty handed.[64] A few months later, however, Royle comes through and offers to lend it to Darwin, to which Darwin replies and promises: "I will take the greatest care of this valuable work and will return it as soon as I can." Darwin also immediately asks whether Royle could provide him with another work, "Blacklock's treatise on sheep" (which Royle had cited in his *Essay on the Productive Resources of India* regarding sheep-breeding in India).[65] Upon returning the *Transactions*, Darwin writes to thank Royle, discussing what he learned of the varieties of domestic animals and plants common in India. He is disappointed to have learned that experiments in introducing new varieties to India are so recent that the effects of climate are unknown. But, in an indication of the moral and scientific value British naturalists placed on the work of Company science in India (as well as Darwin's desire to stay on Royle's good side), he closes by expressing his "delight and astonishment" at "the energetic attempts to do good by such numbers of people & most of them evidently not personally interested in the result. Long may our rule flourish in India. I declare all the labor shown in these Transactions is enough by itself to make one proud of one's countrymen."[66]

Club and society business also brought Darwin and Royle together. On one occasion, Darwin mentions that Royle had discouraged him from accepting the post of secretary of the Geological Society. On another, Darwin solicited Royle's support for the election of his brother Erasmus Darwin into the Athenaeum Club. At the foundation of the reformist Philosophical Club of the Royal Society, Royle invites Darwin to join (Darwin, noting with regret he "cannot often dine out," declines, though he does give his enthusiastic support for the new venture).[67] Behind Royle's back, however, Darwin displays a touch of the elitism sometimes directed at salaried medical men and overseas naturalists employed by the Company. To his friend Joseph Dalton Hooker, the botanist son of Kew director William Jackson Hooker, Darwin writes that he is reading Royle's work on the Himalayas "& I have picked out some things which have interested me, but he strikes me as rather dullish & with all his Materia Medica smells of the

[64] "I had no idea how rare a book the Agricult & Hort Journal Was .... I have long wished to look through some of the papers ... they shall be carefully returned." Darwin to Royle, April 12 and May 17, 1847. Darwin Correspondence Project, "Letter no. 1109," www.darwinproject.ac.uk/letter/DCP-LETT-1109.xml; Darwin Correspondence Project, "Letter no. 1047," www.darwinproject.ac.uk/letter/DCP-LETT-1047.xml.

[65] Darwin to Royle, August 14, 1847. Darwin Correspondence Project, "Letter no. 1108," www.darwinproject.ac.uk/letter/DCP-LETT-1108.xml.

[66] Darwin to Royle, September 1, 1847. Darwin Correspondence Project, "Letter no. 1112," www.darwinproject.ac.uk/letter/DCP-LETT-1112.xml.

[67] Darwin to Royle, c. May 28, 1847. Darwin Correspondence Project, "Letter no. 1047," www.darwinproject.ac.uk/letter/DCP-LETT-1047.xml.

Doctor's shop. I shall ever hate the name of Materia Medica, since hearing [Edinburgh professor Andrew] Duncan's lectures at 8 o'clock in a winter's morning – a whole, cold, breakfastless hour on the properties of rhubarb!"[68]

Hooker, for his part, also often references Royle (and sometimes Horsfield) as sources he might have access to in his correspondence with Darwin.[69] This was especially true, unsurprisingly, during his trip to India in the late 1840s. More surprisingly, Hooker, who, as future director of Kew, often belittled the work of naturalists working in the colonies, displayed some humility with respect to the challenges facing European naturalists in India: "I am perfectly bewildered by the number of facts hourly thrown before me whose importance I can scarce appreciate from my ignorance of Indian Nat-Hist. & all I can do now is to attempt to collect those relating to the larger or more common animals."[70]

From the Great Exhibition to the Great Rebellion

By 1850, the India House museum was displaying all kinds of new material that had arrived in the wake of the wars of the 1830s and 1840s. A London guidebook from 1851 was particularly impressed by some of the plunder from the recent Anglo-Nepalese war: "a copy of the great cyclopedic aggregate of Tibetan literature, contained in upwards of 300 large oblong volumes, printed with wooden blocks on the paper of the country. There is but one other set of this work in Europe – in the National Library of France, both having been procured by Mr. Hodgson when political resident at Nepal." Other newly acquired loot and purchases mentioned by the guide include some of Charles Masson's covert grave-robbed antiquities from Afghanistan ("reliques and curiosities found in the Topes of Afghanistan"). The guide also highlights some of the more spectacular items that had now been on display in India House for nearly fifty years: the huge silk lanterns from China; the large cuneiform tablet from near Baghdad, still undeciphered; a large collection of ancient sculpture from central India ("Hindu idols");

[68] Darwin to J. D. Hooker, April 18, 1847. Darwin Correspondence Project, "Letter no. 1082," www.darwinproject.ac.uk/letter/DCP-LETT-1082.xml.

[69] For example: "You must get Griffith's Journals as soon as you can – ask Royle about them, they are published here and are full of crude facts which I have not yet studied." This is in reference to *The Calcutta Journal of Natural History and Miscellany of the Arts and Sciences in India*, edited by John McClelland, botanical section. J. D. Hooker to Darwin, March 16, 1848. Darwin Correspondence Project, "Letter no. 1158." Regarding Horsfield, and Hooker's discussion of how evidence from Java may relate to theories of polygenism, see J. D. Hooker to Darwin, March 23, 1845. Darwin Correspondence Project, "Letter no. 844."

[70] J. D. Hooker to Darwin, March 16, 1848. Darwin Correspondence Project, "Letter no. 1158."

jade and rock scenery carvings from China ("handsome models of the Chinese *beau ideal* of country village life"); weaponry and musical instruments; and Tipu's tiger, as well as a treasury of rare and valuable Qurans, "some owned by kings, some miniature copies, very ancient ones dated to the seventh century ... many belonged to the library of Tipu Sultan, having been presented to the Company's library by the captors of Seringapatam."

The passages leading between the original library and museum were now lined with dresses, fabrics and ornamentation, "some of Indian, some of Malay or Javanese, and some of Abyssinian origin," as well as "models of boats and instruments of various kinds," not only the ones commissioned from Margaret Tytler, but now many more scale reproductions of scenes, figures and, in the case of the cosmopolitan city of Lahore, an entire city. Into the natural history display rooms, new "Abyssinian material" – collected on the recent military foray into Abyssinia – is also highlighted, as well as the "very extensive and complete" ornithological collection and the "remarkable" entomological collection, which, like the manuscript collection, is judged "unrivalled."[71]

Last but not least was the transformed Pay Office on the ground floor, into which all of the items too large or heavy to be taken upstairs were placed. Here the guide describes "more miscellaneous variety of objects":

> Hindu images and specimens of sculpture, a state palankeen [palanquin] and elephant seat and trappings captured at Bhurtpore, Chinese lanterns, a model of the car of Jagannath, and of one of the bhaulis or large wells of Hindustan, and a well-preserved series of the cases enshrining an Egyptian mummy [and] ... a collection of eastern Mammalia, and one of Indian fishes.

But the main feature of the large new gallery was the collection of fossils from the Siwalik Hills, "presented partly to the British Museum and partly to the museum of the Company ... the most striking object is a cast of the restored shell, upper and lower, of the gigantic tortoise, made up of fossil bones actually found and divided between the Company's and the British Museum."

Despite its growth and expansion, however, in 1851 the library and museum at India House was, for the first time, *not* the place to see the largest collection of materials on Asia in London. For that, the public would have to go to the India Section of the Great Exhibition of the Industry of All Nations, which ran at the purpose-built Crystal Palace from May to October. First conceived by the Society of Arts as a way to boost the quality and range of British manufactures (and to become more

[71] Weale. *London Exhibited in 1851*, pp. 598–600.

competitive with the high-quality production of France, which had a tradition of industrial exhibitions), the idea of an industrial exhibition in London was welcomed by the Board of Trade and the Company, whose chairman, Sir Archibald Galloway, served on the planning commission (as did Francis Edgerton, president of the Royal Asiatic Society).[72]

Royle was the critical point of connection between the two. Royle published prodigiously, but he had a much greater impact as a curator and promoter of industrial exhibitions. He was the head of the India section (General Commissioner and Keeper for India) and, as an authority on botany and *materia medica*, was also given the supervision of two classes of the British section: "Substances Used as Food" and "Vegetable and Animal Substances." Royle's vision for the Exhibition of 1851 was to boost Britain's manufacturing quality and therefore its export trade (see Figure 7.4). The Exhibition, it was hoped, would stimulate innovation by presenting the world of manufacturing materials, techniques and raw materials to the widest possible audience in the most efficient way. From Royle's perspective, the particular challenge of matching British factories with Asian suppliers was of massive political and economic significance, and the Great Exhibition would, he believed, be the start of a new era for the Indian supply of raw goods for British industry.

But Royle's was only one vision for the Great Exhibition, and his aims did not go unchallenged. Priti Joshi has traced the processes through which collections were formed in the colonies to be exported, and revealed yet another part of the Exhibition that took a different direction than the planners had imagined. Most importantly, Joshi has shown how dependent were Royle and the London organizers upon the decisions and designs of the local organizing committees, which in many cases were composed of both Indian and British gentlemen. One Anglo-Indian newspaper closely reported on the preparations, and the coverage reveals the diverging visions of what kinds of materials should be sent to represent a region. Royle had widely distributed what the *Friend*

[72] For a classic postcolonial studies perspective on the Great Exhibition and similar world fairs as sites through which a certain Victorian cultural hegemony was established, see Breckenridge, Carol A. "The Aesthetics and Politics of Colonial Collecting: India at World Fairs." *Comparative Studies in Society and History* 31, no. 2 (1989): 195–216. More recently, the discursive power of the Great Exhibition has been cast as less top down and more multidirectional and contingent. See Joshi, Priti. "Miles Apart: The India Display at the Great Exhibition." *Museum History Journal* 9, no. 2 (2016): 136–152. There have been many histories of the Great Exhibition. See, for example, Auerbach, Jeffrey A. *The Great Exhibition of 1851: A Nation on Display*. Yale University Press, 1999; Auerbach, Jeffrey A. *Britain, the Empire, and the World at the Great Exhibition of 1851*. Ashgate Publishing Ltd, 2008.

Figure 7.4 View of part of the India Section of the Great Exhibition in 1851, showing samples of horn, skins, furs and other materials of interest to manufacturers. Between these displays, in the background, the more famous elephant *howdah* and other spectacular works are visible. From an illustration by Joseph Nash in *Comprehensive Pictures of the Great Exhibition of 1851, Vols. I and II: From the Originals Painted for His Royal Highness Prince Albert*. Dickenson Publishers, 1854. Courtesy of the Cornell University Library.

of India (1850) called an "enormous list of articles."[73] As might be expected from Royle, the list was heavy on raw materials (plant, animal and mineral) and also asked for samples of manufactures produced with those materials. One report praises the list but worries it does not do enough to distinguish "purely Indian" productions. But just what should represent "purely Indian" productions is up for debate, as when, a few months later, one organizing committee in Bengal submitted its items and the same paper reacted with dismay. The committee had selected many luxurious items produced specifically for the exhibition, such as carved-ivory chairs, jewels, cashmere shawls and a cushion embroidered with the names of Victoria and Albert in diamonds and pearls. The *Friend* complained that "none of these articles are exactly calculated to display either the resources or peculiarities of India. Indeed we would rather that these magnificent presents had not been sent, as they will tend to revive the old assertion – not yet extinct – that India is land where gold and diamonds are the most common specimens of the mineral world."[74] The same pattern repeated itself all over British India, from Bombay to Singapore. Organizing committees, usually made up of a mix of government employees, prominent merchants, civic leaders and educators, were hugely successful in collecting materials to be forwarded on to London. The small port city of Singapore, whose organizing committee included prominent merchants Tan Kim Seng and Syed Omar, contributed 663 items, all shipped tariff-free as with all contributions to the Exhibition.[75]

The end result was a much larger and more eclectic collection than perhaps Royle had expected. Virtually all of the categories that Royal had suggested for collection were well represented in the final exhibition (although not always from the regions that Royle had suggested samples be provided from). However, these were not displayed, as had been the original plan, according to type rather than region, so that, for example, samples of raw cotton from all over the world could be compared side by side. Instead, largely due to time constraints, the exhibitions were organized into regional displays. Furthermore, artisanal luxury items such as jewels, thrones and extravagant textiles were also much more strongly represented, and it was this vision of the opulent orient that garnered most of the press and public attention. Still, for Royle, it was a great success, most importantly as the beginning of, as one commentator put it, "a great

[73] Quoted in Joshi. "Miles Apart," p. 140. The list is appendixed to Royle, J. Forbes. *On the Culture and Commerce of Cotton in India, and Elsewhere*, 1851.

[74] Quoted in Joshi. "Miles Apart," p. 140.

[75] See alphabetical entry under Indian Archipelago in *Official Catalogue of the Great Exhibition of the Works of Industry of All Nations, 1851* .... Spicer Brothers, 1851.

transformation in Education." The Exhibition had convinced many, Royle believed, of the social, economic and scientific significance of a new *biogeography*. As he put it:

> I could almost hope that the time is come, or very nearly so, in which knowledge of natural subjects should be considered a part of general education, and that what is called the study of geography be connected with a general knowledge of the soils, the climate, the plants and the animals of the different regions of the globe, and not be confined, as it often is, to boundaries, to the heights of mountains, the lengths of rivers, and to the bare enumeration of places. Some of the improved views, now entertained on such subjects, must be ascribed to the discovery that so many made of their own ignorance at the Great Exhibition of 1851, which in this, as in so many points, will continue to be, as it has already been, of immense benefit both to producers and consumers in all parts of the world.[76]

Whether or not the Great Exhibition had such a wide effect on geographical thinking, it did result in some radical changes to the Company's museum spaces in India House. One set of changes was connected to the fact that the scale of the accumulated India donations went well beyond what Royle had imagined. The Company's warehouses in New Street were modified to enable storage of the articles as they arrived for the Exhibition.[77] And a radically successful new self-funding mode of accumulation was born. After the Exhibition, the materials were to be auctioned to cover expenses and create a fund for future exhibitions. Before the India materials were auctioned, Horsfield and Royle were permitted to make selections from the exhibition material for the India House collections that remained at the New Street Warehouse:[78] "In making this selection I have confined myself . . . to such objects as are in accordance with the primary concerns of the museum, namely: native weaving & apparel; native shipping; figures in clay illustrating the native costumes and trades; native musical instruments; and miscellaneous articles illustrating Indian Ethnology."

[76] In carrying out these ideas Dr. Royle was heading an important national movement in education. From obituary in "Dr. J. Forbes Royle, F.r.s." *The Literary Gazette: A Weekly Journal of Literature, Science, and the Fine Arts* 2138 (January 9, 1858): 38.

[77] BL IOR L/SUR/1/2 ff.144, September 1851.

[78] BL IOR MSS EUR F303/39–41, November 11, 1851, ff. 146. Letter from Horsfield to the court noting "such articles as have not yet been disposed from the exhibition are now in the New Street Warehouse" – he asks to examine them and pick what he sees fit to save for the museum; BL MSS EUR F303/39–541, April 21, 1851, f. 31. Horsfield to "J D D," the assistant secretary. Royle also asks permission to select a few items, and they are equally domestic and artisanal, "a remarkable specimens of cutlery characteristic of different parts of India and the Ivory Carriage as well as the . . . agate work be retained, as also one piece of Dacca Muslin as a specimen for future examination." BL MSS EUR F/303/39–41 F&H, May 5, 1852.

Materials were also donated to a dozen other institutions, and the remainder was auctioned off in June 1852. In that year, the Company also donated over 120 items related to arms, armor and military history to the newly opened Asiatic Room in the Tower of London.[79] It is worth noting that this seems to be one of the first times Horsfield used "Ethnology" in describing the museum's collecting interests. Both textiles and musical instruments already had designated gallery spaces, and ship models had also long been collected, but other scale model collections (aside from those Margaret Tytler had commissioned) were only now starting to take up more museum space. From 1851 onward, the focus of the museum would indeed take a strongly "ethnological" turn, but in a way that was directed as much to British manufacturers, who were interested in items of dress and decoration from a commercial vantage, as to the small but growing profession of ethnological science. It is unclear how much, if any, of this material from the Great Exhibition was put on display at that time. Both storage and display space were tight. Horsfield's request for dry cellar space in the basement of India House to store museum specimens was denied.[80] At the same time, the New Street Warehouse, where many records were stored, was dealing with a serious rat infestation and new record and book storage space was being installed in the attic and the basement.[81]

It is possible that the popularity of the Great Exhibition, and the Company's very visible role in it all, helped smooth Company–state relations in the next charter renewal season of 1853. The subject of raw material supplies for British manufactures received a great deal of attention. Now Royle was called in to testify, giving precise and technical reports on topics such as the difference between long-staple American and short-staple Indian cotton, the problem of adulteration in Indian cotton exports, the state of Indigo farm financing around Bombay and ongoing experiments in growing American varieties of cotton. In addition to cotton and indigo, Royle was also asked to report on the progress of the relatively new tea plantations in the north of India, on the prospects of mass production of chinchona and on the sourcing of rope-making fibers such as hemp and flax from India. These were all areas in which British industrialists and politicians worried about global supply chains. It was often Royle's job to not only report on British India's current ability to supply raw goods need but also defend the Company's historical efforts to supply such materials.

[79] MacGregor. *Company Curiosities*, p. 201.
[80] BL IOR L/SUR/1/2, ff. 151, 153–154, December to January 1853.
[81] BL IOR L/SUR/1/2, ff. 38–44, December 1853 and ff. 20–21, ff. 123–124.

The role of museums in colonial relations was also still very much on the minds of legislators. The state of museums in British India was part of the much larger review of the state of "native education" in the wake of the new education laws passed in 1835 that codified T. B. Macaulay's Anglicist model for the colonies. The tone of the reporting on the state of Indian education is generally self-congratulatory. As a summary from the statistical office at India House put it in 1853: "In its attempts to introduce and extend the pursuit of the higher branches of sound and useful learning the Government may be regarded as completely successful. Every year will add something to the evidence of its success."[82] In the previous two decades, the Anglicist model of education – European literature, philosophy and science taught in English – had taken hold on the subcontinent: "English is now the classical language of India," noted the same report approvingly.[83] Annual expenditure on education had increased from the £10,000 required of the Company in 1813 to "between £70,000 and £80,000 per annum."[84] And a new examination system for native entry into the civil service had been established throughout the presidencies.

During the 1853 charter debates, once again, Haileybury and the directors' control over the patronage came under severe questioning. Macaulay and others, again, advocated for opening British entry to the Indian civil service to competitive examination rather than continuing to allow the directors to hand-pick the incoming classes to Haileybury. At first it seemed that the Company's control of the patronage and the Haileybury system would survive intact. The new charter of 1854 kept Haileybury open. However, instead, six months later, at the request of the Board of Control, Parliament would move, without debate, to close Haileybury for good. Two developments were behind the rapid reversal of Haileybury's fortunes. The first was the successful lobbying by established universities, particularly Oxford, for the charter of 1854 to be amended to include a clause that allowed the Board of Control to decide upon the conditions of admittance to Haileybury. Although the Board might then allow the current director-controlled system to continue, it also opened the door to the possibility of introducing a system of admittance based on competitive examination. And, depending on the nature of the examination, this could be a major boon to British universities. Although there were only about forty Indian civil service entry-level

[82] East India Company. *Statistical Papers Relating to India*. Court of Directors of the East-India Company, 1853, p. 79.
[83] East India Company. *Statistical Papers Relating to India*. Court of Directors of the East-India Company, 1853, p. 78.
[84] East India Company. *Statistical Papers Relating to India*. Court of Directors of the East-India Company, 1853, p. 77.

positions per year, it was expected that many hundreds would apply, and, if an entrance exam required the kind of knowledge obtained at university, all of those applicants would be coming from British universities, giving a significant financial and social boost to the university system overall. As Benjamin Jowett, a tutor at Balliol College Oxford, and one of the campaigners, would put it in a letter to William Gladstone, member for Oxford, in July 1854:[85]

> I cannot conceive a greater boon which could be conferred upon the University than a share in the Indian appointments. The inducements thus offered would open to us a new field of knowledge: it would give us another root striking into a new soil of society: it would provide what we have always wanted, a stimulus reaching beyond the Fellowships, for those not intending to take [religious] orders: it would give an answer to the dreary question which a College Tutor so often hears asked by a B.A. even after obtaining a first Class & a Fellowship: "What line of life shall I chose, with no calling to take orders & no taste for the Bar & no Connexions who are able to put me forward in life?"

Importantly, the *British* civil service was, just at this time, *also* undergoing a review that was considered likely to result in the institution of a competitive examination, and the old universities, in particular, were looking to civil service examination in general as another "root striking into a new soil of society," as Jowett called it.

The second development was a report on Indian civil service training and education, commissioned immediately after the completion of the new charter. T. B. Macaulay led the commission and was the primary author. Not only did Macaulay make the case for competitive examination but he also proposed a curriculum and structure for the examination process that essentially required a liberal arts B.A. as preparation. Here, Macaulay returns to his well-known arguments in favor of a classical liberal education as the best preparation for India service: mainly because of the value for "higher order thinking" of an education that can "open, invigorate, enrich the mind," but also because of what, Macaulay assumes, are the limited educational resources for British officers once in India.[86] Macaulay therefore details a preliminary Indian civil service entrance exam that is heavily devoted to "liberal arts," with only a small portion of the exam covering Sanskrit and Arabic. Jowett was also a member of the

[85] Quoted in Moore, R. J. "The Abolition of Patronage in the Indian Civil Service and the Closure of Haileybury College." *The Historical Journal* 7, no. 2 (1964): 246–257, p. 250.

[86] "The servant of the Company is often stationed during a large part of his life, at a great distance from libraries and from European society." Macaulay, T. B. "Report on the Indian Civil Service," November 1854, reprinted in Lowell, A. Lawrence and H. Morse Stephens. *Colonial Civil Service: The Selection and Training of Colonial Officials in England, Holland, and France*. Macmillan & Co., 1900, appendix A, pp. 78, 83.

commission, and the report also echoes Jowett's giddiness at the thought of how the change could energize British universities:[87]

> It is with much diffidence that we venture to predict the effect of the new system; but we think we can hardly be mistaken in believing that the introduction of that system will be an event scarcely less important to this country than to India . . . . We are inclined to think that the [new civil service exams] will produce an effect which will be felt in every seat of learning throughout the realm.

On Macaulay's plan, the select students who succeed in these exams would then be given "probationary" admittance to the Company. The "probationers" would then be required, within a maximum of two years, to take a second exam dealing explicitly with subjects particular to the post ("Indian History, the science of Jurisprudence, financial and commercial science, the oriental tongues [vernacular languages]").[88] The question then arose as to where the probationary studies would be taken: only at Haileybury, or at both Haileybury and other universities? Macaulay noted that, either way, Haileybury would have to be restructured and reorganized, given that entering students would now be mature postgraduates rather than seventeen-year-olds straight out of grammar school.

The Board of Control wholeheartedly adopted Macaulay's recommendations. Haileybury could still have survived as the main training ground for the probationers. But rather than restructure the entire College, the Board of Control argued, the more economical path would be to close it altogether, devolving the orientalist training to other universities as well.[89] Eventually other schools devoted to the specialized education for the Indian civil service would be established, such as the Indian Institute at the University of Oxford (*f.* 1883) and eventually the School of Oriental Studies (*f.* 1916). With private funding, from the Crown as well as Indian elites and "many old Haileyburians," the Oxford Indian Institute was well established with a new building on Broad Street near the Bodleian Library, complete with library and museum (now dispersed among other Oxford museums), and would be a center for Indian civil service training for many years.[90] Thus, in the face of both utilitarian meritocratic ideals and competition from the old universities, the

[87] Macaulay, T. B. "Report on the Indian Civil Service," November 1854, reprinted in Lowell, A. Lawrence and H. Morse Stephens. *Colonial Civil Service: The Selection and Training of Colonial Officials in England, Holland, and France.* Macmillan Co., 1900, appendix A, p. 80.

[88] Macaulay, T. B. "Report on the Indian Civil Service," p. 94.

[89] Moore. "The Abolition of Patronage," p. 254.

[90] Danvers, Frederick Charles, Sir Monier Monier-Williams, Sir Steuart Colvin Bayley, Percy Wigram and Brand Sapte. *Memorials of Old Haileybury College.* A. Constable, 1894, p. xxii.

Company's lucrative control over access to civil service positions was finally and decisively undone.

In the same year that the closure of Haileybury was ordered, plans for a "New Museum" inside India House were approved by the Court of Directors. The Exhibition of 1851 would greatly accelerate of the Company's move toward explicitly industrial exhibitions and a new understanding of *public* utility as fundamentally related to *commercial* utility. Soon yet another industrial exhibition would bring a new wave of donations. In 1854, the New Street Warehouse was again prepared to receive exhibition goods, this time items from Madras for the Paris Exhibition of 1855.[91] Royle was, again, the head of a major India section. And, again, the curators were able to pick and choose which donations would become India House property. This time, however, construction began on a "New Museum" within India House to house the items from the Paris Exhibition and other industrial exhibitions.[92] This would be a major expansion of the footprint of the display space at India House. With oriental architecture and a focus on models of life in India, samples of arts and crafts, and an extensive display of raw materials, a whole swath of the interior India House was remade in the image of the Great Exhibition (see Table 7.1, Figure 7.5 and Figure 7.6). The design and expense of these new spaces, which materially and symbolically overwrote old unused spaces, such as the Tea Sale Room that had formerly been at the heart of the Company's operations, expressed a confident realignment of Company science at India House with the political and public expectations of imperial exhibition and display.

The new museum opened under politically desperate circumstances. After months of flashpoints of conflict within the ranks of the Company's vast Indian army, the first large-scale troop mutiny began at Lucknow in May 1857. The rebellion roiled the northern and central regions of the subcontinent for over two years; the large battles had ended after the Battle of Gwalior in May 1858, but peace was not formally declared until July 1859. At the height of the conflict, the Company's "New Museum," as it was called, would open its doors. It had been designed and promoted by Royle. It would open sometime between November 1857 and January 1858. Royle only just got to see it finished; he died in January 1858. Horsfield took over the domain of the New Museum as well, but not for long. In August of that year, Queen Victoria signed a bill transferring the administration of British India from the Company to the Crown.

A visitor might begin a tour of the Company's New Museum at the new dedicated public entrance, passing under the portico and turning right.

[91] BL IOR L/SUR/1/2, ff. 103–104, 116. [92] BL IOR L/SUR/1/2, ff. 65–66, 78, 81.

Table 7.1 *Some of the items transferred from the Great Exhibition to India House*

48 Intaglios – Medals representing Nepalese Divinities	1 Brass Padlock. 2 Betel Knives & 4 others	1 Cotton Winder (“)
14 Models of Ships, Boats, & Barges	4 Bullett [mounts] – 2 war sticks – 1 Hookah Snake	1 Plough with Oxen (“)
2 Stone Screens	1 Khole Cap with 5 Discus & 4 Knives	1 [N]urrow with Oxen. (“)
2 Marble Screens	5 Sword Sheaths	1 Native Barber (2 figures) (“)
1 Large Brass Peacock Lamp	1 Case containing Arrows – 15 Steel Shod Arrows	1 Rice bruiser with figures (“)
19 Musical Instruments	1 Cross Bow – 1 Common Bow	1 [M]aillery – Native Court (“)
1 Printed Screen of Wood	1 Writing Case and Stand	1 [Churuck] Poojah or Native Religious Penances (Bengal) (“)
10 Models of Bridges, Temples &c	1 Dagger from Madras ([. . .] to Librarian)	2 Native Females (Age & Youth). (“)
2 ditto of Temples – Copper	1 Double Horn of a Rhinoceros	19 Pieces of Native Pottery various
1 ditto Temples – Wood	2 Fish – Stuffed	5 Models in Clay
1 Model of a bruising machine	1 Crown of the King of Oude	1 Queen Victoria modelled in Clay
1 Model of a Native Court of Justice	1 Cap of His Prime Minister	2 Ornaments in Glass Bottles
1 ditto an Oil Mill	1 Embroidered Cashmere Coat	2 Glass Cups, colored
2 Large Buffalo Skulls with Horns	1 ditto [nach] girls dress	5 Biddery – Goblet, Jug, Stand, Vase and Cover
1 Wild Cows Skull	July 1 1852	1 Hookah
14 Pairs of Deer Horns	Articles from the Exhibition continued from previous entry	2 Water Coolers
1 Ivory State Bo[us/nj]e, elaborately carved	1 Beetle Stamp (broken)	1 Idol (Brass)
1 Ivory Elephant with Howdah	4 Siri or Condiment Boxes	2 Lamps (Sacrificial) (“)
1 Ivory Camel with saddle Cloth	2 Shields (Wood)	2 Trays (“)
1 Ivory Set of Workmen	3 Tharau, or Violins	2 Vases (“)
1 Medal of a J[ous]eme Pirate Boat contributed by Captn. Hawkins R.N.	1 Tharau, metal strings	1 Goblet (“)
7 Figurines in Marble	1 Tom-Tom (Drum)	1 Cup (“)
1 Model of a Machine for pounding rice	1 ditto several small drums hung in circle	1 Jungar or Ferry-Boat from Cochin
1 ditto of Mill for pressing Sugar Cane		1 Water Bottle (Brass)

1 ditto of Hand Mill for grinding Wheat
1 ditto Sugar Came Mill broken
ditto Common Native Still
ditto of the Great Nizam Diamond
2 Mahogany Cases with plate glass (two squares broken) & Chubs Patent locks
1 Hookah snake, ornamented with beads
1 Kushkus basket,1 small box made of horn
1 Horn, carved and polished – 1 Box of sandal Wood
1 Mosaic tray from Agra – 1 Ivory Pen Cameo
1 Agate Pen, tray and Instrument & Agate Cup & Saucer
5 Slabs of Agate
1 Quiver covered with velvet & [gold] facings, containing 22 arrows, no. 10756
1 Double Sword, inlaid Gold handle
1 Sword, containing pearls in the blade no. 10735
1 Double Edged thrusting sword from Madras no. 577
1 Battle Axe, with gold & silver handle & cab[. . .]
1 Dagger, with enameled [sic] sheath & Ivory handle
1 The treble knife sheath
1 Jeweled [sic] dagger with jade handle, velvet & gold sheath, ornamented with 19 rubies & 5 diamonds from Lahore
1 ditto with crystal handle & sheath cont. 4 rubies
1 Kattar dagger, 5 spring blades, sheath, gold handle & [. . .]
1 ditto with single blade & gold handle, from Rajah of Boom[..]

1 Instrument (small gongs hung in circle)
1 Model (copper)
4 Spears various
4 Swords &c
2 Knives
1 Model of an Arab Dow ______________ (Ship)
1 [Human] – Bengal Blacksmith _________ (Wood)
1 ditto Blacksmith (several figures) _______ (Wood)
1 [Sovnan] – Native Goldsmith _________ (Wood)
[the above formatting is continued below, but I dont do it here yet]
1 [Mistree] – Carpenter (Clay)
1 ditto – ditto (Wood)
1 Sawyers (2 figures) (Clay)
1 Travelling Blacksmith (Wood)
1 Calico Printer (")
1 Handkerchief Printer (")
1 Weaver (")
1 Hand Mill (")
2 ditto with figures. (")
1 Clog Maker (")
1 Potter (handwork) (")
1 Potter (Lathe) (")
1 Travellers with Oxen (")
1 Tiger Hunter with Tiger (")
1 Palunquin [sic] & 4 Bearers (")
2 Ploughs (")

1 Sprouted Vessel (")
1 Bottle for Scented Water (Pewter)
3 Cups & Covers (")
1 Water Bottle (White Marble)
1 Vase (do.)
1 Plate (do.)
1 Basin (do.)
1 Dish or Tray (do.)
2 Vases with Covers (do.)
1 Goblet (do)
1 Rhinoceros (do)
1 Model of a Temple ([pillars] in imitation of Ivory)
July 15
Articles from the Exhibition
1 Model of an Oil Mill
3 Fine Vessels
1 Instruments from Singapore (Wood)
1 Model of Bridge over [Frisool] Gunga River
1 ditto [Brishnomuti] River
1 ditto Suspension Bridge
1 ditto ditto
1 ditto
1 ditto Ferry Boat
1 ditto Ruth (carriage) used by the [m . . . athas]
1 ditto Garee (carriage) ditto
1 ditto Bullock Carriage from Lahore
1 ditto Palunquin [sic palanquin]
1 ditto Bedstead

ditto with double blad, gold handle and gold Sheath
8 Silver and one Silver [gils] insignia of Nepal Regiments
1 Machine for sorting Coins with 10 [beams]? capable of sorting 1/4 Rupees – Madras Mint 1840
1 Model of Fishing Waters, with netts [sic], Boats &c
1 Model of Burmese Temple. Sculptured in Stone
1 Camel Saddle with Swivel Gun & Saddle Cloth
11 Skins viz. 1 Bear, 2 Tiger, 2 Leopards 1 Buffalo 2 Deer and 3 Goats
21 Coins viz. seven Gold, eight Silver and five Copper
2 Drawings framed
2 Small Marble Screens (in cases)
1 Garland – 5 matchlocks – with 1 rest
2 Clubs for Gymnastics (Mu[gda]rs)
1 Ancient Sword with Heathen mythology
1Wave, or Amazon Sword
4 Swords various – 1 Dirk [. . .]
3 Pair Singapore Shoes – 1 ditto [. . .]
1 Set of Playing Cards with Box / Burmese
1 Khuskus Basket, containing Toys

4 [H]arrows (“)
1 Female Mendicint [sic] (Clay)
1 Potter (“)
1 Cotton Cleaner (“)
1 Dhobie Washerman (“)
1 ditto (Wood)
2 Native Tailors (“)
1 Mai [keray?] Native Court (“)
1 China man (“)
1 Mill for Crushing Sugar Cane (“)
1 Oil Mill (“)
1 Krishna, 6 figures surrounding a revolving Table (Wood)
1 Group of Opium Smokers (Wood)
1 Native Workman (“)
1 Elephant, Howdah & Figures (“)
1 Warrior on Horseback (“)
1 Native Rider on Camel (“)
1 Cow (“)
1 Goblet, richly lacquered (“)
1 Goblet and [Gower] ditto (“)
1 Paper Weight & 3 Bulls ditto (“)
1 Sugar Crushing Machine, 6 figures (Clay)

1 ditto Kero – Maharatta Carriage
1 ditto Carriage from Lahore
2 ditto of Native Courts
2 ditto of Temples on [. . .]
1 ditto of Crushing Machine
1 Drawing framed, Gymnastic sports

“1852 May 25th Received from Mr Downing the under mentioned articles being portions of the Hon Company’s Collection from the Exhibition for deposit in the Museum as Order of Committee” (BL MSS EUR F/303/7).

Figure 7.5 The new gallery in the old tea sale room, transformed by W. Digby Wyatt in an orientalist style. *Illustrated London News*,

This led to a lofty horseshoe-shaped ground-floor gallery (see Figure 7.6). The ground floor was lined with towering glazed cases, reaching nearly to the double-height ceiling. The cases were "filled to overflowing with models" – most, apparently, still without labels – hundreds if not thousands of miniatures, usually made of wood or clay but sometimes ivory or semiprecious stones. The subjects ranged in every direction: famous buildings, manufacturing machines and implements, tools for agricultural production, public works, temples, towns and cities. There were models of modes of transportation – palanquins, carts, wagons and dozens of different boats. There were also models of people working at dozens of different tasks – planting, spinning, weaving, barbering, cooking, snake-charming – and of marriages, religious ceremonies, legal proceedings and military exercises. There were model households depicting the comfortable lifestyle of the wealthy classes.[93]

To a public now thoroughly familiar with live exhibitions, panoramas, dioramas and other spectacles on offer in the metropolis, this strange archive, these shelves upon shelves of miniature abstractions of Indian society, may have been less than overwhelming.[94] Despite Royle's hopes, the extensive display of raw materials and products made from them – thousands upon thousands of samples – may also not have left much of an impression.

But it wasn't only this utilitarian encyclopedia of economic opportunity that was carried over from the Great Exhibition into the New Museum. The exoticism and spectacle was just across a hallway and through another wide, large set of doors. Upon entering the former sale room (see Figure 6.2), the experience turned dramatic and sensorial. One of the largest open spaces in India House was now transformed by Matthew Digby Wyatt, secretary of the Great Exhibition and the Company's surveyor, into a mock "Indian Court" or "Mosque," the interior remade by rows of delicate columns and archways, the walls lined with innate carvings and stone screens (see Figure 7.5). This space was one of the earliest in a coming wave of "orientalist" architectural design – and Wyatt would become a leader in the genre – so much that an architectural trade journal recommended that any serious student of this style make a pilgrimage to the New Museum.[95] The curators referred to this new

[93] "The New Museum at the East India House." *Illustrated London News*, March 6, 1858, p. 230.

[94] Note that rooms crammed full of models could be seen elsewhere. The Royal United Services Institute Museum was known for its roomfuls of exquisite ship models.

[95] "Those who would study Indian Architecture must go there to do it. Of minute carving and metal work there are some beautiful specimens." "Miscellaneous." Verbatim from *The Builder*, quoted in *The Leader* 8, no. 396 (October 24, 1857): 1017.

Figure 7.6 The transformed secretary's apartment for the new museum at India House. *Illustrated London News*, March 6, 1858.

space as the "sculpture gallery," and in it were placed five large sculptures from Amaravati in southern India that had been collected by Mackenzie and sent back to Britain by Wilson in 1827 (see Figure 7.7).[96] As the gallery opened, ninety additional sculptures and fragments (the so-called Elliot marbles) from the Amaravati tope were in transport from Madras and destined for the new museum. (By the time they arrived, it was clear the Company was on track to being disbanded, so the sculptures were moved to an old shed at Fife House, where they sat unprotected for nearly a decade.)[97] But there were also paintings and furnishings such as thrones and palanquins, and soon Horsfield would begin adding other quite different kinds of material. By the summer the walls were also hung with the ghost-white masklike face casts of living subjects recently made

[96] Cohn. *Colonialism and Its Forms of Knowledge*, p. 86.
[97] Cohn. *Colonialism and Its Forms of Knowledge*, pp. 90–91.

Figure 7.7 Carving from the temple or *stupa* at Amaravati. Sculpted panel in limestone carved with the goddess Cundā. Intended for the expanded museum at India House, the Amaravati materials arrived just after the Company was dissolved, so became part of the collection at Fife House, next to the new India Office. Now at the British Museum. © Trustees of the British Museum (asset number 1880,0709.127).

by the Prussian explorer-adventurers the brothers Schlagintweit during their covert exploration of the northern territories.[98]

Despite the oriental spectacle, however, the real draw to the New Museum was the ongoing rebellion in India. Those periodicals that covered the opening of the New Museum generally didn't fail to point out that, as Christian weekly *Leisure Hour* would remark in 1858, "recent events" had brought new interest to such a visit. The India House museum gained much more attention in the context of the rebellion, and the public gaze as reproduced in the periodical press took a harsher view of both the "India" they were being presented with and the

[98] And into the library went, among other things, forty-three large volumes of manuscript records including maps and a huge magnetic survey. "Asia." *The Journal of the Royal Geographical Society* 28 (1858): clxxxi.

Company under whose watch the rebellion broke out. The *Journal of the Royal Geographical Society* was relatively sanguine when it mused that "Dreadful as the recent much-to-be-deplored events in India have been, they will probably bring great advantages to the human race: India will be more entirely ours, and the progress of Christianity and civilization more certain and rapid."[99] Harsher were, for example, commentaries in *The Builder*, imagining the colossal elephant statues of the Delhi Royal Palace standing at the entrance to a London park, and advocating for a large-scale looting of India along the lines of Napoleon in Egypt ("If ever there was a time when we might *justify* the removal . . . of works of art from India is it not now?").[100] Although those statues remained, the scale of looting and plundering in the Siege of Delhi during the rebellion was indeed unprecedented.[101]

The opening of the New Museum was but one expression of the Company's confidence in its knowledge resources and management in the period just before the rebellion. After the rebellion, administrators and observers were thrown into an information panic, as Christopher Bayly has put it, and the whole rebellion came to be seen as a great failure of political intelligence.[102] As with each of the charter renewal debates, the rebellion spurred an anxious reevaluation of the state of British knowledge of its colonial possessions. But now facing an unprecedented crisis of authority, the anxiety ran much deeper. The underlying assumption was that governance is a problem of knowledge, and the more complete or useful or accurate the knowledge, the stronger and more effective the governance. The cause of the rebellion was often diagnosed as a lack of knowledge on multiple fronts: ignorance on the part of Indians about the aims and intentions of English rulers; ignorance on the part of English people about the beliefs and experiences of Indian people. Part of the expression of crisis now took the form of a reconsideration of just what kind of information is needed by a colonial state such as the Government of India.

Another, perhaps even deeper, concern was the failed promise of the utilitarian focus on spreading "knowledge" among the colonial subjects. The quite explicit plan to generate harmony among the colonized and their foreign rulers through a program of education in modern sciences, English literature, moral philosophy and Anglicist political and economic

[99] "Asia." *The Journal of the Royal Geographical Society* 28 (1858): clxxxi.

[100] "War Trophies." *The Builder*, October 23, 1858, p. 708.

[101] And the wrangling over how to allocate the prize money led to a new round of attempted prize reforms. See Gregorian, Raffi. "Unfit for Service: British Law and Looting in India in the Mid-Nineteenth Century." *South Asia: Journal of South Asian Studies* 13, no. 1 (June 1, 1990): 63–84.

[102] Bayly. *Empire and Information.*

theory had not, it seems, had the intended effect. Sir Syed Ahmed Khan (1817–1898), a high court judge, political writer and supporter of the Empire, wrote in the *Causes of the Indian Mutiny* (1862) that, on the part of the British, "the loss of the acquaintance with the Vernacular which prevailed in the old days" was in part to blame. But Khan, a founding member of the Scientific Society of Aligarh and supporter of the Aligarh University, also argued for an expansion of education in "Western science" among Muslims as an inoculation against future clashes.[103] Critically, however, Khan stressed that if it is to play the intended role of aligning the people of India with their British government, the introduction of Western science and other education reforms cannot be an end in itself. They must be (as the utilitarian promise always held out) a step for Indians toward real, meaningful participation in government. The deepest cause of the mutiny, says Khan, was the barrier to Indian participation in the civil service, and no meaningful route to expanding self-government. In a similar vein, S. C. G. Chukerbutty (1824–1874), a prominent doctor (one of the first to be trained in the UK) and supporter of British rule, called the rebellion a war of "ignorance and fanaticism against knowledge and religious toleration, a war in which the educated native has as great a stake as any European in his country."[104] He noted with pride that very few native doctors had joined the rebellion, but he also warned of the dangers posed at the same time by existing barriers for native entry to the Indian Medical Service and the dominance of "European opinions and interests" at the Calcutta Medical College.[105] *This* – the place of colonial scientific and educational institutions within the constellation of unresolved contradictions of British liberal imperialism – is the sense in which, according to Khan and others, the mutiny revealed a crisis of knowledge. Christopher Bayly's conclusion is that – at least in terms of the broad category of political intelligence – Britain's information-gathering practices had indeed, after the 1830s, changed character, with negative consequences for British rule in India. More was not better, and rationalization did not bring new clarity.[106]

Caught by surprise by the Crown's swift decision to abolish it, the Company organized its defense and John Stuart Mill and his office issued a *Memorandum on the Improvements in the Administration of India during the Last Thirty Years*, together with the petition to Parliament to reverse the Crown's decision.[107] Here it is argued that the liberal imperial project of

[103] Khan Sir Sayyid Aḥmad. *The Causes of the Indian Revolt*. Medical Hall Press, 1873.
[104] Kumar, Deepak. "The 'Culture' of Science and Colonial Culture, India 1820–1920." *The British Journal for the History of Science* 29, no. 2 (1996): 195–209, p. 200.
[105] Kumar. "The 'Culture' of Science and Colonial Culture, India 1820–1920," p. 200.
[106] See Bayly. *Empire and Information*.
[107] Mill, John Stuart. *Memorandum of the Improvements in the Administration of India during the Last Thirty Years, and the Petition of the East-India Company to Parliament*. Printed by order of the Court of Proprietors of the East-India Company by Cox & Wyman, 1858.

the last thirty years has – bar a few bumps such as the ongoing state monopoly on salt production – been more transformative than any other period in the history of the Empire. Mill draws a picture of a rapidly improving colony, covering everything from judicial and land reform to the establishment of police and prisons; the abolition of slavery and forced labor; the protection of oppressed races; the suppression of piracy, extortion, *suttee* and witchcraft; the establishment of medical schools, hospitals and clinics that now serve over half a million patients a year; the introduction of education at all levels of society from villages to the large metropolitan universities, including initiatives to expand female education; the improvement and extension of irrigation canals, transcontinental roads and railways; and the beginnings of the electric telegraph. Mention of the New Museum come near the end under "Miscellaneous Improvements" where the work of the "unrivalled" Dr. Royle is highlighted, and where the new "Industrial Museum" at India House is presented as evidence of "the accelerating" of the improvement of the productions of India.[108] Still, after unrolling this long list of "improvements," Mill is compelled to end with a plea for leniency, given the great informational challenge inherent in being "a Government of foreigners, over a people most difficult to be understood, and still more difficult to be improved – a Government which has had all its knowledge to acquire, by a slow process of study and experience."[109]

*

> We had thought there was really nothing left of the belongings of the old East India Company to be appropriated by the Imperial Government; but we are reminded by a rumor which has recently been in circulation that there is still something left of an Indian character to be fused into the general mass of Imperialism. It is even said that the old Museum and Library of the India-house are to be made over bodily to the gigantic establishment in Great Russell-street, there to become part and parcel of the national collections known by the name of the British Museum.
>
> [Anon.], *The Times*, London, July 11, 1861

The New Museum at India House would have given space and resources for a new science of colonial trade and industry. Instead, in the wake of the rebellion, a much deeper reorganization of (what was formerly) Company science was now underway. Despite the rumors, the Company's collections were not at first transferred to the British Museum. The Government of India Act of 1858 had abolished the Company and transferred administration of British India to the new

[108] Mill. *Memorandum on the Improvements*, p. 85.
[109] Mill. *Memorandum on the Improvements*, p. 94.

India Office, located in Whitehall near the Foreign and Colonial Office. The library and museum were moved to a vacant building in Whitehall, Fife House, which was adjacent to the museum of the Royal United Services Institute, where many of the Company's war trophies had already been donated. The remaining contents of India House were auctioned off – thousands of desks and bookshelves and hundreds of carpets, down to the mantelpieces and lighting fixtures – and in 1860 India House was razed.[110]

The demolition of the Company and its headquarters, and the absorption of the Government of India into the British state, would mark the ideological, formal and material transfer of Company science to the British state. But although British India eventually became folded into Britain's new "general mass of Imperialism," most of the resources of Company science would begin to be spun off into other institutions; and although it was initially thought necessary to keep the centers of colonial science in close physical proximity to the Colonial Office, the razing of India House would also mark the beginning of an even more intense separation of certain forms of science from state administration. The abolition of Haileybury would lead to the absorption of orientalist training, civil service exam preparation and scholarship on Asia into Oxford and other universities. Likewise, the abolition of the Company's library and museum would, over the course of another half-century, in a process deserving a separate study, lead to the absorption of the old Company's knowledge resources and expertise by new, specialized museums and growing university departments. Under a new welding of public science and educational regimes to a consolidated imperial government, Britain's growing dominance of the production of science within its empire would take new forms.[111]

Initially at Whitehall, the small gallery space that Fife House provided was devoted to the material that had made up the new museum: raw materials and finished products. No longer would John Stuart Mill be passing rooms full of stuffed birds and pinned insects as he made his way to budget meetings.[112] Like his father James Mill, John Stuart strongly advocated for the value of centralization and consolidation of knowledge resources within metropolitan centers: "power may be localized," he

[110] BL IOR H/787, no. 35.

[111] The Royal Botanical Gardens at Kew, for example, as Lucile Brockway has shown, would by the 1870s have grown into the most important, and an extremely active, center for economic botany in the empire. See Brockway, Lucile H. *Science and Colonial Expansion: The Role of the British Royal Botanic Gardens*. Yale University Press, 2002.

[112] On J. S. Mill at India House, see Stack, David. "The Pleasures of Office Life: Mill at East India House." *Nineteenth-Century Prose* 47, no. 1 (March 22, 2020): 55–92.

wrote in 1861, "but knowledge, to be most useful, must be centralized."[113] Both across the Empire and within Britain, a new geography of science was emerging out of the rubble of the old India House. By the end of the century, however, the overlapping worlds of science, company and state had disaggregated. Orientalists, historians and naturalists who once worked side by side (if not always in harmony) at the Company had found new, separate homes in university departments that divided the natural and the social sciences. Only the administrator, now in pursuit of a newly redrawn domain of "political intelligence," remained part of government. Likewise, the museum and library, which once contained under one roof materials ranging from historical manuscripts to cultural artifacts to natural history specimens, were broken up across other "national" museums: the British Museum, the Natural History Museum and the Victoria and Albert Museum, as well as the new India Museum in Calcutta. The separation of Company and state, and the remaking of Company science into public science, transformed the landscape of the sciences in Britain. What remained constant, however, was the steady accumulation of Britain's global information resources, and the growing divergence between the scale of Europe's information stores and those of its colonies.

[113] Mill, John Stuart. *Considerations on Representative Government* (1861), reprinted in H. B. Acton, ed. *J.S. Mill: Utilitarianism, Liberty and Representative Government*. Dent, 1972, p. 357.

Conclusion

A Thames Mahal

A vacant plot near the new County Hall on the "Surrey bank" of the Thames suited perfectly. The building plan, commissioned by the East India Association, by the influential architect Robert S. Chisholm, former architect for the Government of Madras, called for 16,000 square feet of exhibition, education and storage space. It would rise on the bank of the Thames in an ambitious oriental style, complete with domes and turrets, announcing to London its purpose of being the center of knowledge of Asia. This was the 1910 proposal for a new India Museum, which would gather together once again all the now-dispersed East India Company collections and, more importantly, provide a dedicated space for the scientific specimens and works of art from India that continued to arrive at London's docks, since "at present India sends her geological and mineralogical products to Jermyn Street, her vegetable products to Kew, and her antiquities to the British Museum."[1]

Chisholm's design was the last, and unsuccessful, iteration of a push for a dedicated space in London to house the India Museum collections that had begun almost as soon as India House was demolished. The first proposal, for a new India Institute, was the work of Royle's successor, John Forbes Watson, in 1874.[2] Forbes Watson had joined the India Office as the Reporter on the Products of India in 1859. For fifteen years, Forbes Watson had managed the museum at Fife House, continued to organize the India sections of international exhibitions, published reports on the natural resources and industries of India, and

[1] Anon. "India in London: A New Scheme." *The Sphere: An Illustrated Newspaper for the Home* 41 (June 25, 1910): 358.

[2] The full title of Forbes Watson's circular proposing the Institute gives a sense of the scale of ambition: Watson, J. Forbes. *On the Establishment in Connection with the India Museum and Library of an Indian Institute for Lecture, Enquiry, and Teaching; Its Influence on the Promotion of Oriental Studies in England, on the Progress of Higher Education among the Natives of India and on the Training of Candidates for the Civil Service of India.* W. H. Allen & Co., 1875.

developed a wide network of correspondence and specimen exchange. Now, in 1874, he and a range of "India interests" were mounting a campaign to make the next home of the Company's collections even grander than the first one at India House had been. Construction could begin as early as 1875. The budget of £50,000 was large but, so it was argued, more than reasonable given the importance of the object, and especially considering that the new India Museum Calcutta was expected to cost twice as much. This was the proposal for a new "India Institute," a combined library, museum, research center and civil service training institute. In a letter in *Nature*, the orientalist Hyde Clarke argued that such a plan was long overdue and admonished government for neglecting its duty to care for and make use of the old Company's collections.[3] In another letter in *Nature*, the naturalist Alfred Russel Wallace avidly supported the idea and tried to increase its appeal to Parliament by suggesting ways to achieve the same scientific and educational results, but with a cheaper price tag than that of Forbes Watson's palatial proposal.[4]

Wallace's instincts were right. In Parliament, the proposal ran aground on the question of who would supply the funds. Whereas the Company had used Indian tax revenue to fund the museum and library, as part of the Home Government, by this time the use of Indian tax revenue for such a project – perhaps because it was now clearly separate from the state administration, perhaps because the national museum movement in India was now going strong or perhaps because Bengal was again in the grips of a terrible famine – was not on the table. Forbes Watson thus had to argue for the use of British public funds to support the India Institute. As he argued, it was about time:[5]

> the whole of the collections has either been purchased by Indian money, or presented by people connected with India ... the cost of maintenance of the Museum, as also that of the Department of the Reporter on the Products of India, from the action of which England derives a benefit fully equal to that of India, is entirely borne by India .... Under such circumstances, it may be held that England has sufficient interest in the undertaking to warrant her taking a share in the cost of erecting a suitable structure for the Museum and Library ... in view of recent circumstances [i.e. the famines] fresh in the memory of everybody, such a course would be only a graceful act on the part of England.

In the end the grand plan would not succeed. No palatial Mughal-style building went up along the Thames, no institution for the literary, artistic

[3] Clarke, Hyde. "East India Museum." *Nature* 8, no. 183 (1873): 5–6.

[4] "It seems to me one of the greatest popular delusions, that specimens of natural history require lofty halls and spacious galleries for their preservation and exhibition in a useful manner." Wallace, Alfred R. "East India Museum." *Nature* 8, no. 183 (1873): 5.

[5] Forbes Watson. *On the Establishment.*

and scientific study of Asia was opened to the British (and, as it was imagined, also Indian) public. This would not become a new center for commercial and cultural exchange between India and Britain to the supposed economic and social benefit of both. Instead, as had begun in the 1880s, the Company's collections would remain divided between the British Museum, Kew Gardens and the new South Kensington museums, including the Natural History Museum and the future Victoria and Albert Museum. Teaching and scholarship became the domain of new specialized university programs such as the Oxford India Institute and the School of Oriental Studies at London University. In this final step, the Company's imperial scientific and educational resources were transformed into a British public resource intended to, among other things, demonstrate the civility and liberality of the imperial project.[6]

In the later part of the Victorian era, state-funded museums and exhibitions were simultaneously places where a new ideal of inclusive public participation in politics and culture was nurtured that depended on using works of art and craft, natural history specimens and antiquities, and even exhibited peoples extracted from colonized regions. Via the institution of the public museum, British cultural and religious chauvinism was hardened into racialist and racist beliefs about who should and should not be granted political rights and economic sovereignty. Today, as the colonial foundations of Britain's public museums have become a subject of heated public discussion and critique, those institutions have become the site for a new round of debate over the tensions between liberalism and empire. The contradictions of liberal imperialism are woven deeply into the fabric of these institutions. That legacy is a huge dilemma for these institutions, which otherwise still carry an aura and a mission devoted to liberal, even progressive, cosmopolitan ideals.

The colonial origins of many of the world's most famous collections had long been largely ignored or irrelevant in the public eye. Clearly that is no longer the case, although natural history museums have so far faced less scrutiny over the provenance of their collections and fewer calls for repatriation. It may be that the only way for museums to begin to extract themselves from the hypocritical bind their colonial history places them in is to fully and openly acknowledge that history. In some areas of the professional museum world, moves are already being made in that direction: some museums have begun a new round of much more in-depth provenance research than ever. In grappling with the role of the Company and its monopoly in the history of public science, I hope to have added a useful new perspective on the debates surrounding postcolonial collections today.

[6] See, for example, Mehta. *Liberalism and Empire*.

In addition, I hope to have made clear how the dilemmas posed by the colonial history of public museums are only one part of an even wider story that includes public university systems, the structure of public–private investment in scientific research and development, and the deep and ongoing structural inequalities in global scientific practice. Scientific research and innovation today involves a sprawling, transregional set of enterprises, many of which are as historically rooted in the "great data divergence" of the imperial era as Europe's national museums.[7] Compared to the postcolonial dilemmas of museums today, the global inequalities in access to scientific and cultural resources (i.e. data, education, instrumentation, expertise) that are maintained by these other institutions are arguably much more severe. It is therefore especially important to recognize the much broader imprint that colonialism has made upon science. The emergence of European museum cultures in the colonial era was a key moment in the history of the making of the modern global political economy of science.[8] In the last fifty years, knowledge resource management in the natural sciences has radically changed, and a new era of corporate collecting has expanded alongside government-funded programs. There has also been a gradual shift from collecting whole specimens to collecting (or buying or renting) genetic data.[9] These developments have been accompanied by a set of regulations and agreements that have both accelerated and restricted accumulation by European and American corporations and states, profoundly shaping the global political economy of science today.[10]

I have tried to give a longer historical view of the making of the global political economy of science, one in which museums, libraries, colleges and other institutions for the accumulation and management of information are key. Equally important is how the history of these institutions can help to clarify the complex interplay between public and private interest and the the fuzzy boundary between "state science" and "corporate science." It is now common to worry about the threat of the "corporatization" of science in, for example, the growing influence of industry upon

[7] See, for example, Krige, John, ed. *How Knowledge Moves: Writing the Transnational History of Science and Technology*. University of Chicago Press, 2019; Tyfield, David, Rebecca Lave, Samuel Randalls and Charles Thorpe. *The Routledge Handbook of the Political Economy of Science*. Routledge, 2017.

[8] See, for example, Strasser, Bruno J. "Collecting Nature: Practices, Styles, and Narratives." *Osiris* 27, no. 1 (2012): 303–340; Strasser, Bruno J. *Collecting Experiments: Making Big Data Biology*. University of Chicago Press, 2019.

[9] Parry, Bronwyn. *Trading the Genome: Investigating the Commodification of Bio-Information*. Columbia University Press, 2004.

[10] On "informationalization," see Schiller, Dan. *How to Think about Information*. University of Illinois Press, 2007. Also see Parry, Bronwyn and Beth Greenhough. *Bioinformation*. John Wiley & Sons, 2017.

publicly funded universities or the corporate sponsorship of many of the biggest exhibitions in public museums. This book has explored both the very long history of corporate engagement with science and the very long history of our preoccupation with that connection. We have also seen the organization of science under a very different form of state, one in which the distinction between "public" and "private" became unsettled exactly when it was applied to science, revealing the historical contingency of our familiar forms of state or public science, and of their dependency on their supposed opposites, private or corporate science. However, if gaining a better understanding of the place of the Company in the making of Britain's "second scientific revolution" has in some ways naturalized ties between corporate and state interests within institutions of science and education, I hope that this case might also allow us to think more clearly about the likely future consequences of allowing those deepening connections to go unchecked.

Finally, in following the rise and decline of the Company and its museum, this book also traces the reorganization of British institutions of science across the nineteenth century through three interlinked arguments. First, it was stimulated by the expanding collections made by states and state-like bodies such as the Company; second, the acceleration of such accumulation was itself conditioned by the political economy of scientific practice under colonial capitalism; and third, it was through this reorganization that the institutional distinctions between "public" and "private" science began to crystalize into their modern forms. Taken together, this is how the sweeping changes across the sciences in nineteenth-century Britain depended upon the advance of colonial capitalism. That political debt was then partly obscured when, in the case of Britain, state science subsequently claimed a different form of monopoly on scientific knowledge.

Bibliography

Primary Sources

Archive and Manuscript Sources

British Library (BL) Add.Or.1967–2007: Drawings of Boats (Canton)
BL IOR/A/2/19 Papers Relating to the Negation with His Majesty ... 1833
BL IOR/E/4/711: Bengal Public Letters
BL IOR/E/4/787: Dispatches to India and Bengal
BL IOR/H/787: Papers Concerning the India Museum and Its Collections and the India Office Site
BL IOR/L/F/2: Finance and Home Committee Minutes
BL MSS EUR D562/16: Extracts of Canton Consultations
BL MSS EUR F303: Records of the East India Company Library

Chambers, Neil (ed.). *The Indian and Pacific Correspondence of Sir Joseph Banks, 1768–1820*. Pickering & Chatto, 2008.
Crowe, Michael J. (ed.). *A Calendar of the Correspondence of Sir John Herschel.* Cambridge University Press, 1998.
Epsilon: https://epsilon.ac.uk
Gupta, P. C. (ed.). *Fort William–India House Correspondence and Other Contemporary Papers Relating Therto* (Public Series). Vol. 13: 1796–1800. National Archives of India, 1959.
Natural History Museum London, Z MSS Horsfield
Natural History Museum London, Z MSS Ind (Documents of the India Museum)
National Museum of Scotland: Margaret Tytler collection
Philips, C. H. (ed.). *The Correspondence of Lord William Cavendish Bentinck*. Vol. 2. Oxford University Press, 1977, pp. 917–920.
Ricardo, David. *Works and Correspondence*. Edited by Piero Sraffa with the collaboration of M. H. Dobb, Cambridge University Press, 1951.
Royal Botanical Gardens Kew: Herbarium Presentations to 1900
The Darwin Correspondence Project: https://darwinproject.ac.uk
Zoological Society Library: Council Minutes

UK Government/East India Company Publications

Arberry, A. J. *The Library of the India Office: A Historical Sketch.* India Office, 1938.

Bombay (Presidency) Government Central Museum and George C. M. Birdwood. *Catalogue of the Economic Products of the Presidency of Bombay: Being a Catalogue of the Government Central Museum, Division 1., Raw Produce (Vegetable)*. Printed at the Education Society's Press, 1862.

East India Company. *Papers Respecting the Negociation with His Majesty's Ministers for a Renewal of the East-India Company's Exclusive Privileges . . . for the Use of the Court of Proprietors.* E. Cox and Son for the East India Company, 1813.

East India Company. *A Catalogue of the Library of the Hon. East-India Company.* J. & H. Cox for the East India Company, 1845.

East India Company. *Statistical Papers Relating to India.* Court of Directors of the East-India Company, 1853.

Griffith, William, and John McClelland. *Posthumous Papers Bequeathed to the Honorable the East India Company, and Printed by Order of the Government of Bengal . . .. Journals of Travels . . . William Griffith.* Bishop's College Press, 1847.

Griffith, William, and John McClelland. *Posthumous Papers Bequeathed to the Honourable the East India Company, and Printed by Order of the Government of Bengal: Palms of British East India.* Printed by C. A. Serrao, 1850.

Horsfield, Thomas. *A Catalogue of the Mammalia in the Museum of the Hon. East-India Company.* Printed for the East India Company by J. & H. Cox, 1851.

Leveson-Gower, Granville George. *Select Committee of House of Lords on Operation of Act for Better Government of H.M. Indian Territories First Report, Minutes of Evidence, Appendix; Second Report; Third Report; Index.* House of Commons Papers, Sessional Papers, March 1852.

Mill, John Stuart. Memorandum of the Improvements in the Administration of India during the Last Thirty Years, and the Petition of the East-India Company to Parliament. Printed by order of the Court of Proprietors of the East-India Company by Cox & Wyman, 1858.

Sainsbury, W. Noel (ed.). Calendar of State Papers Colonial, East Indies, China and Japan. Vol. 2, 1513–1616. Her Majesty's Stationery Office, 1864.

United Kingdom House of Commons. *The Fifth Report from the Select Committee of the House of Commons on the Affairs of the East India Company, Dated 28th July, 1812.* Vol. 3. House of Commons, 1812.

United Kingdom House of Commons. Report from the Select Committee on the Affairs of the East India Company: With Minutes of Evidence in Six Parts, and an Appendix and Index to Each. 734, 735 I–VI. Parliament. H. of C. Reports and Papers, 1832.

United Kingdom House of Commons. Return of Papers in Possession of East India Company, Showing Measures Taken since 1836 to Promote Cultivation of Cotton in India. *19th Century House of Commons Sessional Papers.* Vol. 42, Paper Number 439 (January 1, 1847).

Wilson, Horace Hayman, and Charles Masson. *Ariana Antiqua: A Descriptive Account of the Antiquities and Coins of Afghanistan.* East India Company, 1841.

Periodicals

Alexander's East India and Colonial Magazine (London)
The Asiatic Annual Register or a View of the History of Hindustan and of the Politics, Commerce and Literature of Asia (London)
The Asiatic Journal and Monthly Register for British and Foreign India, China, and Australia (London)
The Athenaeum (London)
The Bombay Times and Journal of Commerce (1838–1859) (Bombay)
The British Friend of India Magazine, and Indian Review (London)
The Builder (London)
The Edinburgh Review (Edinburgh)
The Emerald (Dublin)
The Gentleman's Magazine (London)
The Illustrated London News (London)
The Leader (London)
The Literary Gazette: A Weekly Journal of Literature, Science, and the Fine Arts (London)
London Chronicle (London)
The London General Railway, Steam-Boat, and Omnibus Guide (London)
The London Saturday Journal (London)
Morning Post (London)
Nature (London)
The Oracle (London)
Penny Magazine of the Society for the Diffusion of Useful Knowledge (London)
Singapore Chronicle and Commercial Advertiser (Singapore)
The Sphere (London)
The Times (London)
The Weekly True Sun (London)

Other Printed Primary Sources

[no author]. *The Charter, By-Laws and Regulations of the Zoological Society of London, inc. March 27, 1829*. Waterlow and Sons, 1829.

[no author]. *Official Catalogue of the Great Exhibition of the Works of Industry of All Nations, 1851 ...*. Spicer Brothers, 1851.

Aikin, Arthur. *Illustrations of Arts and Manufactures: Being a Selection from a Series of Papers Read before the Society for the Encouragement of Arts, Manufactures, and Commerce*. J. Van Voorst, 1841.

Atkin, George. *The British and Foreign Homœopathic Medical Directory and Record, 1853*. Aylott & Company, 1853.

Babbage, Charles. *Reflections on the Decline of Science in England, and on Some of Its Causes*. B. Fellowes, 1830.

Beatson, Alexander. *A View of the Origin and Conduct of the War with Tippoo Sultaun Comprising a Narrative of the Operations of the Army under the Command of Lieutenant-General Harris, and of the Siege of Seringapatam*. London, 1800.

Brayley, Edward Wedlake, and Joseph Nightingale. *A Topographical and Historical Description of London and Middlesex ....* Vol. 2. Sherwood, Neely and Jones and G. Cowie, 1814.

Brayley, Edward Westlake, and John Britton. *The Beauties of England and Wales, Or, Delineations, Topographical, Historical, and Descriptive, of Each County ....* Vol. 10. T. Maiden, 1810.

Brown, Samuel, and James Petiver. "An Account of Part of a Collection of Curious Plants and Drugs, Lately Given to the Royal Society by the East India Company." *Philosophical Transactions (1683–1775)* 22 (1700): 579–594.

Buchanan, Francis. *A Journey from Madras Through the Countries of Mysore, Canara and Malabar ....* Vol. 1. Black, Parry and Kingsbury, 1807.

Cantor, Theodore. *Catalogue of Reptiles Inhabiting the Malayan Peninsula and Islands.* Printed by J. Thomas, 1847.

Clarke, Henry Greene. *The East India Museum; a Description of the Museum and Library of the Honourable East India Company, Leadenhall Street.* H. G. Clarke & Co., 1851.

Colebrooke, Henry Thomas (ed.). *On the Valley of the Setlej River in the Himalaya Mountains: from the journal of Captain A. Gerard / with remarks by Henry Thomas Colebrooke.* Cox and Baylis, 1825.

Colebrooke, Henry Thomas. "On Dichotomous and Quinary Arrangements in Natural History." *Zoological Journal* 4 (1828): 43–46.

Colebrooke, Henry Thomas (trans.), Brahmagupta and Bhaskara. *Algebra, with Arithmetic and Mensuration, from the Sanscrit of Brahmegupta and Bhascara.* John Murray, 1817.

Dalrymple, Alexander. *Historical Fragments of the Mogul Empire. Of the Morattoes, and of the English Concerns, in Indostan.* London, 1782.

Dalrymple, Alexander. *An Historical Collection of the Several Voyages and Discoveries in the South Pacific Ocean.* 1770.

Danvers, Frederick Charles, Sir Monier Monier-Williams, Sir Steuart Colvin Bayley, Percy Wigram and Brand Sapte. *Memorials of Old Haileybury College.* A. Constable, London, 1894.

Douglas, John William. "Entomological Localities." *The Zoologist* 10 (1852).

Everest, George. *A Series of Letters Addressed to ... the Duke of Sussex, as President of the Royal Society, Remonstrating against the Conduct of That Learned Body.* London, 1839.

Falconer, Hugh. *Palæontological Memoirs and Notes of the Late Hugh Falconer: With a Biographical Sketch of the Author.* R. Hardwicke, 1868.

Finlayson, George, and Sir Thomas Stamford Raffles. *The Mission to Siam, and Hué the Capital of Cochin China, in the Years 1821–2.* John Murray, 1826.

Foote, Samuel. "The Nabob; a Comedy, in Three Acts. As It Is Performed at the Theatre-Royal in the Haymarket." London, 1778.

Forbes Watson, John. *On the Measures Required for the Efficient Working of the India Museum and Library with Suggestions for the Foundation, in Connection with Them, of an Indian Institute for Enquiry, Lecture, and Teaching.* H. M. Stationery Office, 1874.

Foster, William. *The East India House, Its History and Associations.* John Lane, 1924.

Grant, James. *The Great Metropolis*. T. Foster, 1837.

Gray, John E. "Some Remarks on Natural History Museums." *Analyst* 5 (1836): 273–280.

Hager, Joseph. *A Dissertation on the Newly Discovered Babylonian Inscriptions*. Wilks and Taylor, 1801.

Hakluyt, Richard. *The Discoveries of the World [. . .]*, 1601. Reprinted as Galvano, António. *Discoveries of the World: From Their First Original unto the Year of Our Lord 1555*. Cambridge University Press, 2010.

Halhed, Nathaniel Brassey *A Code of Gentoo Laws*. London, 1776.

Hooker, Joseph Dalton, and Thomas Thomson. *Flora Indica*. W. Pamplin, 1855.

Horsfield, Thomas. "Systematic Arrangement and Description of Birds from the Island of Java." *Transactions of the Linnean Society of London* 13 (1822): 133.

Horsfield, Thomas. *Zoological Researches in Java, and the Neighbouring Islands*. Kingsbury, Parbury, & Allen, 1824.

Horsfield, Thomas. *Descriptive Catalogue of the Lepidopterous Insects Contained in the Museum of the Honourable East-India Company: Illustrated by Coloured Figures of New Species and of the Metamorphosis of Indian Lepidoptera, with Introductory Observations on a General Arrangement of This Order of Insects*. Vol. 1, Parts I–II. Parbury, Allen, 1828.

Horsfield, Thomas. *A Catalogue of the Mammalia in the Museum of the Hon. East-India Company*. J. & H. Cox, 1851.

Horsfield, Thomas, and Frederic Moore. *A Catalogue of the Birds in the Museum of the Honorable East India Company*. Vol. 1. W. H. Allen & Co., 1854.

Horsfield, Thomas, and Frederic Moore. *A Catalogue of the Lepidopterous Insects in the Museum of the Hon. East-India Company*. Vol. 1. W. H. Allen and Co., 1857.

Horsfield, Thomas, John Joseph Bennett and Robert Brown. *Plantae Javanicæ Rariores: Descriptæe Iconibusque Illustrateæ, Quas in Insula Java, Annis 1802–1818, Legit et Investigavit*. G. H. Allen & Company, 1838.

Jervis, Thomas Best. "Address Delivered at the Geographical Section of the British Association, Friday August 26th 1839. Descriptive of the State, Progress and Prospects of the Various Surveys, and Other Scientific Inquiries, Instituted by the Honorable East India Company Throughout Asia." *The Transactions of the Bombay Geographical Society* 4 (August 1840): 157–189.

Jones, William. "The Design of a Treatise on the Plants of India." In *The Works of Sir William Jones*. J. Stockdale and J. Walker, 1807.

Kelly, Patrick. *The Universal Cambist, and Commercial Instructor: Being a Full and Accurate Treatise on the Exchanges, Monies, Weights and Measures of All Trading Nations and Their Colonies; with an Account of Their Banks, Public Funds, and Paper Currencies*. Published by the author, 1821.

Khan, Sir Sayyid Aḥmad. *The Causes of the Indian Revolt*. Medical Hall Press, 1873.

Lowell, A. Lawrence, and H. Morse Stephens. *Colonial Civil Service: The Selection and Training of Colonial Officials in England, Holland, and France*. Macmillan & Co., 1900.

Macleay, William Sharp, and Thomas Horsfield. *Annulosa Javanica Or an Attempt to Illustrate the Natural Affinities and Analogies of the Insects Coll. in Java by Thomas Horsfield*. Kingsbury, 1825.

Magalotti, Lorenzo. *Travels of Cosmo the Third, Grand Duke of Tuscany, through the England during the Reign of King Charles the Second (1669)* .... J. Mawman, 1821.

Malthus, Thomas Robert. *Statements Respecting the East-India College: With an Appeal to Facts, in Refutation of the Charges Lately Brought against It, in the Court of Proprietors*. John Murray, 1817.

Marsden, William. *The History of Sumatra: Containing an Account of the Government, Laws, Customs and Manners of the Native Inhabitants, with a Description of the Natural Productions, and a Relation of the Ancient Political State of That Island*. Printed for the author, 1784.

Marsden, William. *Bibliotheca Marsdeniana Philologica et Orientalis*. London, 1827.

Mill, James. *The History of British India*. Baldwin, Cradock, and Joy, 1817.

Mill, James, and Horace Hayman Wilson. *The History of British India*. J. Madden, 1845.

Mitra, Rajendralal. *Centenary Review of the Asiatic Society of Bengal, from 1784 to 1883. Part I. History of the* Society. Thacker, Spink and Co., 1885.

Moor, Edward. *The Hindu Pantheon*. J. Johnson, 1810.

Moore, Frederic. *A Catalogue of the Birds in the Museum of the Hon. East-India Company*. W. H. Allen and Co., 1854.

Murray, Hugh, Peter Gordon and John Crawfurd. *An Historical and Descriptive Account of China: Its Ancient and Modern History, Language, Literature, Religion, Government, Industry, Manners, and Social State* .... The Edinburgh Cabinet Library, vol. 18–20. Oliver & Boyd, 1843.

Pearce, Robert Rouiere. *Memoirs of the Most Noble Richard Marquess Wellesley*. Richard Bentley, 1847.

Roberts, William. *The Book-Hunter in London: Historical and Other Studies of Collectors and Collecting*. E. Stock, 1895.

Rochefort, Charles-César, comte de. *The History of the Caribby-Islands ... With a Caribbian Vocabulary / Rendered into English by John Davies* .... London, 1666.

Roxburgh, William, Joseph Banks, William Bulmer, D. Mackenzie and G. Nicol. *Plants of the Coast of Coromandel: Selected from Drawings and Descriptions Presented to the Hon. Court of Directors of the East India Company*. Vol. 1. Printed by W. Bulmer and Co. for G. Nicol, Bookseller, 1795.

Royle, John Forbes. *An Essay on the Antiquity of Hindoo Medicine: Including an Introductory Lecture to the Course of Materia Medica and Therapeutics, Delivered at King's College*. Allan & Co., 1837.

Royle, John Forbes. *Illustrations of the Botany and Other Branches of the Natural History of the Himalayan Mountains and of the Flora of Cashmere*. Wm. H. Allen, 1839.

Royle, John Forbes. *Essay on the Productive Resources of India*. W. H. Allen and Co., 1840.

Royle, John Forbes. *On the Production of Isinglass along the Coasts of India with a Notice of Its Fisheries*. W. H. Allen and Co., 1842.

Royle, John Forbes. *A Manual of Materia Medica and Therapeutics: Including the Preparations of the Pharmacopoeias of London, Edinburgh and Dublin*. J. Churchill, 1847.

Royle, John Forbes. *On the Culture and Commerce of Cotton in India, and Elsewhere.* Smith, Elder & Co., 1851.

Royle, John Forbes. *Descriptive Catalogue of the Woods Commonly Employed in This Country for the Mechanical and Ornamental.* Holtzapffel, 1852.

Royle, John Forbes. *The Fibrous Plants of India, Fitted for Cordage, Clothing, and Paper. With an Account of the Cultivation and Preparation of Flax, Hemp, and Their Substitutes.* Smith, Elder & Co., 1855.

Royle, John Forbes. *Review of the Measures Which Have Been Adopted in India for the Improved Culture of Cotto*n. Smith, Elder & Co., 1857.

Smith, Charles Roach. *Illustrations of Roman London.* London, 1859.

Smith, James Edward. "A Review of the Modern State of Botany, with a Particular Reference to the Natural Systems of *Linnaeus* and *Jussieu*," in James Edward Smith and Lady Smith (eds.), *Memoir and Correspondence of the Late Sir James Edward Smith.* Longman, Rees, Orme, Brown, Green, and Longman, Volume II, 1832, pp. 441–591.

Sprenger, Aloys. *A Catalogue of the Arabic, Persian and Hindu'sta'ny Manuscripts, of the Libraries of the King of Oudh.* Thomas, 1854.

Stewart, Charles. *A Descriptive Catalogue of the Oriental Library of the Late Tippoo Sultan of Mysore to which are prefixed Memoirs of Hyder Aly Khan and his son Tippoo Sultan.* Cambridge, 1809.

Taylor, William C. "On the Present State and Future Prospects of Oriental Literature, Viewed in Connection with the Royal Asiatic Society." *Journal of the Royal Asiatic Society of Great Britain and Ireland* 2 (1835).

Watson, John Forbes. *On the Establishment in Connection with the India Museum and Library of an Indian Institute for Lecture, Enquiry, and Teaching; Its Influence on the Promotion of Oriental Studies in England, on the Progress of Higher Education among the Natives of India and on the Training of Candidates for the Civil Service of India.* W. H. Allen & Co., 1875.

Weale, John (ed.). *London Exhibited in 1851: Elucidating Its Natural and Physical Characteristics, Its Antiquity and Architecture; Its Arts, Manufactures, Trade and Organization, Its Social, Literary, and Scientific Institutions and Its Numerous Galleries of Fine Art.* John Weale, 1851.

Wheatstone, Charles. "On the Resonances or Reciprocated Vibrations of Columns of Air." *Quarterly Journal of Science, Literature and Art* 3 (March 1828): 175–183.

Wilkes, John (ed.). *Encyclopaedia Londinensis*, 1815. J. Britton, *Illustrations of the Public Buildings of London*, vol. 2. London, 1828.

Wilkins, Charles, and Anthon Charles (eds.). *The Bhăgavăt-Gēētā, or, Dialogues of Krēēshnă and Ărjŏŏn : In Eighteen Lectures, with Notes.* Printed for C. Nourse, 1785.

Wilson, Horace Hayman. *Mackenzie Collection: A Descriptive Catalogue of the Oriental Manuscripts and Other Articles Illustrative of the Literature, History, Statistics and Antiquities of the South of India, Collected by the Late Lieut.-Col. Colin Mackenzie.* Asiatic Society Press, 1828.

Wilson, Horace Hayman. *Essays Analytical, Critical, and Philological on Subjects Connected with Sanskrit Literature.* London, 1864.

Secondary Sources

Dissertations

Cook, Andrew. "Alexander Dalrymple (1737–1808), Hydrographer to the East India Company and the Admiralty, as Publisher: A Catalogue of Books and Charts." St Andrews, 1993 (Vol .1) and 2012 (Vol. 2).
Thomas, Adrian P. "Calcutta Botanic Garden: Knowledge Formation and the Expectations of Colonial Botany, 1833–1914." Ph.D. dissertation King's College London, 2016.
McCartor, Robert Lynn. "The John Company's College: Haileybury and the British Government's Attempt to Control the Indian Civil Service." Ph.D. thesis submitted to Texas Tech University, 1981.

Online Secondary Resources

Ɛpsilon: https://epsilon.ac.uk.
Oxford University Press: Dictionary of National Biography: https://oxforddnb .com.
The Stanford Encyclopedia of Philosophy: https://plato.stanford.edu.
University College London: Survey of London: www.ucl.ac.uk/bartlett/architec ture/research/survey-london/survey-london-volumes.

Printed Secondary Sources

Acton, H. B. (ed.). *J.S. Mill: Utilitarianism, Liberty and Representative Government.* Dent, 1972.
Alam, Muzaffar, and Sanjay Subrahmanyam. "The Making of a Munshi." *Comparative Studies of South Asia, Africa and the Middle East* 24, no. 2 (2004): 61–72.
Alborn, Timothy L. "Boys to Men: Moral Restraint at Haileybury College." *Malthus, Medicine, & Morality* (January 1, 2000): 33–55.
Altick, Richard Daniel. *The Shows of London*. Harvard University Press, 1978.
Archer, Mildred. *Natural History Drawings in the India Office Library*. Published for the Commonwealth Relations Office by H. M. Stationery Office, 1962.
Archer, Mildred, Christopher Rowell and Robert Skelton. *Treasures from India: The Clive Collection at Powis Castle*. Meredith Press, 1987.
Armitage, David. *The Ideological Origins of the British Empire*. Cambridge University Press, 2000.
Arnold, David. *Colonizing the Body: State Medicine and Epidemic Disease in Nineteenth-Century India*. University of California Press, 1993.
Arnold, David. "Hunger in the Garden of Plenty: The Bengal Famine of 1770." In Alessa Jones (ed.), *Dreadful Visitations: Confronting Natural Catastrophe in the Age of Enlightenment*. Routledge, 1999, pp. 81–111.
Arnold, David. *New Cambridge History of India: Science Technology and Medicine in Colonial India*. Cambridge University Press, 2000.

Arnold, David. "Agriculture and 'Improvement' in Early Colonial India: A Pre-History of Development." *Journal of Agrarian Change* 5, no. 4 (2005): 505–525.

Arnold, David. "Plant Capitalism and Company Science: The Indian Career of Nathaniel Wallich." *Modern Asian Studies* 42, no. 5 (2008): 899–928.

Arnold, Ken. *Cabinets for the Curious: Looking Back at Early English Museums*. Ashgate, 2006.

Auerbach, Jeffrey A. *The Great Exhibition of 1851: A Nation on Display*. Yale University Press, 1999.

Auerbach, Jeffrey A. *Britain, the Empire, and the World at the Great Exhibition of 1851*. Ashgate Publishing Ltd, 2008.

Axelby, Richard. "Calcutta Botanic Garden and the Colonial Re-Ordering of the Indian Environment." *Archives of Natural History* 35, no. 1 (2008): 150–163.

Baber, Zaheer. "Colonizing Nature: Scientific Knowledge, Colonial Power and the Incorporation of India into the Modern World-System." *The British Journal of Sociology* 52, no. 1 (December 15, 2003): 37–58.

Baber, Zaheer. "The Plants of Empire: Botanic Gardens, Colonial Power and Botanical Knowledge." *Journal of Contemporary Asia* 46, no. 4 (October 2016): 659–679.

Baker, Alexi. "'Scientific' Instruments and Networks of Craft and Commerce in Early Modern London." In *Cities and Solidarities: Urban Communities in Pre-Modern Europe*, edited by Justin Colson and Arie van Steensel, Taylor & Francis, 2017, pp. 245–274.

Barrera, Antonio. "Local Herbs, Global Medicines: Commerce, Knowledge, and Commodities in Spanish America," in Pamela H. Smith and Paula Findlen (eds.), *Merchants and Marvels: Commerce, Science, and Art in Early Modern Europe*. Routledge, 2002, pp. 163–182.

Barrera-Osorio, Antonio. *Experiencing Nature: The Spanish American Empire and the Early Scientific Revolution*. University of Texas Press, 2006.

Bastin, John Sturgus. "The Geological Researches of Dr. Thomas Horsfield in Indonesia 1801–1819," *Bulletin of the British Museum* 10 (1982): 75–115.

Bastin, John Sturgus. *The Natural History Researches of Dr. Thomas Horsfield (1773 – 1859): First American Naturalist of Indonesia*. Oxford University Press, 1990.

Bastin, John Sturgus, and Chong Guan Kwa. *Natural History Drawings: The Complete William Farquhar Collection – Malay Peninsula, 1803–1818*. Editions Didier Millet, 2010.

Bastin, John Sturgus, and Thomas Horsfield. *Zoological Researches in Java, and the Neighbouring Islands*. Oxford University Press, 1990.

Bayly, Christopher A. *Empire and Information Intelligence Gathering and Social Communication in India, 1780–1870*. Cambridge University Press, 1996.

Bearce, George D. "Lord William Bentinck: The Application of Liberalism to India." *The Journal of Modern History* 28, no. 3 (1956): 234–246.

Beckert, Sven. *Empire of Cotton: A Global History*. Vintage Books, 2015.

Behrendt, Kurt. "Charles Masson and the Buddhist Sites of Afghanistan: Explorations, Excavations, Collections 1832–1835." *South Asian Studies* 36, no. 1 (January 2, 2020): 107–109.

Bellenoit, Hayden J. *The Formation of the Colonial State in India: Scribes, Paper and Taxes, 1760–1860*. Routledge, 2017.

Bennett, Jim, and Rebekah Higgitt. "London 1600–1800: Communities of Natural Knowledge and Artificial Practice." *The British Journal for the History of Science* 52, no. 2 (June 2019): 183–196.

Bennett, Tony. *Pasts beyond Memory: Evolution, Museums, Colonialism*. Taylor & Francis Group, 2004.

Binnema, Theodore. *Enlightened Zeal: The Hudson's Bay Company and Scientific Networks, 1670–1870*. University of Toronto Press, 2014.

Blagden, Charles Otto. *Catalogue of Manuscripts in European Languages Belonging to the Library of the India Office . . .: The Mackenzie Collections. Pt. I. The 1822 Collection & the Private Collection*. Oxford University Press, 1916.

Blake, David M. "Colin Mackenzie: Collector Extraordinary." *The British Library Journal* 17, no. 2 (1991): 128–150.

Boot, H. M. "Real Incomes of the British Middle Class, 1760–1850: The Experience of Clerks at the East India Company." *The Economic History* Review 52, no. 4 (1999): 638–68. p. 639.

Bowen, Huw V. *Revenue and Reform: The Indian Problem in British Politics 1757–1773*. Cambridge University Press, 2002.

Bowen, Huw V. *The Business of Empire: The East India Company and Imperial Britain, 1756–1833*. Cambridge University Press, 2006.

Bowen, Huw V. "Trading with the Enemy: British Private Trade and Supply of Arms to India, 1750–1850," in Richard Harding and Sergio Solbas Ferri (eds.), *The Contractor State and Its Implications, 1659–1815*. Universidad de Las Palmas de Gran Canaria, Servicio de Publicaciones, 2012, pp. 32–53.

Bowen, John. "The East India Company's Education of Its Own Servants." *Journal of the Royal Asiatic Society of Great Britain and Ireland* 3, no. 4 (1955): 105–123.

Breckenridge, Carol A. "The Aesthetics and Politics of Colonial Collecting: India at World Fairs." *Comparative Studies in Society and History* 31, no. 2 (1989): 195–216.

Brewer, John. *The Sinews of Power: War, Money, and the English State, 1688–1783*. Knopf, 1989.

Briggs, Asa. *The Age of Improvement, 1783–1867*. Longmans, 1960.

Brittlebank, Kate. "Accessing the Unseen Realm: The Historical and Textual Contexts of Tipu Sultan's Dream Register." *Journal of the Royal Asiatic Society* 21, no. 2 (2011): 159–175.

Brock, C. Helen. "The Happiness of Riches," in W. F. Bynum and Roy Porter (eds.), *William Hunter and the Eighteenth-Century Medical World*. Cambridge University Press, 1985, pp. 35–56.

Brockway, Lucile H. *Science and Colonial Expansion: The Role of the British Royal Botanic Gardens*. Yale University Press, 2002.

Browne, Janet. "Biogeography and Empire," in Nicholas Jardine, James A. Secord and E. C. Spary (eds.), *Cultures of Natural History*. Cambridge University Press, 1996, pp. 306–320.

Burke, Peter. *A Social History of Knowledge II: From the Encyclopaedia to Wikipedia*. Polity Press, 2012.

Cannon, Susan Faye. *Science in Culture: The Early Victorian Period.* Science History Publications, 1978.

Carey, Daniel, and Claire Jowitt (eds.). *Richard Hakluyt and Travel Writing in Early Modern Europe.* Routledge, 2016.

Carroll, Diana J. "William Marsden, the Scholar behind the History of Sumatra." *Indonesia and the Malay World* 47, no. 137 (2019): 66–89.

Carson, Penelope. "Grant, Charles (1746–1823), Director of the East India Company and Philanthropist." *Oxford Dictionary of National Biography.* Oxford University Press.

Chakrabarti, Pratik, and Jaydeep Sen. "'The World Rests on the Back of a Tortoise': Science and Mythology in Indian History." *Modern Asian Studies* (January 1, 2016): 808–840.

Chakrabarty, Dipesh. *Provincializing Europe: Postcolonial Thought and Historical Difference.* Princeton University Press, 2009.

Chambers, Neil. *Joseph Banks and the British Museum: The World of Collecting, 1770–1830.* Taylor & Francis Group, 2007.

Chatterjee, Kumkum. *Merchants, Politics, and Society in Early Modern India: Bihar, 1733–1820.* Brill, 1996.

Chaudhuri, K.N. "The English East India Company's Shipping (c. 1670–1760)," in Jaap Bruijn and Femme Gaastra (eds.), *Ships, Sailors and Spices: East India Companies and Their Shipping in the 16th, 17th and 18th Centuries.* NEHA, 1993, pp. 49–80.

Chen, Songchuan. "An Information War Waged by Merchants and Missionaries at Canton: The Society for the Diffusion of Useful Knowledge in China, 1834–1839." *Modern Asian Studies; Cambridge* 46, no. 6 (November 2012): 1705–1735.

Choi, Ja Yun. "A 'Most Interesting Subject for the Investigation of the Philosopher': Conjectural History in John Barrow's Travels in China." *Journal for Eighteenth-Century Studies* 42, no. 3 (2019): 303–320.

Ciepley, David. "Beyond Public and Private: Toward a Political Theory of the Corporation." *American Political Science Review* 107, no. 1 (2013): 139–158.

Cohn, Bernard S. *Colonialism and Its Forms of Knowledge: The British in India.* Princeton University Press, 1996.

Colpitts, George. "Knowing Nature in the Business Records of the Hudson's Bay Company, 1670–1840." *Business History* 59, no. 7 (October 3, 2017): 1054–1080.

Cook, Harold J. *Matters of Exchange: Commerce, Medicine, and Science in the Dutch Golden Age.* Yale University Press, 2007.

Cooper, Tompson. "Bridge, Bewick, (1767–1833), Mathematician." *Oxford Dictionary of National Biography.* Oxford University Press.

Cornish, Caroline, and Felix Driver. "'Specimens Distributed': The Circulation of Objects from Kew's Museum of Economic Botany, 1847–1914." *Journal of the History of Collections* 32, no. 2 (August 8, 2020): 327–340.

Courtright, Paul B. "Wilson, Horace Hayman (1786–1860), Sanskritist." *Oxford Dictionary of National Biography.* Oxford University Press.

Crawford, D. G. *A History of the Indian Medical Service: 1600–1913.* W. Thacker, 1914.

Curthoys, Mark. "Dealtry, William (1775–1847), Church of England Clergyman." *Oxford Dictionary of National Biography*. Oxford University Press.

Dalrymple, William. *The Anarchy: The East India Company, Corporate Violence, and the Pillage of an Empire*. Bloomsbury Publishing, 2019.

Dalrymple, William. "Opinion: The Original Evil Corporation." *The New York Times*, September 4, 2019.

Damodaran, Vinita. "Famine in Bengal: A Comparison of the 1770 Famine in Bengal and the 1897 Famine in Chotanagpur." *The Medieval History Journal* 10, no. 1–2 (October 1, 2006): 143–181.

Damodaran, Vinita. "The East India Company, Famine and Ecological Conditions in Eighteenth-Century Bengal," in Vinita Damodaran, Anna Winterbottom and Alan Lester (eds.), *The East India Company and the Natural World*. Palgrave Macmillan UK, 2015, pp. 80–101.

Damodaran, Vinita, Anna Winterbottom and Alan Lester (eds.). *The East India Company and the Natural World*. Palgrave Macmillan, 2015.

Davids, Karel. "Public Knowledge and Common Secrets. Secrecy and Its Limits in the Early-Modern Netherlands." *Early Science and Medicine* 10, no. 3 (2005): 411–427.

Davies, James Q. "Instruments of Empire," in James Q. Davies and Ellen Lockhart (eds.), *Sound Knowledge: Music and Science in London 1789–1851*. University of Chicago Press, 2017, pp. 145–174.

Davis, Richard H. "Wilkins, Kasinatha, Hastings, and the First English Bhagavad Gītā." *International Journal of Hindu Studies* 19, no. 1–2 (2015): 39–57.

Delbourgo, James. *Collecting the World: Hans Sloane and the Origins of the British Museum*. The Belknap Press of Harvard University Press, 2017.

Delgoda, Sinharaja Tammita. "'Nabob, Historian and Orientalist.' Robert Orme: The Life and Career of an East India Company Servant (1728–1801)." *Journal of the Royal Asiatic Society* 2, no. 3 (1992): 363–376.

Desmond, Adrian. "The Making of Institutional Zoology in London 1822–1836: Part I." *History of Science* 23, no. 2 (June 1, 1985): 153–185.

Desmond, Ray. *The India Museum, 1801–79*. H. M. Stationery Office, 1982.

Desmond, Ray. *The European Discovery of the Indian Flora*. Royal Botanic Gardens, 1992.

Dirks, Nicholas B. "Colonial Histories and Native Informants: Biography of an Archive," in Carol A. Breckenridge and Peter van der Veer (eds.), *Orientalism and the Postcolonial Predicament: Perspectives on South Asia*. University of Pennsylvania Press, 1993, pp. 279–313.

Driver, Felix, and Sonia Ashmore. "The Mobile Museum: Collecting and Circulating Indian Textiles in Victorian Britain." *Victorian Studies* 52, no. 3 (2010): 353–385.

Driver, Felix, and Luciana Martins. *Tropical Visions in an Age of Empire*. University of Chicago Press, 2005.

Eastwood, David. "'Amplifying the Province of the Legislature': The Flow of Information and the English State in the Early Nineteenth Century." *Historical Research* 62, no. 149 (1989): 276–294.

Edney, Matthew H. "The Atlas of India 1823–1947/The Natural History of a Topographic Map Series." *Cartographica: The International Journal for Geographic Information and Geovisualization* 28, no. 4 (December 1991): 59–91.

Edney, Matthew H. *Mapping an Empire: The Geographical Construction of British India, 1765–1843*. University of Chicago Press, 1997.

Ehrlich, Joshua. "Empire and Enlightenment in Three Letters from Sir William Jones to Governor-General John Macpherson." *The Historical Journal* 62, no. 2 (2019): 541–555.

Ehrlich, Joshua. *The East India Company and the Politics of Knowledge*. Cambridge University Press, 2023.

Endersby, Jim. "Lumpers and Splitters: Darwin, Hooker, and the Search for Order." *Science* 326, no. 5959 (December 11, 2009): 1496–1499.

Endersby, Jim. *Imperial Nature: Joseph Hooker and the Practices of Victorian Science*. University of Chicago Press, 2020.

Engberg-Pedersen, Anders. *Empire of Chance: The Napoleonic Wars and the Disorder of Things*. Harvard University Press, 2015.

Erikson, Emily. *Between Monopoly and Free Trade: The English East India Company, 1600–1757*. Princeton University Press, 2014.

Erikson, Emily, and Sampsa Samila. "Networks, Institutions, and Uncertainty: Information Exchange in Early-Modern Markets." *The Journal of Economic History*, no. 4 (2018): 1034.

Erikson, Emily, and Peter Bearman. "Malfeasance and the Foundations for Global Trade: The Structure of English Trade in the East Indies, 1601–1833." *American Journal of Sociology* 112, no. 1 (2006): 195–230.

Estève, Marie-Hélène "Introduction," in Jonville, Eudelin de. *Quelques Notions Sur l'Isle de Ceylon*. Ginkgo Editeur, 2012.

Fan, Fa-ti. *British Naturalists in Qing China: Science, Empire, and Cultural Encounter*. Harvard University Press, 2004.

Fan, Fa-ti. "Science in Cultural Borderlands: Methodological Reflections on the Study of Science, European Imperialism, and Cultural Encounter." *East Asian Science, Technology and Society* 1 (2007): 213–231.

Fennell, Shailaja. "Malthus, Statistics and the State of Indian Agriculture." *The Historical Journal* 63, no. 1 (February 2020): 159–185.

Finn, Margot C. "Material Turns in British History I: Loot." *Transactions of the Royal Historical Society* 28 (December 2018): 5–32.

Finn, Margot, and Kate Smith. *The East India Company at Home, 1757–1857*. UCL Press, 2018.

Fleetwood, Lachlan. "Science and War at the Limit of Empire: William Griffith with the Army of the Indus." *Notes and Records: The Royal Society Journal of the History of Science* 75, no. 3 (April 2020).

Freudenthal, Gideon, and Peter McLaughlin. "Classical Marxist Historiography of Science: The Hessen-Grossmann-Thesis," in Gideon Freudenthal and Peter McLaughlin (eds.), *The Social and Economic Roots of the Scientific Revolution: Texts by Boris Hessen and Henryk Grossmann*. Boston Studies in the Philosophy of Science. Springer, 2009, pp. 1–40.

Gahtan, Maia, Alessandro Nova and Eva-Maria Troelenberg (eds.). *Collecting and Empires: The Impact of Empires on Collections and Museums from Antiquity to the Present*. Harvey Miller Publishers, 2019.

Gascoigne, John. *Science in the Service of Empire: Joseph Banks, the British State and the Uses of Science in the Age of Revolution*. Cambridge University Press, 1998.

Gascoigne, John. *Science and the State: From the Scientific Revolution to World War II*. Cambridge University Press, 2019.

Gasso Miracle, M. Eulalia. "On Whose Authority? Temminck's Debates on Zoological Classification and Nomenclature: 1820–1850." *Journal of the History of Biology* 44, no. 3 (January 1, 2011): 445–481.

Goss, Andrew. *The Routledge Handbook of Science and Empire*. Routledge, Taylor & Francis Group, 2023.

Gregorian, Raffi. "Unfit for Service: British Law and Looting in India in the Mid-Nineteenth Century." *South Asia: Journal of South Asian Studies* 13, no. 1 (June 1, 1990): 63–84.

Grigson, Caroline. *Menagerie: The History of Exotic Animals in England, 1100–1837*. Oxford University Press, 2016.

Grove, A. T. "St Helena as a Microcosm of the East India Company World," in Vinita Damodaran, Anna Winterbottom and Alan Lester (eds.), *The East India Company and the Natural World*. Palgrave Macmillan, 2015, 249–269.

Grove, Richard H. *Green Imperialism: Colonial Expansion, Tropical Island Edens and the Origins of Environmentalism, 1600–1860*. Cambridge University Press, 1996.

Guha, Ranajit. *Dominance without Hegemony: History and Power in Colonial India*. Harvard University Press, 1997.

Guha, Ranajit. *A Rule of Property for Bengal: An Essay on the Idea of Permanent Settlement*. Orient BlackSwan and Permanent Black, 2016 [1963].

Hadden, Richard W. *On the Shoulders of Merchants: Exchange and the Mathematical Conception of Nature in Early Modern Europe*. SUNY Press, 1994.

Hall, Marie Boas. *All Scientists Now: The Royal Society in the Nineteenth Century*. Cambridge University Press, 1984.

Harding, Richard, and Sergio Solbes Ferri. *The Contractor State and Its Implications, 1659–1815*. Universidad de Las Palmas de Gran Canaria, 2012.

Harling, Philip, and Peter Mandler. "From 'Fiscal-Military' State to Laissez-Faire State, 1760–1850." *Journal of British Studies* 32, no. 1 (1993): 44–70.

Harris, Steven J. "Long-Distance Corporations, Big Sciences, and the Geography of Knowledge." *Configurations* 6, no. 2 (1998): 269–304.

Harrison, Mark. "Science and the British Empire." *Isis* 96, no. 1 (March 2005): 56–63.

Harrison, Mark. "Russell, Patrick (1727–1805), Physician and Naturalist." *Oxford Dictionary of National Biography*. Oxford University Press. https://doi.org/10.1093/ref:odnb/24334.

Highton, Hester. "Thomas Hood (bap. 1566, d. 1620)." *Oxford Dictionary of National Biography*. Oxford University Press.

Hill, Christopher. *Intellectual Origins of the English Revolution – Revisited*. Clarendon Press, 1997.

Hilts, Victor L. "Aliis Exterendum, or, the Origins of the Statistical Society of London." *Isis* 69, no. 1 (1978): 21–43.

Hodges, Ian. "Western Science in Siam." *Osiris* 13 (1998): 80–95.

Hoock, Holger. "The British State and the Anglo-French Wars over Antiquities, 1798–1858." *The Historical Journal* 50, no. 1 (March 2007): 49–72.

Hoquet, Thierry, "Botanical Authority: Benjamin Delessert's Collections between Travelers and Candolle's Natural Method (1803–1847)." *Isis* 105, no. 3 (2012): 508–539.

Howes, Jennifer. "Colin Mackenzie and the Stupa at Amaravati." *South Asian Studies* 18, no. 1 (2002): 53–65.

Ince, Onur Ulas. *Colonial Capitalism and the Dilemmas of Liberalism*. Oxford University Press, 2018.

Jain, Nalini. "Colonial Circuits of Power, Indian Raw Materials: An Analysis of Nathaniel Brassey Halhed's A Code of Gentoo Laws (1776)." *South Asian Review* 26, no. 2 (December 1, 2005): 3–20.

Jansari, Sushma. "Roman Coins from the Masson and Mackenzie Collections in the British Museum." *South Asian Studies* 29, no. 2 (2013): 177–193.

Jasanoff, Maya. "Collectors of Empire: Objects, Conquests and Imperial Self-Fashioning." *Past & Present* 184 (2004): 109–135.

Jasanoff, Maya. *Edge of Empire: Lives, Culture, and Conquest in the East, 1750–1850*. Knopf Doubleday Publishing Group, 2007.

Johnston, Stephen. "Mathematical Practitioners and Instruments in Elizabethan England." *Annals of Science* 48 (1991): 319–344.

Jonsson, Fredrik Albritton. "Rival Ecologies of Global Commerce: Adam Smith and the Natural Historians." *American Historical Review* 115, no. 5 (2010): 1342–1363.

Joshi, Priti. "Miles Apart: The India Display at the Great Exhibition." *Museum History Journal* 9, no. 2 (2016): 136–152.

Kathirithamby-Wells, Jeyamalar. "Peninsular Malaysia in the Context of Natural History and Colonial Science." *New Zealand Journal of Asian Studies* 11, no. 1 (2009): 337–374.

Koerner, Lisbet. *Linnaeus: Nature and Nation*. Harvard University Press, 2009.

Kriegel, Lara. *Grand Designs: Labor, Empire, and the Museum in Victorian Culture*. Duke University Press, 2008.

Krige, John (ed.). *How Knowledge Moves: Writing the Transnational History of Science and Technology*. University of Chicago Press, 2019.

Kumar, Deepak. "The 'Culture' of Science and Colonial Culture, India 1820–1920." *The British Journal for the History of Science* 29, no. 2 (1996): 195–209.

Lamond, J. M. "Afghanistan Collections of William Griffith." *Notes of the Royal Botanical Gardens Edinburgh* 30 (1970): 159–175.

Latour, Bruno. *Science in Action: How to Follow Scientists and Engineers through Society*. Harvard University Press, 1987.

Lawrence, Jonathan. "Building a Library: The Arabic and Persian Manuscript Collection of Sir William Jones." *Journal of the Royal Asiatic Society* 31, no. 1 (January 2021): 1–70.

Lidchi, Henrietta, and Stuart Allan. *Dividing the Spoils: Perspectives on Military Collections and the British Empire*. Manchester University Press, 2022.

Lipkowitz, Elise S. "Seized Natural-History Collections and the Redefinition of Scientific Cosmopolitanism in the Era of the French Revolution." *The British Journal for the History of Science* 47, no. 1 (March 2014): 15–41.

Liscombe, R. Windsor. "Wilkins, William (1778–1839), Architect and Antiquary." *Oxford Dictionary of National Biography*. Oxford University Press.

Lowther, David. "Preliminary Analysis of the Hodgson Collection at the Zoological Society of London." *Archives of Natural History* 43, no. 1 (April 1, 2016): 90–94.

Lowther, David A. "The Art of Classification: Brian Houghton Hodgson and the 'Zoology of Nipal'." *Archives of Natural History* 46, no. 1 (April 1, 2019): 1–23.

MacGregor, Arthur. *Company Curiosities: Nature, Culture and the East India Company, 1600–1874*. Reaktion Books, 2018.

Mackenzie, John M. "Empire and Metropolitan Cultures," in Andrew Porter and Wm Roger Louis (eds.), *The Oxford History of the British Empire: Volume III – The Nineteenth Century*. Oxford University Press, 1999, pp. 270–293.

MacKenzie, John M. *Museums and Empire: Natural History, Human Cultures and Colonial Identities*. Manchester University Press, 2009.

Majeed, Javed. *Ungoverned Imaginings: James Mill's The History of British India and Orientalism*. Clarendon Press, 1992.

Makepeace, Margaret. *The East India Company's London Workers: Management of the Warehouse Labourers, 1800–1858*. Boydell Press, 2010.

Mantena, Rama Sundari. *The Origins of Modern Historiography in India: Antiquarianism and Philology, 1780–1880*. Palgrave Macmillan, 2012.

Markham, Clements R., *Narratives of the Mission of George Bogle to Tibet, and of the Journey of Thomas Manning to Lhasa*. Trübner, 1876.

Marshall, P. J. "The Making of an Imperial Icon: The Case of Warren Hastings." *The Journal of Imperial and Commonwealth History* 27, no. 3 (1999): 1–16.

Marx, Karl, and Friedrich Engels. *The German Ideology* (written 1845–6, published 1935). Reprinted in Tucker, *The Marx-Engels Reader*. W. W. Norton, 1978.

Mathur, Saloni. *India by Design: Colonial History and Cultural Display*. University of California Press, 2007.

May, William Edward "The Log-Books Used by Ships of the East India Company." *The Journal of Navigation* 27, no. 1 (January 1974): 116–118.

Mazzucato, Mariana. *The Entrepreneurial State: Debunking Public vs. Private Sector Myths*. Public Affairs, 2015.

McAleer, John. "Exhibiting 'The Strangest of All Empires': The East India Company, East India House, and Britain's Asian Empire," in Stephanie Barczewski and Martin Farr (eds.), *The MacKenzie Moment and Imperial History: Essays in Honour of John M. MacKenzie*. Springer International Publishing, 2019, pp. 25–45.

McClellan, James E., and François Regourd. "The Colonial Machine: French Science and Colonization in the *Ancien Regime*." *Osiris* 15 (2000): 31–50.

McClellan, James E., and François Regourd. *The Colonial Machine: French Science and Overseas Expansion in the Old Regime*. Brepols, 2011.

McNair, James B. "Thomas Horsfield: American Naturalist and Explorer." *Torreya* 42, no. 1 (1942): 1–9.

Mcouat, Gordon. "Cataloguing Power: Delineating 'Competent Naturalists' and the Meaning of Species in the British Museum." *The British Journal for the History of Science* 34, no. 1 (2001): 1–28.

Mehta, Uday Singh. *Liberalism and Empire: A Study in Nineteenth-Century British Liberal Thought*. University of Chicago Press, 2018.

Meiring, Henry-James. "Thomas Robert Malthus, Naturalist of the Mind." *Annals of Science* 77, no. 4 (October 2020): 495–523.

Menon, Minakshi. "Medicine, Money, and the Making of the East India Company State: William Roxburgh in Madras, c. 1790," in Anna Winterbottom and Facil Tesfaye (eds.), *Histories of Medicine and Healing in the Indian Ocean World: The Medieval and Early Modern Period*. Palgrave Macmillan US, 2016, pp. 151–178.

Menon, Minakshi. "Transferrable Surveys: Natural History from the Hebrides to South India." *Journal of Scottish Historical Studies* 38, no. 1 (May 1, 2018): 143–159.

Menon, Minakshi. "Indigenous Knowledges and Colonial Sciences in South Asia." *South Asian History and Culture* 13, no. 1 (January 2, 2022): 1–18.

Menon, Minakshi. "What's in a Name? William Jones, 'Philological Empiricism' and Botanical Knowledge Making in Eighteenth-Century India." *South Asian History and Culture* 13, no. 1 (January 2, 2022): 87–111.

Mercer, Malcolm. "Collecting Oriental and Asiatic Arms and Armour: The Activities of British and East India Company Officers, c.1800–18501." *Arms & Armour* 15, no. 1 (2018): 1–21.

Milford, Lionel Sumner. *Haileybury College, Past and Present*. T. F. Unwin, 1909.

Miller, David Philip. "Longitude Networks on Land and Sea: The East India Company and Longitude Measurement 'in the Wild', 1770–1840," in Rebekah Higgitt and Richard Dunn (eds.), *Navigational Enterprises in Europe and Its Empires, 1730–1850*. Palgrave Macmillan, 2016, pp. 223–247.

Moore, R. J. "The Abolition of Patronage in the Indian Civil Service and the Closure of Hailybury College." *The Historical Journal* 7, no. 2 (1964): 246–257.

Moriarty, G. P., and John D. Haigh. "Henley, Samuel (1740–1815), Church of England Clergyman and Writer." *Oxford Dictionary of National Biography*. Oxford University Press.

Morrell, Jack, and Arnold Thackray. *Gentlemen of Science: The Early Years of the British Association for the Advancement of Science*. Oxford University Press, 1981.

Morus, Iwan Rhys. "Manufacturing Nature: Science, Technology and Victorian Consumer Culture." *The British Journal for the History of Science* 29, no. 4 (December 1996): 403–434.

Nair, Savithri Preetha. "'. . . Of Real Use to the People': The Tanjore Printing Press and the Spread of Useful Knowledge." *The Indian Economic and Social History Review* 48, no. 4 (2011): 497–529.

Nayar, Pramod K. (ed.). *Colonial Education in India, 1781–1945*. Vol. 1. Routledge, 2019.

Nechtman, Tillman W. "Nabobs Revisited: A Cultural History of British Imperialism and the Indian Question in Late-Eighteenth-Century Britain." *History Compass* 4, no. 4 (July 2006): 645–667.

Nechtman, Tillman W. "A Jewel in the Crown? Indian Wealth in Domestic Britain in the Late Eighteenth Century." *Eighteenth-Century Studies* 41, no. 1 (2007): 71–86.

Novick, Aaron. "On the Origins of the Quinarian System of Classification." *Journal of the History of Biology* 49, no. 1 (2016): 95–133.

Ogborn, Miles. *Indian Ink Script and Print in the Making of the English East India Company*. University of Chicago Press, 2007.

Ong, Seng P. "Jurisdictional Politics in Canton and the First English Translation of the Qing Penal Code (1810)." *Journal of the Royal Asiatic Society* 20, no. 2 (2010): 141–165.

Parry, Bronwyn. *Trading the Genome: Investigating the Commodification of Bio-Information*. Columbia University Press, 2004.

Payne, Anthony. "Hakluyt, Richard (1552?–1616), Geographer." *Oxford Dictionary of National Biography*. Oxford University Press. https://doi.org/10.1093/ref:odnb/11892.

Pedley, Mary Sponberg. *The Commerce of Cartography: Making and Marketing Maps in Eighteenth-Century France and England*. University of Chicago Press, 2005.

Pelling, Madeleine. "Collecting the World: Female Friendship and Domestic Craft at Bulstrode Park." *Journal for Eighteenth-Century Studies* 41, no. 1 (2018): 101–120.

Pitts, Jennifer. *A Turn to Empire: The Rise of Imperial Liberalism in Britain and France*. Princeton University Press, 2009.

Prakash, Gyan, *Another Reason: Science and the Imagination in Modern India*. Princeton University Press, 1999.

Pratt, Louise. *Imperial Eyes: Travel Writing and Transculturation*. Routledge, 2007.

Puga, Rogério Miguel. "The First Museum in China: The British Museum of Macao (1829–1834) and Its Contribution to Nineteenth-Century British Natural Science." *Journal of the Royal Asiatic Society* 22 (2012): 575–586.

Pullen, John M. "Malthus, (Thomas) Robert (1766–1834), Political Economist." *Oxford Dictionary of National Biography*. Oxford University Press.

Quilley, Geoff. *British Art and the East India Company*. Boydell Press, 2020.

Qureshi, Sadiah. *Peoples on Parade: Exhibitions, Empire, and Anthropology in Nineteenth Century Britain*. University of Chicago Press, 2011.

Raina, Dhruv. "The European Construction of 'Hindu Astronomy' (1700–1900)," in Knut A. Jacobsen and Ferdinando Sardella (eds.), *Handbook of Hinduism in Europe*, vol. 2. Brill, 2020, pp. 123–151.

Raj, Kapil. *Relocating Modern Science: Circulation and the Construction of Knowledge in South Asia and Europe, 1650–1900*. Springer, 2007.

Raman, Bhavani. *Document Raj: Writing and Scribes in Early Colonial South India*. University of Chicago Press, 2012.

Ratcliff, Jessica. "The Great (Data) Divergence: Global History of Science within Global Economic History," in Patrick Manning and Daniel Rood (eds.), *Global Scientific Practice in an Age of Revolutions, 1750–1850*. University of Pittsburgh Press, 2016, pp. 237–254.

Ratcliff, Jessica. "Travancore's Magnetic Crusade: Geomagnetism and the Geography of Scientific Production in a Princely State." *British Journal for the History of Science; Norwich* 49, no. 3 (September 2016): 325–352.

Ratcliff, Jessica. "Hand-in-Hand with the Survey: Surveying and the Accumulation of Knowledge Capital at India House during the Napoleonic Wars." *Notes and Records: The Royal Society Journal of the History of Science* 73, no. 2 (June 20, 2019): 149–166.

Rehbock, Philip F. *The Philosophical Naturalists: Themes in Early Nineteenth-Century British Biology*. University of Wisconsin Press, 1983.

Rendall, Jane. "Scottish Orientalism: From Robertson to James Mill." *The Historical Journal* 25, no. 1 (1982): 43–69.

Retford, Kate, and Susanna Avery-Quash (eds.). *The Georgian London Town House: Building, Collecting and Display*. Bloomsbury Visual Arts, 2021.

Richards, Thomas. *The Imperial Archive: Knowledge and the Fantasy of Empire*. Verso, 1993.

Robb, Peter. "Completing 'Our Stock of Geography', or an Object 'Still More Sublime': Colin Mackenzie's Survey of Mysore, 1799–1810." *Journal of the Royal Asiatic Society* 8, no. 2 (1998): 181–206.

Roberts, Lissa. "Accumulation and Management in Global Historical Perspective: An Introduction." *History of Science* 52, no. 3 (September 1, 2014): 227–246.

Robins, Nick. *The Corporation That Changed the World: How the East India Company Shaped the Modern Multinational*. Pluto Press, 2012.

Rocher, Rosane. "Hamilton, Alexander (1762–1824), Orientalist." *Oxford Dictionary of National Biography*. Oxford University Press.

Rocher, Rosane. *Orientalism, Poetry, and the Millennium: The Checkered Life of Nathaniel Brassey Halhed, 1751–1830*. Motilal Banarsidass, 1983.

Rocher, Rosane, and Ludo Rocher. *The Making of Western Indology: Henry Thomas Colebrooke and the East India Company*. Routledge, 2012.

Rodrigues, Louiza. "Commercialisation of Forests, Timber Extraction and Deforestation of Malabar: Early Nineteenth Century." *Proceedings of the Indian History Congress* 73 (2012): 809–819.

Routledge, David. "The History of the Philippine Islands in the Late Eighteenth Century: Problems and Prospects." *Philippine Studies* 23, no. 1/2 (1975): 36–52.

Sanchez, Rafael Torres. *Military Entrepreneurs and the Spanish Contractor State in the Eighteenth Century*. Oxford University Press, 2016.

Sandholtz, Wayne. *Prohibiting Plunder: How Norms Change*. Oxford University Press, 2007.

Sangwan, Satpal. "Natural History in Colonial Context: Profit or Pursuit? British Botanical Enterprise in India 1778–1820," in Patrick Petitjean, Catherine Jami and Anne Marie Moulin (eds.), *Science and Empires*. Boston Studies in the Philosophy of Science. Springer, 1992, pp. 281–298.

Saville-Smith, Kay J. *Provincial Society and Empire: The Cumbrian Counties and the East Indies, 1680–1829*. Boydell Press, 2018.

Schabas, Margaret. "John Stuart Mill: Evolutionary Economics and Liberalism." *Journal of Bioeconomics; New York* 17, no. 1 (2015): 97–111.

Schaffer, Simon, Lissa L. Roberts, Kapil Raj and James Delbourgo. *The Brokered World: Go-Betweens and Global Intelligence, 1770–1820*. Science History Publications, 2009.

Schiller, Dan. *How to Think about Information*. University of Illinois Press, 2007.
Schroeder, Ralph, and Richard Swedberg. "Weberian Perspectives on Science, Technology and the Economy." *The British Journal of Sociology* 53, no. 3 (2002): 383–401.
Schultz, Bart, and Georgios Varouxakis. *Utilitarianism and Empire*. Lexington Books, 2005.
Secord, Anne. "Corresponding Interests: Artisans and Gentlemen in Nineteenth-Century Natural History." *The British Journal for the History of Science* 27, no. 4 (December 1994): 383–408.
Seth, Sanjay. *Subject Lessons: The Western Education of Colonial India*. Duke University Press, 2007.
Sheets-Pyenson, Susan. *Cathedrals of Science: The Development of Colonial Natural History Museums during the Late Nineteenth Century*. McGill-Queen's University Press, 1988.
Siegel, Jonah. *The Emergence of the Modern Museum: An Anthology of Nineteenth-Century Sources*. Oxford University Press, 2008.
Sisir, Kumar Das. *Sahibs and Munshis: An Account of the College of Fort William*. Orion Publications, 1960.
Sivalingam, S. "Bibliography on Pearl Oysters." *Bulletin of the Fisheries Research Station* 13 (1962): 1–21.
Sivasundaram, Sujit. "Trading Knowledge: The East India Company's Elephants in India and Britain." *The Historical Journal* 48, no. 1 (March 2005): 27–63.
Sivasundaram, Sujit. "'A Christian Benares': Orientalism, Science and the Serampore Mission of Bengal." *The Indian Economic & Social History Review* 44, no. 2 (April 2007): 111–145.
Sivasundaram, Sujit. "Tales of the Land: British Geography and Kandyan Resistance in Sri Lanka, c. 1803–1850." *Modern Asian Studies* 41, no. 5 (2007): 925–965.
Sivasundaram, Sujit. "Sciences and the Global: On Methods, Questions, and Theory." *Isis* 101, no. 1 (March 2010): 146–158.
Sivasundaram, Sujit. *Islanded: Britain, Sri Lanka, and the Bounds of an Indian Ocean Colony*. University of Chicago Press, 2013.
Smith, Andrew, and Daniel Simeone. "Learning to Use the Past: The Development of a Rhetorical History Strategy by the London Headquarters of the Hudson's Bay Company." *Management & Organizational History* 12, no. 4 (October 2, 2017): 334–356.
Smith, Charlotte H. F., and Michelle Stevenson. "Modeling Cultures: 19th Century Indian Clay Figures." *Museum Anthropology* 33, no. 1 (2010): 37–48.
Srinivasachariar, C. S. "Robert Orme and Colin Mackenzie: Two Early Collectors of Manuscripts and Records." *Proceedings of the Indian Historical Records Commission* 6 (January 1924): 84.
Stack, David. "The Pleasures of Office Life: Mill at East India House." *Nineteenth-Century Prose* 47, no. 1 (March 22, 2020): 55–92.
Standaert, Nicolas. "Jean-François Foucquet's Contribution to the Establishment of Chinese Book Collections in European Libraries: Circulation of Chinese Books." *Monumenta Serica* 63, no. 2 (July 3, 2015): 361–424.

Stark, Ulrike. *Empire of Books, An: The Naval Kishore Press and the Diffusion of the Printed Word in Colonial India*. Orient Blackswan, 2009.

Stern, Philip J. *The Company-State: Corporate Sovereignty and the Early Modern Foundations of the British Empire in India*. Oxford University Press, 2011.

Stewart, Larry. "Other Centres of Calculation, Or, Where the Royal Society Didn't Count: Commerce, Coffee-Houses and Natural Philosophy in Early Modern London." *British Journal for the History of Science* 32, no. 2 (1999): 133–153.

Stewart, Larry, and Kelly J. Whitmer. "Expectations and Utility in Eighteenth-Century Knowledge Economies Notes and Records Special Issue Introduction." *Notes and Records of the Royal Society of London* 72, no. 2 (2018): 111–117.

Stoler, Ann Laura. *Along the Archival Grain: Epistemic Anxieties and Colonial Common Sense*. Princeton University Press, 2009.

Strasser, Bruno J. "Collecting Nature: Practices, Styles, and Narratives." *Osiris* 27, no. 1 (2012): 303–340.

Strasser, Bruno J. *Collecting Experiments: Making Big Data Biology*. University of Chicago Press, 2019.

Subrahmanyam, Sanjay. "Frank Submissions: The Company and the Mughals between Sir Thomas Roe and Sir William Norris," in H. V. Bowen, Lincoln Margarette and Nigel Rigby (eds.), *The Worlds of the East India Company*. D. S. Brewer, 2002, pp. 69–96.

Subrahmanyam, Sanjay. *Europe's India: Words, People, Empires, 1500–1800*. Harvard University Press, 2017.

Subramanian, S. "A Brief History of the Organisation of Official Statistics in India during the British Period." *Sankhyā: The Indian Journal of Statistics (1933–1960)* 22, no. 1/2 (1960): 85–118.

Sutton, Elizabeth A. *Capitalism and Cartography in the Dutch Golden Age*. University of Chicago Press, 2015.

Sutton, Jean. *Lords of the East: The East India Company and Its Ships*. Conway Maritime Press, 1981.

Talbot, Philip A. "Colonel William Henry Sykes: His Contribution to Statistical Accounting." *Accounting History* 15, no. 2 (May 1, 2010): 253–276.

Tavolacci, Laura. "Vegetable Gardens versus Cash Crops: Science and Political Economy in the Agricultural and Horticultural Society of India, 1820–40." *History Workshop Journal* 88 (October 1, 2019): 24–46.

Taylor, Eva Germaine Rimington. *The Mathematical Practitioners of Tudor and Stuart England*. Institute of Navigation at the Cambridge University Press, 1970.

Terrall, Mary, and Adriana Craciun (eds.). *Curious Encounters: Voyaging, Collecting, and Making Knowledge in the Long Eighteenth Century*. University of Toronto Press, 2019.

Thackray, Arnold. "Natural Knowledge in Cultural Context: The Manchester Mode." *The American Historical Review* 79, no. 3 (1974): 672–709.

Thell, Anne M. *Minds in Motion: Imagining Empiricism in Eighteenth-Century British Travel Literature*. Bucknell University Press, 2017.

Théodoridès, Jean. "Humboldt and England." *The British Journal for the History of Science* 3, no. 1 (June 1966): 39–55.

Thomas, Adrian P. "The Establishment of Calcutta Botanic Garden: Plant Transfer, Science and the East India Company, 1786–1806." *Journal of the Royal Asiatic Society* 16, no. 2 (2006): 165–177.

Thomas, Jennifer. "Compiling 'God's Great Book [of] Universal Nature': The Royal Society's Collecting Strategies." *Journal of the History of Collections* 23, no. 1 (May 1, 2011): 1–13.

Tiffin, Sarah. *Southeast Asia in Ruins: Art and Empire in the Early 19th Century.* NUS Press, 2016.

Trautmann, Thomas R. (ed.). *The Madras School of Orientalism: Producing Knowledge in Colonial South India.* Oxford University Press, 2009.

Travers, Robert. *Ideology and Empire in Eighteenth-Century India: The British in Bengal.* Cambridge University Press, 2007.

Travers, Robert. *Empires of Complaints: Mughal Law and the Making of British India, 1765–1793.* Cambridge University Press, 2022.

Tribe, Keith. "Professors Malthus and Jones: Political Economy at the East India College 1806–1858." *The European Journal of the History of Economic Thought* 2, no. 2 (1995): 327–354.

Turner, Ian M. "Plant Species Described by William Griffith in 'Some Account of the Botanical Collection Brought from the Eastward by Dr. Cantor.'" *Edinburgh Journal of Botany* 72, no. 3 (November 2015): 413–421.

Turner, Ian M. "Natural History Publications Arising from Theodore Cantor's Visit to Chusan, China, in 1840." *Archives of Natural History* 43, no. 1 (April 1, 2016): 30–40.

Tyfield, David, Rebecca Lave, Samuel Randalls and Charles Thorpe. *The Routledge Handbook of the Political Economy of Science.* Routledge, 2017.

Verner, Coolie. "John Seller and the Chart Trade in Seventeenth-Century England," in Norman Thrower and Joseph William (eds.), *The Compleat Plattmaker: Essays on Chart, Map, and Globe Making in England in the Seventeenth and Eighteenth Centuries.* University of California Press, 1978, pp. 127–158.

Vicziany, Marika. "Imperialism, Botany and Statistics in Early Nineteenth-Century India: The Surveys of Francis Buchanan (1762–1829)." *Modern Asian Studies* 20, no. 4 (1986): 625–660.

Watson, Mark F., and Henry J. Noltie. "Career, Collections, Reports and Publications of Dr Francis Buchanan (Later Hamilton), 1762–1829: Natural History Studies in Nepal, Burma (Myanmar), Bangladesh and India. Part 1." *Annals of Science* 73, no. 4 (October 1, 2016): 392–424.

Watt, James. *British Orientalisms, 1759–1835.* Cambridge University Press, 2019.

Webster, Anthony. "The Political Economy of Trade Liberalization: The East India Company Charter Act of 1813." *The Economic History Review* 43, no. 3 (1990): 404–419.

Webster, Anthony. "The Strategies and Limits of Gentlemanly Capitalism: The London East India Agency Houses, Provincial Commercial Interests, and the Evolution of British Economic Policy in South and South East Asia 1800–50." *The Economic History Review* 59, no. 4 (November 1, 2006): 743–764.

White, Daniel E. *From Little London to Little Bengal: Religion, Print, and Modernity in Early British India, 1793–1835*. Johns Hopkins University Press, 2013.

Wimsatt, William K. "Foote and a Friend of Boswell's: A Note on the Nabob." *Modern Language Notes* 57, no. 5 (1942): 325–335.

Winichakul, Thongchai. *Siam Mapped: A History of the Geo-Body of a Nation*. University of Hawaii Press, 1994.

Winterbottom, Anna. "Producing and Using the Historical Relation of Ceylon: Robert Knox, the East India Company and the Royal Society." *The British Journal for the History of Science* 42, no. 4 (December 2009): 515–538.

Winterbottom, Anna. *Hybrid Knowledge in the Early East India Company World*. Palgrave Macmillan, 2015.

Winterbottom, Anna. "Science," in William A. Pettigrew and David Veever (eds.), *The Corporation as a Protagonist in Global History, c. 1550–1750*. Brill, 2019, pp. 232–254.

Yazdani, Kaveh. "Haidar 'Ali and Tipu Sultan: Mysore's Eighteenth-Century Rulers in Transition." *Itinerario* 38, no. 2 (August 2014): 101–120.

Zastoupil, Lynn. *Rammohun Roy and the Making of Victorian Britain*. Palgrave Macmillan US, 2010.

Zastoupil, Lynn, and M. Moir (eds.). *The Great Indian Education Debate: Documents Relating to the Orientalist-Anglicist controversy, 1781–1843*. Curzon, 1999.

Zuidervaart, Huib J., and Rob H. van Gent. "'A Bare Outpost of Learned European Culture on the Edge of the Jungles of Java': Johan Maurits Mohr (1716–1775) and the Emergence of Instrumental and Institutional Science in Dutch Colonial Indonesia." *Isis* 95, no. 1 (2004): 1–33.

Index

Printed in the United States
by Baker & Taylor Publisher Services